Electrical Science: An Introduction

Electrical Science: An Introduction

Craig Cain

Willford Press
www.willfordpress.com

Published by Willford Press,
118-35 Queens Blvd., Suite 400,
Forest Hills, NY 11375, USA

ISBN: 978-1-68285-740-3

Cataloging-in-Publication Data

Electrical science : an introduction / Craig Cain.
p. cm.
Includes bibliographical references and index.
ISBN 978-1-68285-740-3
1. Electrical engineering. 2. Electricity. 3. Electric power. I. Cain, Craig.
TK153 .E44 2019
621.3--dc23

For information on all Willford Press publications
visit our website at www.willfordpress.com

Willford Press

Printed in China.

Contents

Preface VII

Chapter 1 **Electricity and Electric Current** **1**

i. Electric Current 15
ii. Electrical Units of Measure 22
iii. Electromotive Force 25

Chapter 2 **Electrical Circuits: Analysis Techniques** **45**

i. Electrical Circuit 45
ii. Circuit Elements 58
iii. Circuit Diagram 61
iv. Phasor Diagrams 63
v. Electrical Impedance 70
vi. Network Analysis 73
vii. Ohm's Law 80
viii. Kirchhoff's Circuit Laws 85
ix. Source Transformation 90
x. Substitution Theorem 92
xi. Norton's Theorem 94
xii. Compensation Theorem 97
xiii. Superposition Theorem 99
xiv. Thevenin's Theorem 102

Chapter 3 **Diodes and Transistors** **107**

i. Diode 107
ii. Zener diode 111
iii. Bridge Rectifier 117
iv. Wave-shaping Circuits 121
v. Diode Clipping Circuits 131
vi. Transistor 141
vii. Bipolar Junction Transistor 146
viii. Field-effect Transistor 152
ix. Transistor Biasing 158
x. Transistor Amplifiers 163

Chapter 4 **Capacitance and Capacitors** 170
i. Capacitance 170
ii. Mutual and Self Capacitance 173
iii. Parasitic Capacitance 176
iv. Quantum Capacitance 182
v. Capacitor 184

Chapter 5 **Magnetism and Electromagnetism** 223
i. Magnetism 223
ii. Magnetic Moment 231
iii. Gauss's Law for Magnetism 235
iv. Electromagnetism 238
v. Oersted's Law 242
vi. Electromagnetic Induction 242
vii. Faraday's Law of Induction 246
viii. Maxwell's Equations 250

Permissions

Index

Preface

The study and application of electromagnetism, electricity and electronics is under the scope of a field of engineering called electrical engineering. It has a number of significant subfields such as electronic engineering, control engineering, power engineering, telecommunications, radio-frequency engineering, signal processing, etc. It has applications in the fields of hardware engineering, microwave engineering, nanotechnology, mechatronics and electrochemistry, besides others. Some of the most important innovations of electrical science in history are the development of the telegraph, telephone and electrical power generation and distribution. This textbook is a valuable compilation of topics, ranging from the basic to the most complex theories and principles in the field of electrical science. Most of the topics introduced herein cover new techniques and applications of electrical science in a comprehensive manner. For all those who are interested in this discipline, this book can prove to be an essential guide.

To facilitate a deeper understanding of the contents of this book a short introduction of every chapter is written below:

Chapter 1- The electric charge is a basic property of matter. All physical phenomena that are associated with the motion and presence of electric charges together constitute electricity. Some of the vital topics central to the understanding of electricity, such as electric current, units of measurement and electromotive force, have been extensively discussed in this chapter.

Chapter 2- Electrical circuits are electrical networks that consist of electrical components in a closed loop thus giving a return path for the passing of the electric current. Finding voltages and currents through each component in the network is achieved through certain network analysis techniques. The aim of this chapter is to provide a detailed examination of the major concepts of network analysis and significant analysis techniques, such as circuit elements, circuit diagram, phasor diagram, Kirchhoff's circuit laws, Thévenin's theorem, Norton's theorem, etc.

Chapter 3- A diode is an electrical component with two terminals, which conducts current in only one direction. A transistor is any semiconductor device that is used for amplifying electronic signals and electrical power. This chapter has been carefully written to provide an easy understanding of the varied aspects of diodes and transistors. It discusses in detail topics like Zener diode, bridge rectifier, field-effect transistor, bipolar junction transistor, transistor biasing, etc.

Chapter 4- Any change in an electric charge relative to a change in the electric potential is expressed in terms of a ratio called the capacitance. A passive electrical component, which stores electrical potential energy is known as a capacitor. The topics elaborated in this chapter on the fundamental concepts of capacitors and capacitance include mutual and self capacitance, parasitic capacitance and quantum capacitance, which will help in developing a comprehensive understanding of this physical quantity and its measurement.

Chapter 5- Magnetism refers to the class of physical phenomena that occurs as a result of magnetic fields. Electromagnetism is a branch of physics, which is concerned with the study of the electromagnetic force that is produced between charged particles. In order to completely understand the phenomena of magnetism and electromagnetism, it is necessary to understand the concepts of magnetic moment, Gauss's law for magnetism, Oersted's law, electromagnetic induction, Faraday's law of induction and Maxwell's equations, which have been carefully analyzed in this chapter.

Finally, I would like to thank the entire team involved in the inception of this book for their valuable time and contribution. This book would not have been possible without their efforts. I would also like to thank my friends and family for their constant support.

Craig Cain

1

Electricity and Electric Current

The electric charge is a basic property of matter. All physical phenomena that are associated with the motion and presence of electric charges together constitute electricity. Some of the vital topics central to the understanding of electricity, such as electric current, units of measurement and electromotive force, have been extensively discussed in this chapter.

Electricity is the physical flow of electrons, referred to as an electrical current. Electricity is an energy carrier that efficiently delivers the energy found in primary sources to end users, who in turn convert it into energy services.

Electricity can be created in three ways:

- The most common is through electro-magnetic conversion, where electricity is generated by moving an electric conductor, such as a wire, inside a magnetic field. The most practical example of this method is a generator connected to a turbine. The turbine provides the motion required to move the conductor in the generator. This energy for motion can come from various technologies, for example wind turbines, hydropower, or the steam created from heat produced in nuclear fission or coal combustion.
- Electricity can also be created through a chemical reaction, for example in a battery or fuel cell.
- Finally, electricity can be created through solid-state conversion, wherein electricity is generated using the structure and properties of a solid. The specially constructed solid consists of different molecules packed closely together that create an electric current when stimulated. An example of a technology that utilizes solid state conversion is a solar PV cell.

It is important to note that electricity is the same, regardless of how it is produced. Thus the electricity generated from an electromagnetic generator is the same as that from a battery. Once electricity is created it is the most versatile form of energy.

The rate at which electricity is produced is referred to as a Watt. The quantity of energy used over a certain period of time is referred to as kWh.

Types of Electricity

Electricity is of two type: static electricity and current electricity.

Static Electricity

When electricity gathers in one place it is known as static electricity (the word static means something that does not move).

Static electricity often happens when you rub things together. If you rub a balloon against your pullover 20 or 30 times, you'll find the balloon sticks to you. This happens because rubbing the balloon gives it an electric charge (a small amount of electricity). The charge makes it stick to your pullover like a magnet, because your pullover gains an opposite electric charge. So your pullover and the balloon attract one another like the opposite ends of two magnets.

Lightning happens when static electricity (built up in one place) turns to current electricity

Have you ever walked across a nylon rug or carpet and felt a slight tingling sensation? Then touched something metal, like a door knob or a faucet (tap), and felt a sharp pain in your hand? That is an example of an electric shock. When you walk across the rug, your feet are rubbing against it. Your body gradually builds up an electric charge, which is the tingling you can sense. When you touch metal, the charge runs instantly to Earth—and that's the shock you feel.

Lightning is also caused by static electricity. As rain clouds move through the sky, they rub against the air around them. This makes them build up a huge electric charge. Eventually, when the charge is big enough, it leaps to Earth as a bolt of lightning. You can often feel the tingling in the air when a storm is brewing nearby. This is the electricity in the air around you.

Working of Static Electricity

Electricity is caused by electrons, the tiny particles that "orbit" around the edges of atoms, from which everything is made. Each electron has a small negative charge. An atom normally has an equal number of electrons and protons (positively charged particles in its nucleus or center), so atoms have no overall electrical charge. A piece of rubber is made from large collections of atoms called molecules. Since the atoms have no electrical charge, the molecules have no charge either—and nor does the rubber.

Suppose you rub a balloon on your pullover over and over again. As you move the balloon back and forward, you give it energy. The energy from your hand makes the balloon move. As it rubs against the wool in your pullover, some of the electrons in the rubber molecules are knocked free and gather on your body. This leaves the balloon with slightly too few electrons. Since electrons are negatively charged, having too few electrons makes the balloon slightly positively charged. Your pullover meanwhile gains these extra electrons and becomes negatively charged. Your pullover is negatively charged, and the balloon is positively charged. Opposite charges attract, so your pull-over sticks to the balloon.

Current Electricity

When electrons move, they carry electrical energy from one place to another. This is called current electricity or an electric current. A lightning bolt is one example of an electric current, although it does not last very long. Electric currents are also involved in powering all the electrical appliances that you use, from washing machines to flashlights and from telephones to MP3 players. These electric currents last much longer.

Have you heard of the terms potential energy and kinetic energy? Potential energy means energy that is stored somehow for use in the future. A car at the top of a hill has potential energy, because it has the potential (or ability) to roll down the hill in future. When it's rolling down the hill, its potential energy is gradually converted into kinetic energy (the energy something has because it's moving).

Static electricity and current electricity are like potential energy and kinetic energy. When electricity gathers in one place, it has the potential to do something in the future. Electricity stored in a battery is an example of electrical potential energy. You can use the energy in the battery to power a flashlight, for example. When you switch on a flashlight, the battery inside begins to supply electrical energy to the lamp, making it give off light. All the time the light is switched on, energy is flowing from the battery to the lamp. Over time, the energy stored in the battery is gradually turned into light (and heat) in the lamp. This is why the battery runs flat.

A battery like this stores electrical potential energy in a chemical form. When the battery is flat, it means you've used up all the stored energy inside by converting it into other forms

Electric Circuits

For an electric current to happen, there must be a circuit. A circuit is a closed path or loop around which an electric current flows. A circuit is usually made by linking electrical components together with pieces of wire cable. Thus, in a flashlight, there is a simple circuit with a switch, a lamp, and a battery linked together by a few short pieces of copper wire. When you turn the switch on, electricity flows around the circuit. If there is a break anywhere in the circuit, electricity cannot flow. If one of the wires is broken, for example, the lamp will not light. Similarly, if the switch is turned off, no electricity can flow. This is why a switch is sometimes called a circuit breaker.

You don't always need wires to make a circuit, however. There is a circuit formed between a storm cloud and the Earth by the air in between. Normally air does not conduct electricity. However, if there is a big enough electrical charge in the cloud, it can create charged particles in the air called ions (atoms that have lost or gained some electrons). The ions work like an invisible cable linking the cloud above and the air below. Lightning flows through the air between the ions.

Moving of Electricity in a Circuit

Materials such as copper metal that conduct electricity (allow it to flow freely) are called conductors. Materials that don't allow electricity to pass through them so readily, such as rubber and plastic, are called insulators. What makes copper a conductor and rubber an insulator?

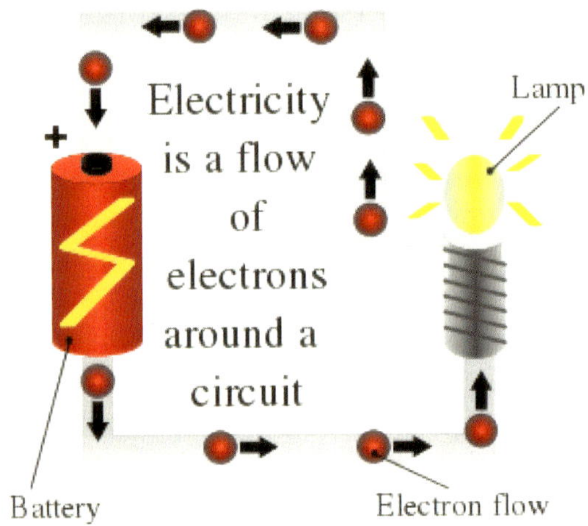

A current of electricity is a steady flow of electrons. When electrons move from one place to another, round a circuit, they carry electrical energy from place to place like marching ants carrying leaves. Instead of carrying leaves, electrons carry a tiny amount of electric charge.

Electricity can travel through something when its structure allows electrons to move through it easily. Metals like copper have "free" electrons that are not bound tightly to their parent atoms. These electrons flow freely throughout the structure of copper and this is what enables an electric current to flow. In rubber, the electrons are more tightly bound. There are no "free" electrons and, as a result, electricity does not really flow through rubber at all. Conductors that let electricity flow freely are said to have a high conductance and a low resistance; insulators that do not allow electricity to flow are the opposite: they have a low conductance and a high resistance.

For electricity to flow, there has to be something to push the electrons along. This is called an electromotive force (EMF). A battery or power outlet creates the electromotive force that makes a current of electrons flow. An electromotive force is better known as a voltage.

Direct Current and Alternating Current

Electricity can move around a circuit in two different ways. In the big picture up above, you can see electrons racing around a loop like race cars on a track, always going in the same direction. This type of electricity is called direct current (DC) and most toys and small gadgets have circuits that work this way.

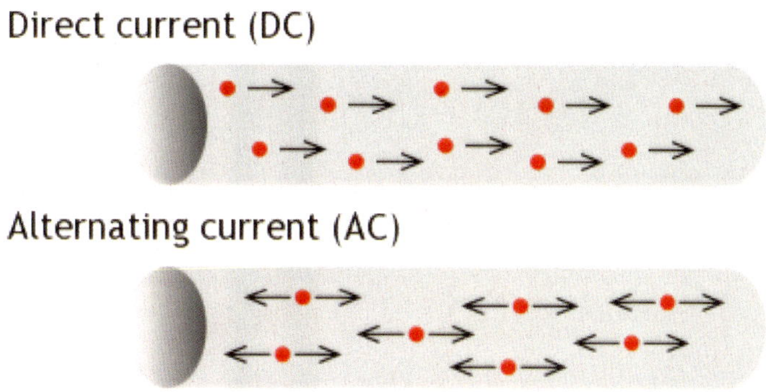

Top: In a direct current (DC) circuit, electrons always flow in the same direction. Bottom: In an alternating current (AC) circuit, the electrons reverse direction many times each second.

The bigger appliances in your home use a different kind of electricity called alternating current (AC). Instead of always flowing the same way, the electrons constantly reverse direction—about 50–60 times every second. Although you might think that makes it impossible for energy to be carried round a circuit, it doesn't. Take the flashlight bulb in the circuit above. With direct current, new electrons keep streaming through the filament (a thin piece of wire inside the bulb), making it heat up and give off light. With alternating current, the same old electrons whiz back and forth in the filament. You can think of them running on the spot, heating up the filament so it still makes bright light we can see. So both types of current can make the lamp work even though they flow in different ways. Most other electric appliances can also work using either direct or alternating current, though some circuits do need AC to be changed to DC (or vice versa) to work correctly.

Making Electricity

Just as electricity can make magnetism, so magnetism can make electricity. A dynamo is a bit like an electric motor inside. When we pedal our bicycle, the dynamo clipped to the wheel spins around. Inside the dynamo, there is a heavy core made from iron wire wrapped tightly around—much like the inside of a motor. The core spins freely inside some large fixed magnets. As we pedal, the core rotates inside these outer magnets and generates electricity. The electricity flows out from the dynamo and powers our bicycle lamp.

The electric generators used in power plants work in exactly the same way, only on a much bigger scale. Instead of being powered by someone's legs, pedaling furiously, these large generators are driven by steam. The steam is made by burning fuels or by nuclear reactions. Power plants can make enormous amounts of electricity, but they waste quite a lot of the energy they produce. The energy has to flow from the plant, where it is made, to the homes, offices, and factories where it is used down many miles of electric power cable. Making electricity in a power plant and delivering it to a distant building can waste up to two thirds of the energy that was originally present in the fuel.

Another problem with power plants is that they make electricity by burning "fossil fuels" such as coal, gas, or oil. This creates pollution and adds to the problem known as global warming (the way Earth is steadily heating up because of the energy people are using). Another problem with fossil fuels is that supplies are limited and they are steadily running out.

Making clean, renewable energy from the wind. Each of these turbines contains an electricity generator in the top section, just behind the spinning rotors

There are other ways to make energy that are more efficient, less polluting, and do not contribute to global warming. These types of energy are called renewable, because they can last indefinitely. Examples of renewable energy include wind turbines and solar power. Unlike huge electric power plants, they are often much more efficient ways of making electricity. Because they can be sited closer to where the electricity is used, less energy is wasted transmitting power down the wires.

Wind turbines are effectively just electric generators with a propeller on the front. The wind turns the propeller, which spins the generator inside, and makes a study current of electricity.

Unlike virtually every other way of making electricity, solar cells (like the ones on calculators and digital watches) do not work using electricity generators and magnetism. When light falls on a solar cell, the material it is made from (silicon) captures the light's energy and turns it directly into electricity. Potentially, this means solar cells are an extremely efficient way to make electricity. A home with solar electric panels on the roof might be able to make most of its own electricity.

The Power of Electricity

Before the invention of electricity, people had to make energy wherever and whenever they needed it. Thus, they had to make wood or coal fires to heat their homes or cook food. The invention of electricity changed all that. It meant energy could be made in one place then supplied over long distances to wherever it was needed. People no longer had to worry about making energy for heating or cooking: all they had to do was plug in and switch on and the energy was there as soon as they wanted it.

Another good thing about electricity is that it's like a common "language" that all modern appliances can "speak." We can run a car using the energy in gasoline, or we can cook food on a barbecue in our garden using charcoal, though we can't run our car on charcoal or cook food with gasoline. But electricity is quite different. We can cook with it, run cars on it, heat our home with it, and charge our cellphone with it. This is the great beauty and the power of electricity: it's energy for everyone, everywhere, and always.

Measuring Electricity

We can measure electricity in a number of different ways, but a few measurements are particularly important.

We can use a digital multimeter like this to measure voltage, current, and resistance.

Voltage

The voltage is a kind of electrical force that makes electricity move through a wire and we measure it in volts. The bigger the voltage, the more current will tend to flow. So a 12-volt car battery will generally produce more current than a 1.5-volt flashlight battery.

Current

Voltage does not, itself, go anywhere: it's quite wrong to talk about voltage "flowing through" things. What moves through the wire in a circuit is electrical current: a steady flow of electrons, measured in amperes (or amps).

Power

Together, voltage and current give we electrical power. The bigger the voltage and the bigger the current, the more electrical power we have. We measure electric power in units called watts. Something that uses 1 watt uses 1 joule of energy each second.

The electric power in a circuit is equal to the voltage × the current (in other words: watts = volts × amps). So if we have a 100-watt (100 W) light and we know our electricity supply is rated as 120 volts (typical household voltage in the United States), the current flowing must be 100/120 = 0.8 amps. If we're in Europe, our household voltage is more likely 230 volts. So if we use the same 100-watt light, the current flowing is 100/230 = 0.4 amps. The light burns just as brightly in both countries and uses the same amount of power in each case; in Europe it uses a higher voltage and lower current; in the States, there's a lower voltage and higher current. (One quick note: 120 volts and 230 volts are the "nominal" or standard household voltages—the voltages we're supposed to have, in theory. In practice, our home might have more or less voltage than this, for all sorts of reasons, but mainly because of how far we are from our local power plant or power supply).

Energy

Power is a measurement of how much energy we're using each second. To find out the total amount of energy an electric appliance uses, we have to multiply the power it uses per second by the total number of seconds we use it for. The result we get is measured in units of power × time, often converted into a standard unit called the kilowatt hour (kWh). If we used an electric toaster rated at 1000 watts (1 kilowatt) for a whole hour, we'd use 1 kilowatt hour of energy; we'd use the same amount of energy burning a 2000 watt toaster for 0.5 hours or a 100-watt lamp for 10 hours.

Electricity meters show the total number of kilowatt hours of electricity we've used. 1 kilowatt hour is equal to 3.6 million joules (J) of energy (or 3.6 megajoules if we prefer).

Electric Charge

Electric charge is defined as: "An electrical property of matter that exists because of an access or a deficiency of electrons". There are two types of electric charge, positive charges and negative charges.

Its formula is given as,

$$Q = I \times t$$

Where, Q is the total charge, I is the amount of current and t is the time.

Properties of Electrical Charge

- Charge is a basic property of a material body due to which attracts or repels another object.
- Friction produces two different types of charge on different materials (such as glass or plastic).
- Like charges always repel each other.
- Unlike charges always attract each other.
- Repulsion is the sure test of charge on a body.

Positive Charge

When a material losses electrons then the number of protons increase in the material ,there is a net positive electrical charge .

The positive and negative labels for electric charge are due to Benjamin Franklin (1706-1790) who, among many other accomplishments, was a scientist of international reputation. It has even been said that Franklin's triumphs in diplomacy in France during the American War of independence may have been made possible because he was so highly regarded as a scientist.

Electrical forces between charged bodies have many industrial applications, among them being electrostatic paint spraying and powder coating, fly ash precipitation, non impact ink jet printing, and photocopying. For example, shows a tiny carrier bead in a photocopying machine, covered with particles of black powder called toner, that stick to the carrier bead by electrostatic forces. These negatively charged toner particles are eventually attracted from their carrier beads to a positive charge latent image of the document to be copied, which is formed on a routing drum. A charged sheet of paper then attracts the toner particles from the drum to itself, after which they are heat fused in place to make the final copy.

Negative Charge

When a material gains electrons then the number of electrons increase in the material, there is a net negative electrical charge.

Electron has negative charge, while proton has positive charge.in a material number of protons an electrons are equal in numbers. Electrical charge is a, which is a fundamental characteristics of electrons and protons, is symbolized by Q.

Static electricity is the presence of net positive and negative charge in a material. Everyone has experienced the effects of static electricity from time to time, for example when attempting to touch a metal surface or another person or when a clothes in the dryer cling together. electric charge

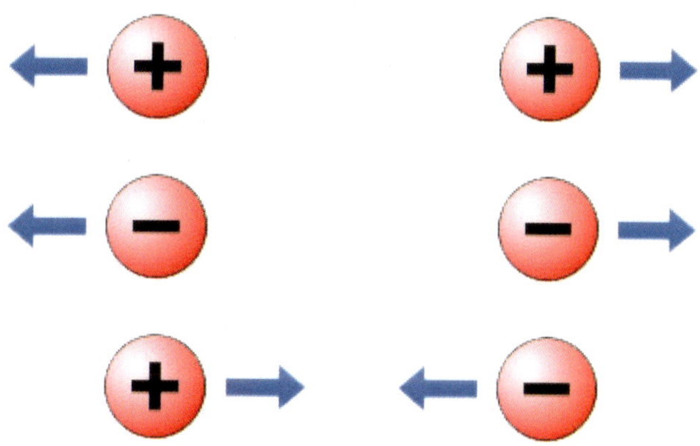

Like charges repel each other and unlike charges attract each other, A force acts between charges, as evidenced by the attraction or repulsion. This force, called an electric field, consists of invisible lines of force.

Electroscopes Detection of Charge

The gold leaf electroscope is a sensitive instrument for detecting charges. It consists of a brass rod with a brass disk at the top and two thin leaves of gold foil hanging at the bottom in Figure. The rod passes through an insulator that keeps the rod in place. Charges can move freely from the disk to the leaves through the rod. A thin aluminium foil is attached on the lower portion of the inside of the jar. Usually, the aluminium foil is grounded by connecting a copper wire. This protects the leaves from the external electrical disturbances.

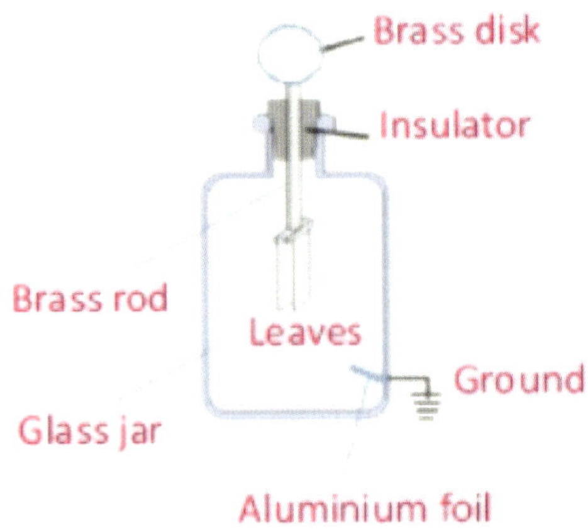

Detecting the Presence of Charge

In order to detect the presence of charge on anybody, bring the body near the disk of an uncharged electroscope. If the body neutral there will be no deflection of the leaves in figure:

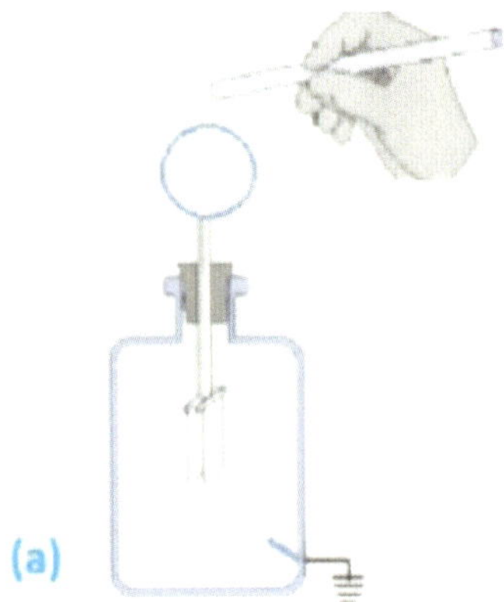

But if the body is positively or negatively charged, the leaves of the electroscope diverge. For example, if the body is negatively charged then due to electrostatic induction, positively charge will appear on the disk while negative charge will appear on the leave in figure:

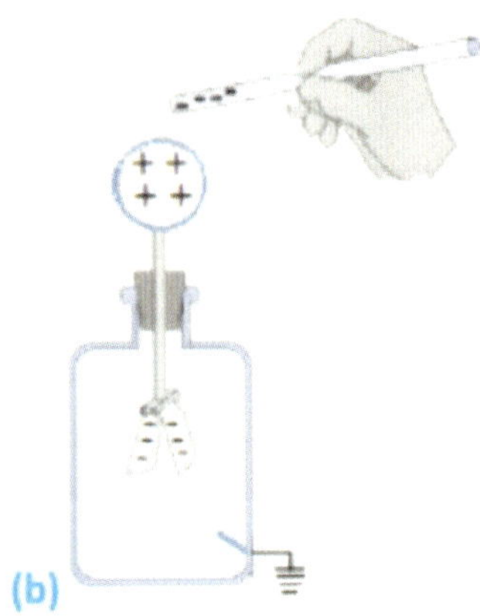

The leaves of the electroscope repel each other and diverge because each leave gets similar charge. The divergence of leaves will depend on the amount of charge.

Charging the Electroscope by Electrostatic Induction

Electroscope can be charged by the process of electrostatic induction. In order to produce positive charge on the electroscope, bring a negatively charged body near the disk of the electroscope in figure:

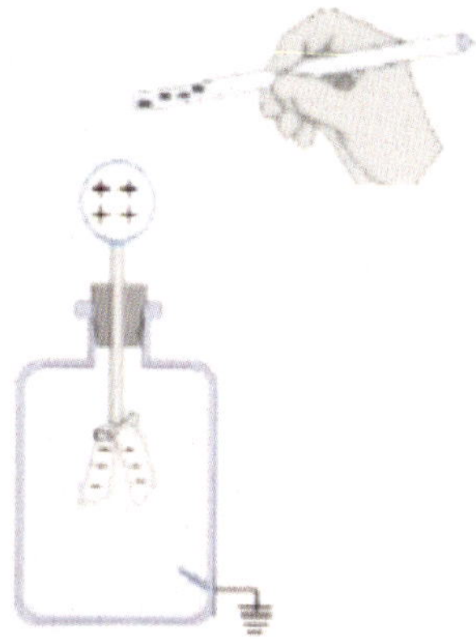

Positive charge will appear on the disk of the electroscope while negative charges will shift to the leaves. Now connect the disk of electroscope to the earthed aluminium foil by a conducting wire in figure:

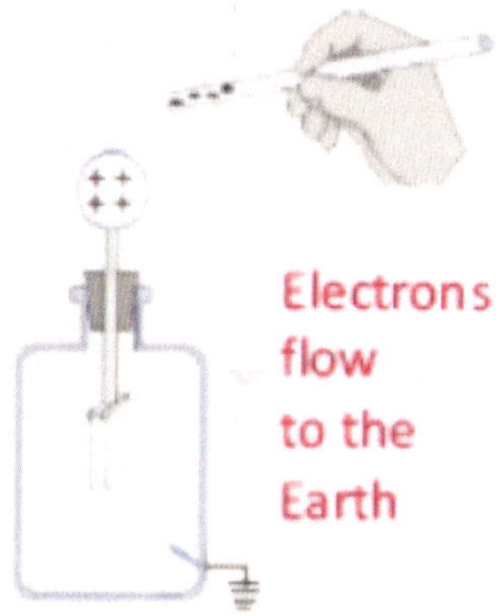

Charge of the leaves will flow to the Earth through the wire. Now if we first break the Earth connection and then remove the rod,the electroscope will be left with positive charge in figure:

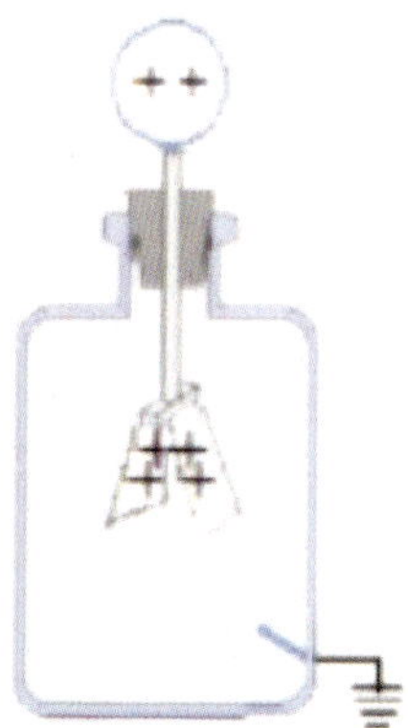

Similarly, electroscope can be charged negatively with the help of a positively charged rod.

Electroscope can also be changed by the process of conduction. Touch a negatively charged rod with the disk of a neutral electroscope. Negative charge from the rod will transfer to the electroscope and will cause it leaves to diverge.

Detecting the Type of Charge

For the detection of type of charge on a body, electroscope is first charged either positively. Suppose the electroscope is positively charged as explain before and are shown in figure:

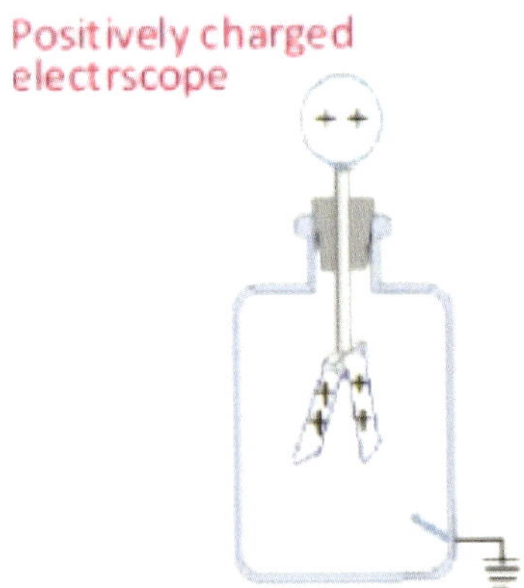

Now in order to detect the type of charge on a body, bring the charged body near the disk of the positively charged electroscope. If the divergence of the leaves increases, the body carries positively charge are shown in figure:

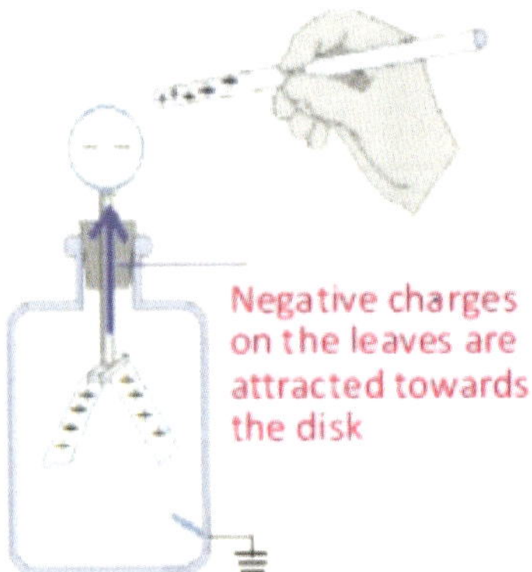

On the other hand if the divergence decreases, the body has negative charge are sown in figure:

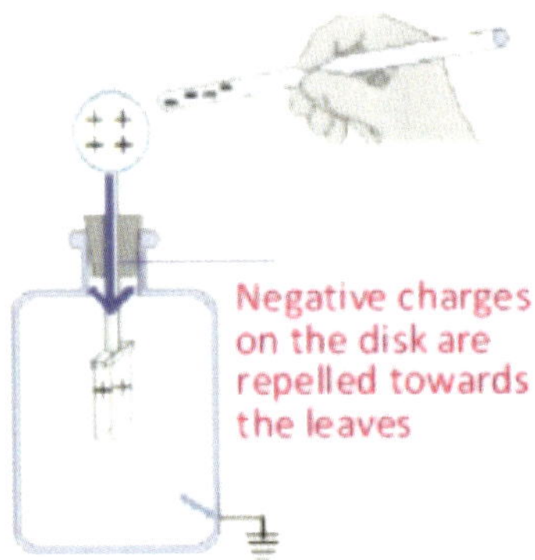

Identifying Conductors and Insulators

Electroscope can also be used to distinguish between insulators and conductors. Touch the disk of a charged electroscope with material under test. If the leaves collapse from their diverged position, the body would be a good conductor. If there is no change in the divergence of the leaves, it will show that the body under test in an insulator.

Electric charge(Q) is measured in Coulombs, symbolized by C.

Charge is conserved:

- In an isolated system charge can neither be created nor destroyed;
- Total charge in the universe is constant;
- Charge can be created or destroyed but in equal and opposite pairs, for instance, a γ-ray can split be an electron and a positron, i.e.

  If $E\gamma$ *(≥ 1.02 MeV)*

  $\gamma \rightarrow e- + e^{+}$

This process is called "pair production". The electron and positron can combine to form γ-ray again. Such a process is termed as pair annihilation.

$e- + e^{+} \rightarrow \gamma(E\gamma = 1.02\ MeV)$

Charge is Quantised

Charge on a body can be integral multiple of electronic charge. i.e. $Q=\pm\ n\ e$. If a body gains electrons, it is said to be negatively charged and if it loses electrons, it is said to be positively charged. Though there are particles called "quarks" which may have charge e/3 or 2e/3.Since these are generated during disintegration of nucleus (neutron, proton and so on) these cannot be transferred. Charge on an electron is 1.6×10^{-19} C.

Electric Field

Electric field is an electric property associated with each point in space when charge is present in any form. The magnitude and direction of the electric field are expressed by the value of E, called electric field strength or electric field intensity or simply the electric field. Knowledge of the value of the electric field at a point, without any specific knowledge of what produced the field, is all that is needed to determine what will happen to electric charges close to that particular point.

Instead of considering the electric force as a direct interaction of two electric charges at a distance from each other, one charge is considered the source of an electric field that extends outward into the surrounding space, and the force exerted on a second charge in this space is considered as a direct interaction between the electric field and the second charge. The strength of an electric field E at any point may be defined as the electric, or Coulomb, force F exerted per unit positive electric charge q at that point, or simply $E = F/q$. If the second, or test, charge is twice as great, the resultant force is doubled; but their quotient, the measure of the electric field

E, remains the same at any given point. The strength of the electric field depends on the source charge, not on the test charge. Strictly speaking, the introduction of a small test charge, which itself has an electric field, slightly modifies the existing field. The electric field may be thought of as the force per unit positive charge that would be exerted before the field is disturbed by the presence of the test charge.

The direction of the force that is exerted on a negative charge is opposite that which is exerted on a positive charge. Because an electric field has both magnitude and direction, the direction of the force on a positive charge is chosen arbitrarily as the direction of the electric field. Because positive charges repel each other, the electric field around an isolated positive charge is oriented radially outward. When they are represented by lines of force, or field lines, electric fields are depicted as starting on positive charges and terminating on negative charges. A line tangent to a field line indicates the direction of the electric field at that point. Where the field lines are close together, the electric field is stronger than where they are farther apart. The magnitude of the electric field around an electric charge, considered as source of the electric field, depends on how the charge is distributed in space. For a charge concentrated nearly at a point, the electric field is directly proportional to the amount of charge; it is inversely proportional to the square of the distance radially away from the centre of the source charge and depends also upon the nature of the medium. The presence of a material medium always diminishes the electric field below the value it has in a vacuum.

At times the electric field itself may become detached from the source charge and form closed loops, as in the case of charges accelerating up and down the transmitting antenna of a television station. The electric field with an accompanying magnetic field is propagated through space as a radiated wave at the same speed as that of light. Such electromagnetic waves indicate that electric fields are generated not only from electric charges but also from changing magnetic fields.

The value of the electric field has dimensions of force per unit charge. In the metre-kilogram-second and SI systems, the appropriate units are newtons per coulomb, equivalent to volts per metre. In the centimetre-gram-second system, the electric field is expressed in units of dynes per electrostatic unit (esu), equivalent to statvolts per centimetre.

Types of Electric Field

The electric fields are of four types:

- Uniform electric field;
- Non-uniform electric field;
- Static electric field;
- Time varying electric field.

Uniform Electric Field

Uniform electric field is an electric field in which at every point within the electric field has constant electric field strength.

Non-uniform Electric Field

Non-uniform electric field is an electric field in which at any two points within the electric field does not have constant electric field strength.

Static Electric Field

The strength of the electric field that does not vary with respect to time is called static electric field. This static electric field is produced by static electric charges.

Time Varying Electric Field

The strength of the electric field that varies with respect to time is called time varying electric field.

Electric Current

An electric current is a flow of electric charge in a circuit. More specifically, the electric current is the rate of charge flow past a given point in an electric circuit. The charge can be negatively charged electrons or positive charge carriers including protons, positive ions or holes.

The basic concept of current is that it is the movement of electrons within a substance. Electrons are minute particles that exist as part of the molecular structure of materials. Sometimes these electrons are held tightly within the molecules and other times they are held loosely and they are able to move around the structure relatively freely.

One very important point to note about the electrons is that they are charged particles - they carry a negative charge. If they move then an amount of charge moves and this is called current.

It is also worth noting that the number of electrons that able to move governs the ability of a particular substance to conduct electricity. Some materials allow current to move better than others.

The motion of the free electrons is normally very haphazard - it is random - as many electrons move in one direction as in another and as a result there is no overall movement of charge.

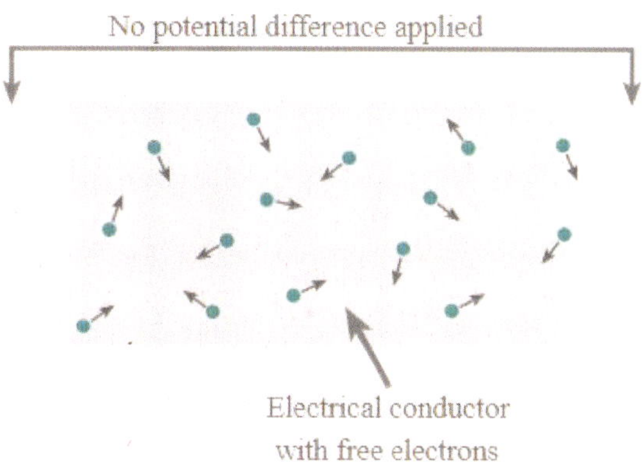

Random electron movement in a conductor with free electrons.

If a force acts on the electrons to move them in a particular direction, then they will all drift in the same direction, although still in a somewhat haphazard fashion, but there is an overall movement in one direction.

The force that acts on the electrons is called and electromotive force, or EMF, and its quantity is voltage measured in volts.

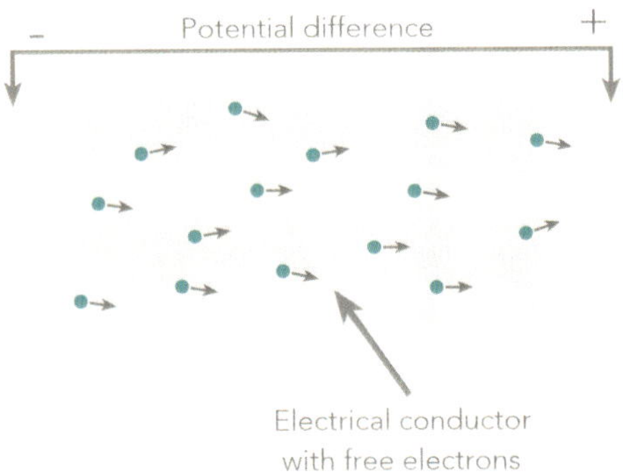

Electron flow under the action of applied electro-motive force.

Electric Current Symbol

The electric current is represented by a symbol I. The symbol I was used by the French physicist "Andre-Marie Ampere". The unit of electric current (ampere) is named after him.

Generation of Electric Current

Atoms are the basic building blocks of matter. Every object in the universe is made up of atoms. Atoms are the tiny particles. Their size is in nanometers.

Each atom consists of subatomic particles such as electrons, protons, and neutrons. These sub-atomic particles are smaller than the atom.

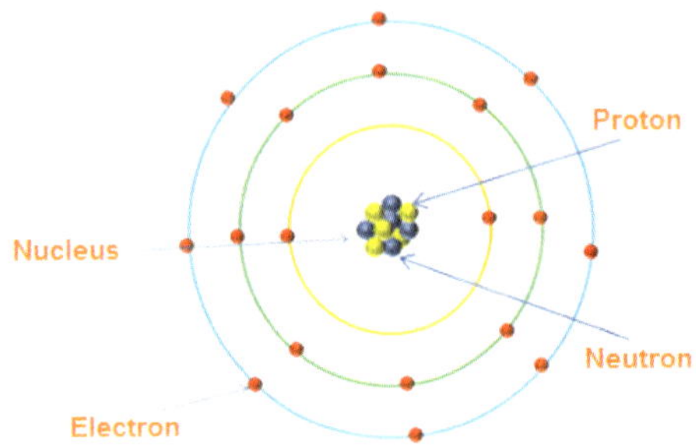

Electrons are the negatively charged particles, protons are the positively charged particles, and neutrons are the neutral particles (no charge).

Protons and neutrons are much heavier than electrons. So the protons and neutrons always reside at the center of the atom. The strong nuclear force between the protons and neutrons make them always stick together.

Protons have positive charge and neutrons have no charge. So the overall charge of the nucleus is positive.

Electrons always revolve around the nucleus because of the electrostatic force of attraction between them.

The electrons revolve around the nucleus in different orbits. Each orbit has an energy level associated with it.

The electrons revolving at a close distance from the nucleus have very low energy. On the other hand, the electrons revolving at a greater distance from the nucleus have very high energy.

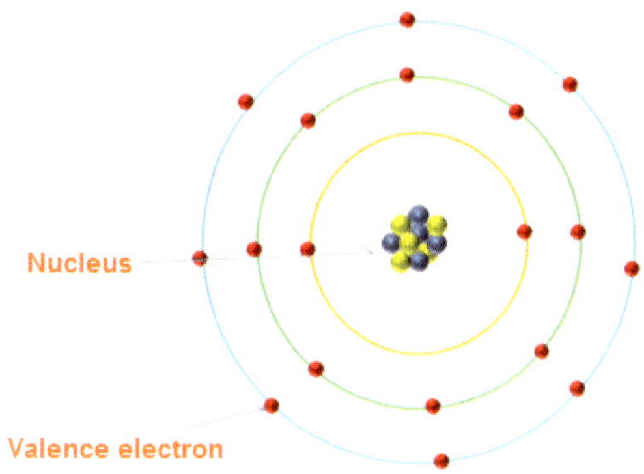

The electrons in the outermost orbit of an atom are called valence electrons. These electrons are very loosely attached to the parent atom. So applying a small amount of energy is enough to make them free from the parent atom.

When a small amount of energy in the form of heat, light, or electric field is supplied to the valence electrons, they gain sufficient energy and then separated from the parent atom.

The electrons that are separated from the parent atom are known as free electrons. These electrons move freely from one place to another place.

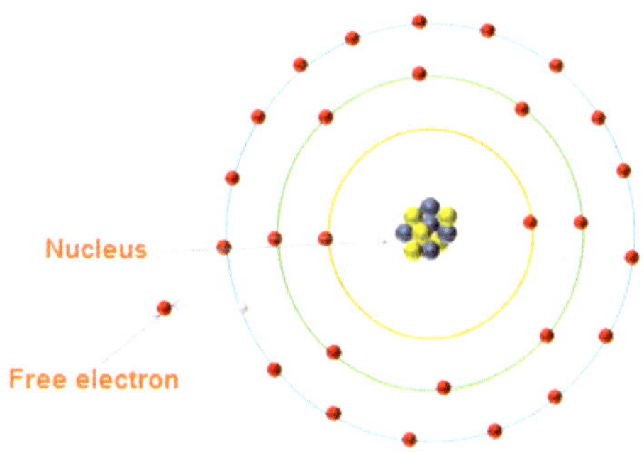

We know that electrons have a negative charge. So the free electrons carry negative charge from one place to another place.

We know that electric current means a flow of charge. So the electrons moving freely from one place to another place will conduct electric current.

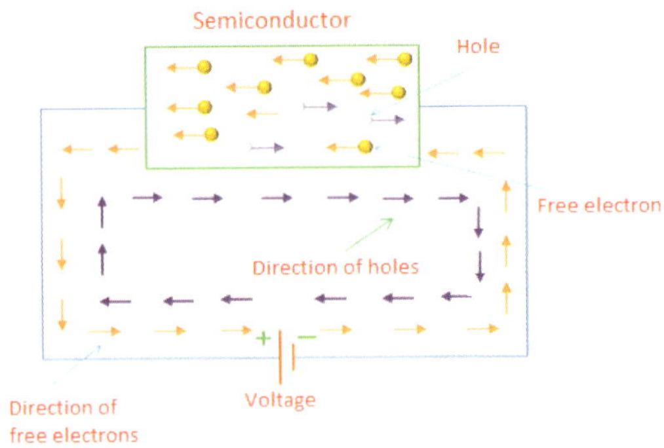

In semiconductors, both free electrons and holes are present. Free electrons are the negatively charged particles. So they carry a negative charge (electric current). Holes are the positively charged particles. So they carry a positive charge (electric current).

Thus, both free electrons and holes conduct electric current in semiconductors.

In conductors, holes are negligible. So the free electrons conduct electric current.

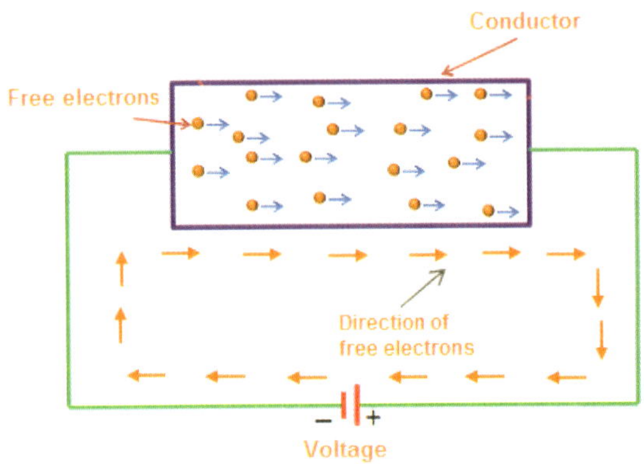

Protons also have the ability to conduct electric current. However, protons cannot move freely from one place to another place like electrons. They always held in a fixed position. So the protons do not conduct electric current.

SI unit of Electric Current

The SI unit of electric current is ampere, which is named after the French physicist Andre-Marie Ampere. The electric current flowing in a conductor or semiconductor is measured in amperes. Ampere is also sometimes referred to as amps or A.

The current flowing through an electronic component (eg: diode) in a circuit is measured by using a device called an ammeter.

Current Direction

When a voltage is applied to a conductor or semiconductor, electric current starts flowing.

In conductors, positively charged protons are held in a fixed position and the negatively charged electrons move from one place to another place by carrying the charge. Thus, electrons conduct electric current in conductors.

In semiconductors, both free electrons and holes carry charge from one place to another place. Thus, electrons and holes conduct electric current in semiconductors.

When voltage is applied, the electrons (negative charges) move from negative end of the battery to the positive end of the battery. So the electrons (negative charges) current direction is from negative to positive.

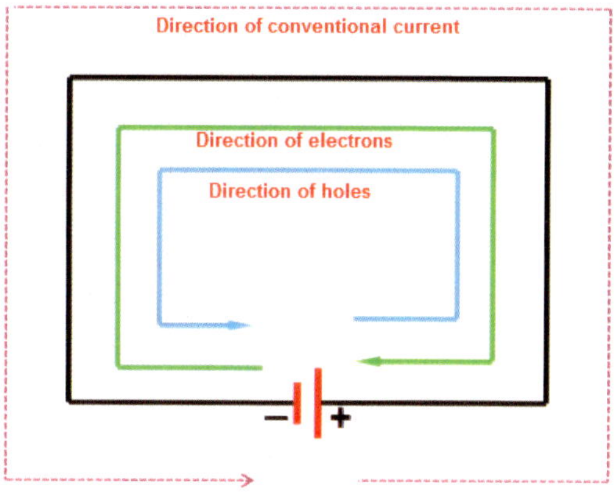

On the other hand, holes (positive charges) move from positive end of the battery to the negative end of the battery. So the holes (positive charges) current direction is from positive to negative.

The conventional current direction is from positive to negative (same as the current direction of positive charges).

The charge of a positively charged particle (hole) is equal to the charge of a negatively charged particle (free electron) but opposite in polarity.

A flow of negative charges in a circuit will produce the current same as the flow of positive charges produce. So it does not matter whether the current is flowing from positive to negative or negative to positive, the generated current will be same.

Types of Current

There are only two types of electrical current, direct current and alternating current.

We abbreviate them as DC and AC respectively. Concept of DC was developed before AC. But AC becomes most popular means of generating, transmitting and distributing of electric power. The direction of the flow of direct current is unidirectional, means this current does not alter its direction during flowing. Most common examples of DC in our daily life, are the current that we get from all kinds of battery system. But most popular form of electrical current is alternating current or AC. AC does have some advantages over DC for generating, transmitting and distributing and that is why the current we get from our electric supply companies, is normally alternating current.

Alternating Current

The current whose flow is not unidirectional moreover it alternates at a frequency, is called alternating current. In other words, the direction of the current continuously changes from forward to backward and then backward to forward in the circuit. The number of times, this direction changes from forward to backward or from backward to forward per second, is referred as frequency of the current. The current produced in an alternator is always an alternating current. The shape of the waveform of an alternating current is usually sinusoidal. But square, triangular and other types of waveform are also available for attending current.

Conventional Direction of Alternating Current

As direct current, alternating current is denoted with arrow. An AC has both forward and backward direction of flow. The arrow head always indicates the forward direction of the current. In different point of view, when the current has a positive valve, the direction of current is same as the reference arrow and when the current gets negative value; its direction is just opposite of the reference arrow.

Mathematical Formulation

Mathematically, it is represented by,

$$I = Q/t$$

Electric-current-equation-1On the basis of the flow of electric charge the current is mainly classified into two types, i.e. alternating current and direct current. In direct current, the charges flow through unidirectional whereas in alternating current the charges flows in both the direction.

Effects of Electric Current

The flow of electric current causes the following effects in a conductor:

- Heating effect;
- Chemical effect;
- Magnetic effect.

Heating Effect

Electric current flowing in an electric iron produces heat. We use this heat for ironing the clothes. An electric current, therefore, produces a heating effect. Whenever an electric current flows through a wire (conductor), a small amount of electrical energy is converted into heat energy. This is called the heating effect of electric current. Heating effect is due to the resistance of the conductor. The conducting wires made of certain substances produce relatively more heat than others of the same dimension depending on the resistance. This effect of electric current is utilised in appliances such as electric stove, boiler, electric fuse, electric iron, electric bulb, water heater, etc.

The heat produced in a conductor, when a current flows through it, is found to depend directly on the following factors:

- Strength of the current;
- Resistance of the conductor;
- Time for which the current flows.

Mathematical expression for the heat produced is given by,

$$Q = VIt$$

$$= I^2Rt$$

$$= \frac{V^2t}{R}$$

The expression $U = VIt$ is called Joule and Lenz's law.

Magnetic Effect of Electric Current

In 1820, Oersted performed an experiment which demonstrated the magnetic effect of electric current. He placed a compass needle parallel to a straight conductor carrying current and found that the needle was deflected. On reversing the current, the needle deflected in the opposite direction. Thus, Oersted concluded that a magnetic field is created around a conductor carrying the electric current. This is known as the magnetic effect of electric current.

Instruments that work based on magnetic effect of electric current are galvanometer, ammeter, voltmeter, etc. The commonly used appliances such as electric fan, telephone, microphone, etc., have components that work on magnetic effect of the electric current.

The direction of the magnetic field due to a straight current-carrying conductor is given by:

- Right-hand clasp rule;
- Maxwell's cork screw rule;
- Ampere's swimming rule.

Mechanical Effect of an Electric Current

A moving charge or a current-carrying conductor placed in a magnetic field experiences a mechanical force. This is referred to as the mechanical effect of an electric current.

If a charge q moves at an angle θ with the direction of a magnetic field of induction B with a speed V, then the magnitude of the mechanical force on it is given by the equation $F = Bq\,V\,sin_{\theta}$.

Chemical Effect of an Electric Current

We know that in electric cells, chemical energy gets converted into electrical energy. The opposite activity is also possible. This means that electrical energy can be used to produce chemical reactions. Some chemical substances in a molten state or in their solution form allow the current to pass through them.

Some examples are sodium chloride solution, copper sulphate solution, sulphuric acid, etc. Such substances are called electrolytes. When an electric current is passed through such chemical compounds, they get decomposed into ions. This process is called electrolysis. Hence, electric current decomposes a few compounds in a molten state or in their aqueous form. This shows that the electric current produces chemical effects.

Electrical Units of Measure

The standard units of electrical measurement used for the expression of voltage, current and resistance are the Volt [V], Ampere [A] and Ohm [Ω] respectively.

These electrical units of measurement are based on the International (metric) System, also known as the SI System with other commonly used electrical units being derived from SI base units.

Sometimes in electrical or electronic circuits and systems it is necessary to use multiples or sub-multiples (fractions) of these standard electrical measuring units when the quantities being measured are very large or very small.

The following table gives a list of some of the standard electrical units of measure used in electrical formulas and component values.

Standard Electrical Units of Measure

Electrical Parameter	Measuring Unit	Symbol	Description
Voltage	Volt	V or E	Unit of Electrical Potential $V = I \times R$
Current	Ampere	I or i	Unit of Electrical Current $I = V \div R$
Resistance	Ohm	R or Ω	Unit of DC Resistance $R = V \div I$
Conductance	Siemen	G or ℧	Reciprocal of Resistance $G = 1 \div R$
Capacitance	Farad	C	Unit of Capacitance $C = Q \div V$

Charge	Coulomb	Q	Unit of Electrical Charge $Q = C \times V$
Inductance	Henry	L or H	Unit of Inductance $V_L = -L(di/dt)$
Power	Watts	W	Unit of Power $P = V \times I$ or $I^2 \times R$
Impedance	Ohm	Z	Unit of AC Resistance $Z^2 = R^2 + X^2$
Frequency	Hertz	Hz	Unit of Frequency $f = 1 \div T$

Multiples and Sub-multiples

There is a huge range of values encountered in electrical and electronic engineering between a maximum value and a minimum value of a standard electrical unit. For example, resistance can be lower than 0.01 Ω or higher than 1,000,000 Ω. By using multiples and submultiple's of the standard unit we can avoid having to write too many zero's to define the position of the decimal point. The table below gives their names and abbreviations.

Prefix	Symbol	Multiplier	Power of Ten
Terra	T	1,000,000,000,000	10^{12}
Giga	G	1,000,000,000	10^{9}
Mega	M	1,000,000	10^{6}
kilo	k	1,000	10^{3}
none	none	1	10^{0}
centi	c	1/100	10^{-2}
milli	m	1/1,000	10^{-3}
micro	µ	1/1,000,000	10^{-6}
nano	n	1/1,000,000,000	10^{-9}
pico	p	1/1,000,000,000,000	10^{-12}

So to display the units or multiples of units for either Resistance, Current or Voltage we would use as an example:

1kV = 1 kilo-volt – which is equal to 1,000 Volts.

1mA = 1 milli-amp – which is equal to one thousandths (1/1000) of an Ampere.

47kΩ = 47 kilo-ohms – which is equal to 47 thousand Ohms.

100uF = 100 micro-farads – which is equal to 100 millionths (100/1,000,000) of a Farad.

1kW = 1 kilo-watt – which is equal to 1,000 Watts.

1MHz = 1 mega-hertz – which is equal to one million Hertz.

To convert from one prefix to another it is necessary to either multiply or divide by the difference between the two values. For example, convert 1MHz into kHz.

Well we know from above that 1MHz is equal to one million (1,000,000) hertz and that 1kHz is equal to one thousand (1,000) hertz, so one 1MHz is one thousand times bigger than 1kHz. Then to convert Mega-hertz into Kilo-hertz we need to multiply mega-hertz by one thousand, as 1MHz is equal to 1000 kHz.

Likewise, if we needed to convert kilo-hertz into mega-hertz we would need to divide by one thousand. A much simpler and quicker method would be to move the decimal point either left or right depending upon whether you need to multiply or divide.

As well as the "Standard" electrical units of measure shown above, other units are also used in electrical engineering to denote other values and quantities such as:

- Wh – The Watt-Hour, The amount of electrical energy consumed by a circuit over a period of time. Eg, a light bulb consumes one hundred watts of electrical power for one hour. It is commonly used in the form of: Wh (watt-hours), kWh (Kilowatt-hour) which is 1,000 watt-hours or MWh (Megawatt-hour) which is 1,000,000 watt-hours.
- dB – The Decibel, The decibel is a one tenth unit of the Bel (symbol B) and is used to represent gain either in voltage, current or power. It is a logarithmic unit expressed in dB and is commonly used to represent the ratio of input to output in amplifier, audio circuits or loudspeaker systems.
- For example, the dB ratio of an input voltage (V_{IN}) to an output voltage (V_{OUT}) is expressed as $20\log_{10}(V_{out}/V_{in})$. The value in dB can be either positive (20dB) representing gain or negative (-20dB) representing loss with unity, ie input = output expressed as 0dB.
- θ – Phase Angle, The Phase Angle is the difference in degrees between the voltage waveform and the current waveform having the same periodic time. It is a time difference or time shift and depending upon the circuit element can have a "leading" or "lagging" value. The phase angle of a waveform is measured in degrees or radians.
- ω – Angular Frequency, Another unit which is mainly used in a.c. circuits to represent the Phasor Relationship between two or more waveforms is called Angular Frequency, symbol ω. This is a rotational unit of angular frequency $2\pi f$ with units in *radians per second*, rads/s. The complete revolution of one cycle is 360° or 2π, therefore, half a revolution is given as 180° or π rad.
- τ – Time Constant, The Time Constant of an impedance circuit or linear first-order system is the time it takes for the output to reach 63.7% of its maximum or minimum output value when subjected to a Step Response input. It is a measure of reaction time.

Electromotive Force

Electromotive force is the electrical potential energy that causes current to flow through a circuit. It results from the difference in charge between two points in a material. This difference is created when an external source such as a battery makes electrons move to both points so that there is an excess at the first point and a deficiency at the other. The volt is the unit that describes the energy that the source of electric power supplies that drives the unit charge within or around the circuit. Electromotive force was first discovered through the study of static electricity

The equation that represents electromotive force is E = V + Ir. In this equation, V is the terminal voltage, I is the circuit amperes, and r is the internal resistance measured in ohms. By definition, a terminal's total voltage is less than the circuit's total electromotive force.

Interpretation

Consider a region of space in which an electromagnetic (e/m) field exists. In the most general sense, any *closed* path *C* (*or loop*) within this region will be called a "circuit" (whether or not the whole or parts of *C* consist of material objects such as wires, resistors, capacitors, batteries, or any other elements whose presence may contribute to the e/m field).

We arbitrarily assign a positive direction of traversing the loop *C*, and we consider an element $\overline{dl}$ of *C* oriented in the positive direction. Imagine now a test charge q located at the position of $\overline{dl}$, and let $\overline{F}$ be the force on *q* at time *t* :

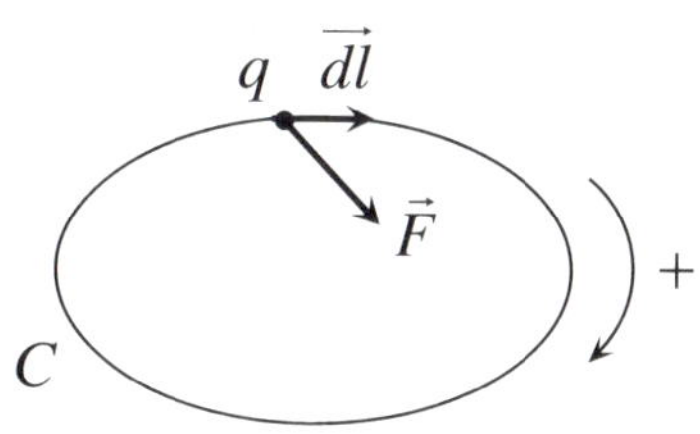

This force is exerted by the e/m field itself, as well as, possibly, by additional energy sources (e.g., batteries) that can interact electrically with *q*. The force per unit charge at the position of $\overline{dl}$ at time *t*, is,

$$\overline{f} = \frac{\overline{F}}{q}$$

Note that $\overline{f}$ is independent of q, since the force by the e/m field and the sources on q is proportional to the charge. In particular, reversing the sign of q will have no effect on $\overline{f}$ (although it will change the direction of $\overline{F}$).

We now define the electromotive force (*emf*) of the circuit C at time *t* as the line integral of along C, taken in the positive sense of C :

$$\varepsilon = \oint_C \overline{f}.\overline{dl}$$

Note that the sign of the emf is dependent upon our choice of the positive direction of circulation of C : by changing this convention, the sign of ε is reversed.

We remark that, in the non-relativistic limit, the emf of a circuit C is the same for all inertial observers since at this limit the force $\vec{F}$ is invariant under a change of frame of reference.

Emf of a Circuit Containing a Battery and a Resistor

Consider a circuit consisting of an ideal battery (i.e., one with no internal resistance) connected to an external resistor. As shown below, the emf of the circuit in the direction of the current is equal to the voltage V of the battery. Moreover, the emf in this case represents the work per unit charge done by the source (battery).

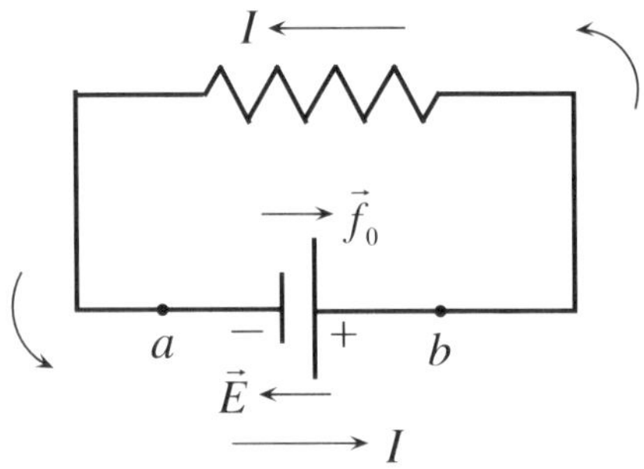

We recall that, in general, the emf of a circuit C at time t is equal to the integral,

$$\varepsilon = \oint_c \vec{f}.\vec{dl}$$

Where, $\vec{f} = \vec{F}/q$ is the force per unit charge at the location of the element $\vec{dl}$ of the circuit, at time t. In essence, we assume that in every element $\vec{dl}$ we have placed a test charge q (this could be, e.g., a free electron of the conducting part of the circuit). The force $\vec{F}$ on each q is then measured simultaneously for all charges at time t. Since here we are dealing with a static (time-independent) situation, however, we can treat the problem somewhat differently: The measurements of the forces $\vec{F}$ on the charges q need not be made at the same instant, given that nothing changes with time, anyway. So, instead of placing several charges q around the circuit and measuring the forces $\vec{F}$ on each of them at a particular instant, we imagine a single charge q making a complete tour around the loop C. We may assume, e.g., that the charge q is one of the (conventionally positive) free electrons taking part in the constant current I flowing in the circuit. We then measure the force $\vec{F}$ on q at each point of C.

We thus assume that q is a positive charge moving in the direction of the current I. We also assume that the direction of circulation of C is the same as the direction of the current (counterclockwise in the figure). During its motion, q is subject to two forces:

(1) the force $\vec{F}_0$ by the source (battery) that carries q from the negative pole a to the positive pole b through the source;

(2) the electrostatic force $\vec{F}_e = q\vec{F}$ due to the electrostatic field $\vec{E}$ at each point of the circuit C (both inside and outside the source).

The total force on q is,

$$\vec{F} = \vec{F_0} + \vec{F_e} = \vec{F_0} + q\vec{E} \Rightarrow \vec{f} = \frac{\vec{F}}{q} = \frac{\vec{F_0}}{q} + \vec{E} \equiv \vec{f}_0 + \vec{E}$$

Then,

$$\varepsilon = \oint_c \vec{f}.\vec{dl} = \oint_c \vec{f_0}.\vec{dl} + \oint_c \vec{E}.\vec{dl} = \oint_c \vec{f_0}.\vec{dl}$$

since, $\oint_c \vec{E}.\vec{dl} = 0$ for an electrostatic field. However, the action of the source on q is limited to the region between the poles of the battery, that is, the section of the circuit from a to b. Hence, $\vec{f_0} = 0$ outside the source, so that above equation reduces to,

$$\varepsilon = \int_a^b \vec{F_0}.\vec{dl}$$

Now, since the current I is constant, the charge q moves at constant speed along the circuit. This means that the total force on q in the direction of the path C is zero. In the interior of the resistor, the electrostatic force $\vec{F_e} = q\vec{E}$ is counterbalanced by the force on q due to the collisions of the charge with the positive ions of the metal. In the interior of the (ideal) battery, however, where there is no resistance, the electrostatic force $\vec{F_e}$ must be counterbalanced by the opposing force $\vec{f_0}$ exerted by the source.

Thus, in the section of the circuit between a and b,

$$\vec{F} = \vec{F_0} + \vec{F_e} = 0 \Rightarrow \vec{f} = \frac{\vec{F}}{q} = \vec{f}_0 + \vec{E} = 0 \Rightarrow \vec{f}_0 = -\vec{E}$$

Equation above then takes the final form,

$$\varepsilon = -\int_a^b \vec{E}.\vec{dl} = V_b - V_a = V$$

where, V_a and V_b are the electrostatic potentials at *a* and *b*, respectively. This is, of course, what every student knows from elementary *e/m courses*.

The work done by the source on *q* upon transferring the charge from *a* to *b* is,

$$W = \int_a^b \vec{F}_0.\vec{dl} = q\int_a^b \vec{f}_0.\vec{dl} = q\varepsilon$$

So, the work of the source per unit charge is $W/q = \varepsilon$. This work is converted into heat in the resistor, so that the source must again supply energy in order to carry the charges once more from *a* to *b*.

Emf and Ohm's Law

Consider a closed wire C inside an e/m field. The circuit may contain sources (e.g., a battery) and may also be in motion relative to our inertial frame of reference. Let q be a test charge at the location of the element $\vec{dl}$ of C, and let $\vec{F}$ be the total force on q (due to the e/m field and/or the sources) at time t. The force per unit charge at the location of $\vec{dl}$ at time *t*, then, is f $\vec{f} = \vec{F}/q$ According to our general definition, the emf of the circuit at time t is,

$$\varepsilon = \oint_c \vec{f}.\vec{dl}$$

Now, if σ is the conductivity of the wire, then, by Ohm's law in its general form we have:

$$\vec{J} = \sigma \vec{f}$$

where, $\vec{J}$ is the volume current density at the location of $\vec{dl}$ at time t. (Note that the more common expression $\vec{J} = \sigma \vec{E}$, found in most textbooks, is a special case of th above formula. Note also that $\vec{J}$ is measured relative to the wire, thus is the same for all inertial observers.) By combining the above two equations we get:

$$\varepsilon = \frac{1}{\sigma}\oint_c \vec{J}.\vec{dl}$$

Taking into account that $\vec{J}$ is in the direction of $\vec{dl}$ at each point of *C*, we write:

$$\vec{J}.\vec{dl} = J\,dl = \frac{I}{S}dl$$

where, *S* is the constant cross-sectional area of the wire. If we make the additional assumption that, at each instant *t*, the current *I* is constant around the circuit (although *I* may vary with time), we finally get:

$$\varepsilon = \frac{1}{\sigma S}I = \frac{\rho l}{S}I = IR$$

where, l is the total length of the wire, $\rho = 1/\sigma$ is the resistivity of the material, and *R* is the total resistance of the circuit. Equation above is the familiar special form of Ohm's law.

Sources of Electromotive Force

A few of the Sources of electromotive force include electrochemical cells, thermoelectric devices, solar cells, electrical generators, transformers, etc.

Electrochemical Cells

An electrochemical cell is a device that produces an electric current from energy released by a spontaneous redox reaction. This kind of cell includes the galvanic, or voltaic, cell, named after Luigi Galvani and Alessandro Volta. These scientists conducted several experiments on chemical reactions and electric current during the late 18th century.

Electrochemical cells have two conductive electrodes, called the anode and the cathode. The anode is defined as the electrode where oxidation occurs. The cathode is the electrode where reduction takes place. Electrodes can be made from any sufficiently conductive materials, such as metals, semiconductors, graphite, and even conductive polymers. In between these electrodes is the electrolyte, which contains ions that can freely move.

The voltaic cell uses two different metal electrodes, each in an electrolyte solution. The anode will undergo oxidation and the cathode will undergo reduction. The metal of the anode will oxidize, going from an oxidation state of 0 (in the solid form) to a positive oxidation state, and it will become an ion. At the cathode, the metal ion in the solution will accept one or more electrons from the cathode, and the ion's oxidation state will reduce to 0. This forms a solid metal that deposits on the cathode. The two electrodes must be electrically connected to each other, allowing for a flow of electrons that leave the metal of the anode and flow through this connection to the ions at the surface of the cathode. This flow of electrons is an electrical current that can be used to do work, such as turn a motor or power a light.

Daniel Cell

An electrochemical cell of zinc and copper metals is known as Daniel cell. It is represented as below:

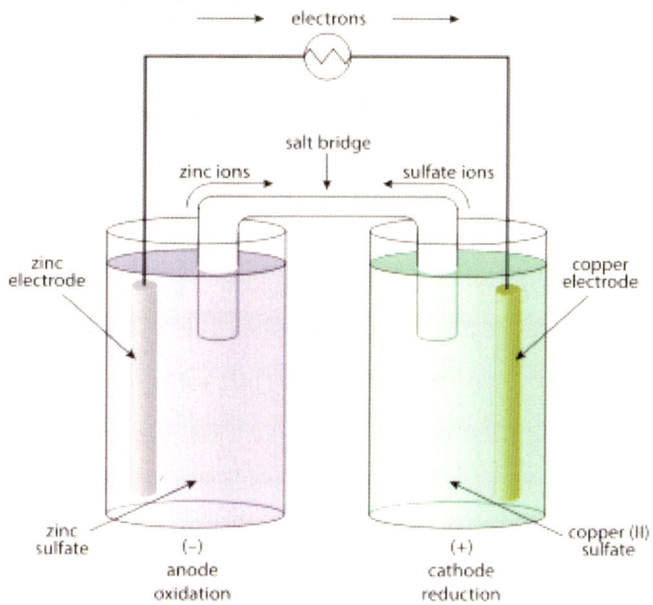

By convention, the Cathode is represented on the RHS and Anode on the LHS.

Functions of Salt Bridge

- Salt bridge allows the flow of current by completing the circuit.
- It assures electrical neutrality on both sides.
- The salt bridge usually comprises of the solution of strong electrolyte known as KNO_3, KCL etc. In most cases, KCI is preferred because the transport numbers of $K^+$and Cl^- are almost same.

Transport Number or Transference Number

The current that passes through an electrolytic solution is carried via the ions. Now, the degree or the fraction of the current that is carried by the ion is known as the transport number or transference number. So,

Transport number of the cation n_c = (current carried by cation/total current)

Transport number of the anion n_a = (current carried by anion/total current)

Evidently $n_c + n_a = 1$

Electrolytic Cells

Electrolysis uses electrical energy to induce a chemical reaction, which then takes place in an electrolytic cell.

In chemistry and manufacturing, electrolysis is a method of using a direct electric current (DC) to drive an otherwise non-spontaneous chemical reaction. Electrolysis is commercially important as a stage in the process of separating elements from naturally occurring sources such as ore.

Electrolysis is the passage of a direct electric current through an ionic substance that is either molten or dissolved in a suitable solvent, resulting in chemical reactions at the electrodes and separation of the materials.

Electrolysis can sometimes be thought of as running a non-spontaneous galvanic cell. Depending on how freely elements give up electrons (oxidation) and how energetically favorable it is for elements to receive electrons (reduction), the reaction may not be spontaneous. By externally supplying the energy to overcome the energy barrier to spontaneous reaction, the desired reaction is "allowed" to run under special circumstances.

The main components required to perform electrolysis are:

- An electrolyte: A substance containing free ions that carry electric current. If the ions are not mobile, as in a solid salt, then electrolysis cannot occur.
- A direct current (DC) supply: Provides the energy necessary to create or discharge the ions in the electrolyte. Electric current is carried by electrons in the external circuit.
- Two electrodes: An electrical conductor that provides the physical interface between the electrical circuit providing the energy and the electrolyte.
- A typical electrolysis cell: A cell used in elementary chemical experiments to produce gas as a reaction product and to measure its volume.

Electrodes of metal, graphite, and semiconductor material are widely used. Choosing a suitable electrode depends on the chemical reactivity between the electrode and electrolyte, and the cost of manufacture.

Other systems that utilize the electrolytic process are used to produce metallic sodium and potassium, chlorine gas, sodium hydroxide, and potassium and sodium chlorate.

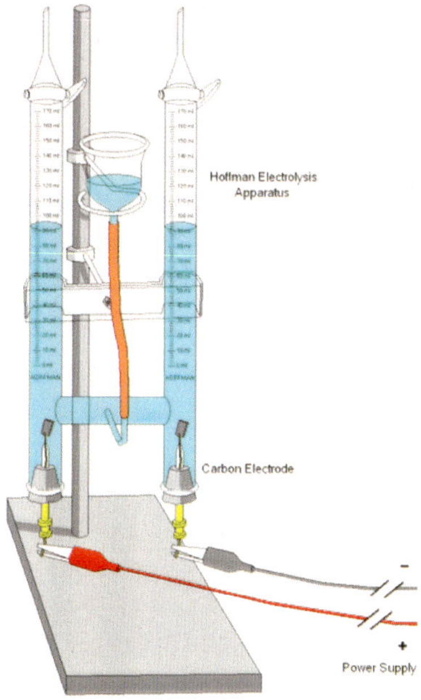

Electrical Generators

Electric generator is a machine that converts mechanical energy to electricity for transmission and distribution over power lines to domestic, commercial, and industrial customers. It is also called dynamo. Generators also produce the electrical power required for automobiles, aircraft, ships, and trains.

The mechanical power for an electric generator is usually obtained from a rotating shaft and is equal to the shaft torque multiplied by the rotational, or angular, velocity. The mechanical power may come from a number of sources: hydraulic turbines at dams or waterfalls; wind turbines; steam turbines using steam produced with heat from the combustion of fossil fuels or from nuclear fission; gas turbines burning gas directly in the turbine; or gasoline and diesel engines. The construction and the speed of the generator may vary considerably depending on the characteristics of the mechanical prime mover.

Nearly all generators used to supply electric power networks generate alternating current, which reverses polarity at a fixed frequency (usually 50 or 60 cycles, or double reversals, per second). Since a number of generators are connected into a power network, they must operate at the same frequency for simultaneous generation. They are therefore known as synchronous generators or, in some contexts, alternators.

Main Components of a Generator

The main components of an electric generator can be broadly classified as follows:

(1) Engine;

(2) Alternator;

(3) Fuel System;

(4) Voltage Regulator;

(5) Cooling and Exhaust Systems;

(6) Lubrication System;

(7) Battery Charger;

(8) Control Panel;

(9) Main Assembly/Frame.

A description of the main components of a generator is given below:

Engine

The engine is the source of the input mechanical energy to the generator. The size of the engine is directly proportional to the maximum power output the generator can supply. There are several factors that you need to keep in mind while assessing the engine of your generator. The manufacturer of the engine should be consulted to obtain full engine operation specifications and maintenance schedules.

- Type of fuel used: Generator engines operate on a variety of fuels such as diesel, gasoline, propane (in liquefied or gaseous form), or natural gas. Smaller engines usually operate on gasoline while larger engines run on diesel, liquid propane, propane gas, or natural gas. Certain engines can also operate on a dual feed of both diesel and gas in a bi-fuel operation mode.
- Overhead Valve (OHV) Engines versus non-OHV Engines: OHV engines differ from other engines in that the intake and exhaust valves of the engine are located in the head of the engine's cylinder as opposed to being mounted on the engine block. OHV engines have several advantages over other engines such as:
    - Compact design
    - Simpler operation mechanism

    - Durability
    - User-friendly in operations
    - Low noise during operations
    - Low emission levels

    However, OHV-engines are also more expensive than other engines.

- Cast Iron Sleeve (CIS) in engine cylinder: The CIS is a lining in the cylinder of the engine. It reduces wear and tear, and ensures durability of the engine. Most OHV-engines are equipped with CIS but it is essential to check for this feature in the engine of a generator. The CIS is not an expensive feature but it plays an important role in engine durability especially if you need to use your generator often or for long durations.

Alternator

The alternator, also known as the 'genhead', is the part of the generator that produces the electrical output from the mechanical input supplied by the engine. It contains an assembly of stationary and moving parts encased in a housing. The components work together to cause relative movement between the magnetic and electric fields, which in turn generates electricity.

- Stator: This is the stationary component. It contains a set of electrical conductors wound in coils over an iron core.
- Rotor/Armature: This is the moving component that produces a rotating magnetic field in any one of the following three ways:
    - By induction: These are known as brushless alternators and are usually used in large generators.
    - By permanent magnets: This is common in small alternator units.
    - By using an exciter: An exciter is a small source of direct current (DC) that energizes the rotor through an assembly of conducting slip rings and brushes.

The rotor generates a moving magnetic field around the stator, which induces a voltage difference between the windings of the stator. This produces the alternating current (AC) output of the generator.

The following are the factors that you need to keep in mind while assessing the alternator of a generator:

- Metal versus Plastic Housing: An all-metal design ensures durability of the alternator. Plastic housings get deformed with time and cause the moving parts of the alternator to be exposed. This increases wear and tear and more importantly, is hazardous to the user.
- Ball Bearings versus Needle Bearings: Ball bearings are preferred and last longer.
- Brushless Design: An alternator that does not use brushes requires less maintenance and also produces cleaner power.

Fuel System

The fuel tank usually has sufficient capacity to keep the generator operational for 6 to 8 hours on an average. In the case of small generator units, the fuel tank is a part of the generator's skid base or is mounted on top of the generator frame. For commercial applications, it may be necessary to erect and install an external fuel tank. All such installations are subject to the approval of the City Planning Division. Common features of the fuel system include the following:

- Pipe connection from fuel tank to engine: The supply line directs fuel from the tank to the engine and the return line directs fuel from the engine to the tank.
- Ventilation pipe for fuel tank: The fuel tank has a ventilation pipe to prevent the build-up of pressure or vacuum during refilling and drainage of the tank. When you refill the fuel tank, ensure metal-to-metal contact between the filler nozzle and the fuel tank to avoid sparks.
- Overflow connection from fuel tank to the drain pipe: This is required so that any overflow during refilling of the tank does not cause spillage of the liquid on the generator set.
- Fuel pump: This transfers fuel from the main storage tank to the day tank. The fuel pump is typically electrically operated.
- Fuel Water Separator/Fuel Filter: This separates water and foreign matter from the liquid fuel to protect other components of the generator from corrosion and contamination.
- Fuel Injector: This atomizes the liquid fuel and sprays the required amount of fuel into the combustion chamber of the engine.

Voltage Regulator

As the name implies, this component regulates the output voltage of the generator. The mechanism is described below against each component that plays a part in the cyclical process of voltage regulation.

- Voltage Regulator: Conversion of AC Voltage to DC Current – The voltage regulator takes up a small portion of the generator's output of AC voltage and converts it into DC current. The voltage regulator then feeds this DC current to a set of secondary windings in the stator, known as exciter windings.
- Exciter Windings: Conversion of DC Current to AC Current – The exciter windings now function similar to the primary stator windings and generate a small AC current. The exciter windings are connected to units known as rotating rectifiers.
- Rotating Rectifiers: Conversion of AC Current to DC Current – These rectify the AC current generated by the exciter windings and convert it to DC current. This DC current is fed to the rotor/armature to create an electromagnetic field in addition to the rotating magnetic field of the rotor/armature.
- Rotor/Armature: Conversion of DC Current to AC Voltage – The rotor/armature now induces a larger AC voltage across the windings of the stator, which the generator now produces as a larger output AC voltage.

This cycle continues till the generator begins to produce output voltage equivalent to its full operating capacity. As the output of the generator increases, the voltage regulator produces less DC current. Once the generator reaches full operating capacity, the voltage regulator attains a state of equilibrium and produces just enough DC current to maintain the generator's output at full operating level.

When you add a load to a generator, its output voltage dips a little. This prompts the voltage regulator into action and the above cycle begins. The cycle continues till the generator output ramps up to its original full operating capacity.

Cooling and Exhaust Systems

(a) Cooling System

Continuous usage of the generator causes its various components to get heated up. It is essential to have a cooling and ventilation system to withdraw heat produced in the process.

Raw/fresh water is sometimes used as a coolant for generators, but these are mostly limited to specific situations like small generators in city applications or very large units over 2250 kW and above. Hydrogen is sometimes used as a coolant for the stator windings of large generator units since it is more efficient at absorbing heat than other coolants. Hydrogen removes heat from the generator and transfers it through a heat exchanger into a secondary cooling circuit that contains de-mineralized water as a coolant. This is why very large generators and small power plants often have large cooling towers next to them. For all other common applications, both residential and industrial, a standard radiator and fan is mounted on the generator and works as the primary cooling system.

It is essential to check the coolant levels of the generator on a daily basis. The cooling system and raw water pump should be flushed after every 600 hours and the heat exchanger should be cleaned after every 2,400 hours of generator operation. The generator should be placed in an open and ventilated area that has adequate supply of fresh air. The National Electric Code (NEC) mandates that a minimum space of 3 feet should be allowed on all sides of the generator to ensure free flow of cooling air.

(b) Exhaust System

Exhaust fumes emitted by a generator are just like exhaust from any other diesel or gasonline engine and contain highly toxic chemicals that need to be properly managed. Hence, it is essential to install an adequate exhaust system to dispose of the exhaust gases. This point can not be emphasized enough as carbon monoxide poisoning remains one of the most common causes for death in post hurricane affected areas because people tend to not even think about it until it's too late.

Exhaust pipes are usually made of cast iron, wrought iron, or steel. These need to be freestanding and should not be supported by the engine of the generator. Exhaust pipes are usually attached to the engine using flexible connectors to minimize vibrations and prevent damage to the generator's exhaust system. The exhaust pipe terminates outdoors and leads away from doors, windows and other openings to the house or building. You must ensure that the exhaust system of your generator is not connected to that of any other equipment. You should also consult the local city ordinances to determine whether your generator operation will need to obtain an approval from the local authorities to ensure you are conforming to local laws a protect against fines and other penalties.

Lubricating System

Since the generator comprises moving parts in its engine, it requires lubrication to ensure durability and smooth operations for a long period of time. The generator's engine is lubricated by oil stored in a pump. You should check the level of lubricating oil every 8 hours of generator operation. You should also check for any leakages of lubricant and change the lubricating oil every 500 hours of generator operation.

Battery Charger

The start function of a generator is battery-operated. The battery charger keeps the generator battery charged by supplying it with a precise 'float' voltage. If the float voltage is very low, the battery will remain undercharged. If the float voltage is very high, it will shorten the life of the battery. Battery chargers are usually made of stainless steel to prevent corrosion. They are also fully automatic and do not require any adjustments to be made or any settings to be changed. The DC output voltage of the battery charger is set at 2.33 Volts per cell, which is the precise float voltage for lead acid batteries. The battery charger has an isolated DC voltage output that does interfere with the normal functioning of the generator.

Control Panel

This is the user interface of the generator and contains provisions for electrical outlets and controls. Different manufacturers have varied features to offer in the control panels of their units. Some of these are mentioned below:

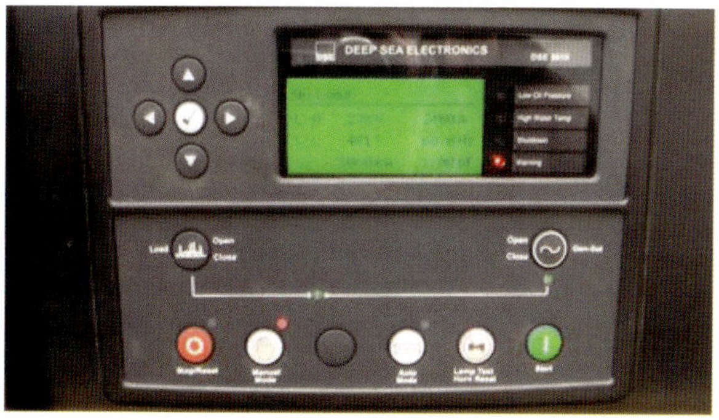

- Electric start and shut-down: Auto start control panels automatically start your generator during a power outage, monitor the generator while in operation, and automatically shut down the unit when no longer required.
- Engine gauges: Different gauges indicate important parameters such as oil pressure, temperature of coolant, battery voltage, engine rotation speed, and duration of operation. Constant measurement and monitoring of these parameters enables built-in shut down of the generator when any of these cross their respective threshold levels.
- Generator gauges: The control panel also has meters for the measurement of output current and voltage, and operating frequency.
- Other controls: Phase selector switch, frequency switch, and engine control switch (manual mode, auto mode) among others.

Main Assembly/Frame

All generators, portable or stationary, have customized housings that provide a structural base support. The frame also allows for the generated to be earthed for safety.

Working of Generators

Generators are basically coils of electric conductors, normally copper wire, that are tightly wound onto a metal core and are mounted to turn around inside an exhibit of large magnets. An electric conductor moves through a magnetic field, the magnetism will interface with the electrons in the conductor to induce a flow of electrical current inside it.

The conductor coil and its core are called the armature, connecting the armature to the shaft of a mechanical power source, for example an motor, the copper conductor can turn at exceptionally increased speed over the magnetic field.

The point when the generator armature first starts to turn, then there is a weak magnetic field in the iron pole shoes. As the armature turns, it starts to raise voltage. Some of this voltage is making on the field windings through the generator regulator. This impressed voltage builds up stronger winding current, raises the strength of the magnetic field. The expanded field produces more voltage in the armature. This, in turn, make more current in the field windings, with a resultant higher armature voltage. At this time the signs of the shoes depended on the direction of flow of current in the field winding. The opposite signs will give current to flow in wrong direction.

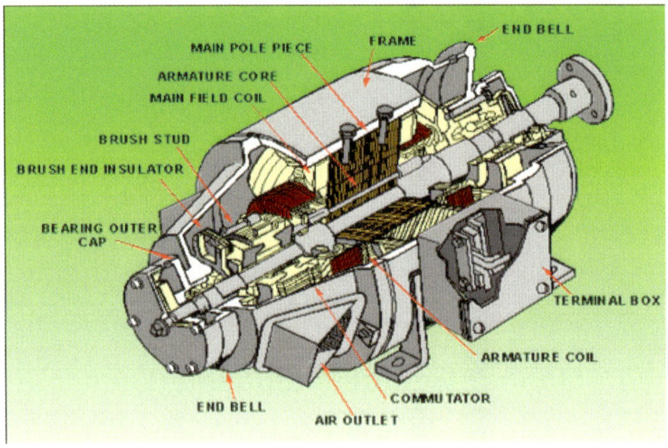

Types of Generators

The generators are classified into types:

- AC generators
- DC generators

AC Generators

These are also called as alternators. It is the most important means of producing electrical power in many of the places since now days all the consumers are using AC. It works based on principle of the electromagnetic induction. These are of two types one is induction generator and other one is synchronous generator. The induction generator requires no separate DC excitation, regulator controls, frequency control or governor. This concept takes place when conductor coils turn in a magnetic field actuating a current and a voltage. The generators should run at a consistent speed to convey a stable AC voltage, even no load is accessible.

Synchronous generators are large size generators mainly used in power plants. These may be rotating field type or rotating armature type. In rotating armature type, armature is at rotor and field is at stator. Rotor armature current is taken through slip rings and brushes. These are limited due to high wind losses. These are used for low power output applications. Rotating field type of alternator is widely used because of high power generation capability and absence of slip rings and brushes.

It can be either 3 phase or two phase generators. A two-phase alternator produces two completely separate voltages. Each voltage may be considered as a single-phase voltage. Each is generated voltage completely independent of the other. The three-phase alternator has three single-phase windings spaced such that the voltage induced in any one phase is displaced by 120° from the other two. These can be connected either delta or wye connections. In Delta Connection each coil end is connected together to form a closed loop. A Delta Connection appears like the Greek Letter Delta (Δ). In Wye Connection one end of each coil connected together and the other end of each coil left open for external connections. A Wye Connection appears as the letter Y.

These generators are packaged with an engine or turbine to be used as a motor-generator set and used in applications like naval, oil and gas extraction, mining machinery, wind power plants etc.

Advantages of AC Generator:

- These Generators are generally maintenance free, because of absence of brushes.
- Easily step up and step down through transformers.
- Transmission link size might be thinner because of step up feature.
- Size of the generator relatively smaller than DC machine.
- Losses are relatively less than DC machine.
- These Generator breakers are relatively smaller than DC breakers.

DC Generators

DC generator is typically found in off-grid applications. These generators give a seamless power supply directly into electric storage devices and DC power grids without novel equipment. The stored power is carries to loads through dc-ac converters. The DC generators could be controlled back to an unmoving speed as batteries tend to be stimulating to recover considerably more fuel.

Classification of DC Generators

D.C Generators are classified according to the way their magnetic field is developed in the stator of the machine:

- Permanent-magnet DC generators
- Separately-excite DC generators
- Self-excited DC generators

Permanent magnet DC generators do not require external field excitation because it has permanent magnets to produce the flux. These are used for low power applications like dynamos. Separately-excite DC generators requires external field excitation to produce the magnetic flux. We can also vary the excitation to get variable output power. These are used in electro plating and electro refining applications. Due to residual magnetism present in the poles of the stator self-excited DC generators can able to produce their own magnetic field ones it is started. These are simple in design and no need to have the external circuit to vary the field excitation. Again these self-excited DC generators are classified into shunt, series, and compound generators.

These are used in applications like battery charging, welding, ordinary lightening applications etc.

Advantages of DC Generator

- Mainly DC machines have the wide variety of operating characteristics which can be obtained by selection of the method of excitation of the field windings.
- The output voltage can be smoothed by regularly arranging the coils around the armature. This leads to less fluctations which is desirable for some steady state applications.
- No shielding need for radiation so cable cost will be less as compared to AC.

Thermoelectric Devices

Thermoelectricity is the direct conversion of temperature gradient to electrical potential difference. phenomenon is described by three effects. The Seebeck effect-discovered by Thomas Seebeck- in 1821 is the conversion of temperature difference into electrical voltage. The Peltier effect -discovered by Jean Peltier in 1934- is the conversion of voltage into temperature gradient. And the Thomson effect–discovered by William Thomson in 1851- is the heat flow in a conductor with terminals held at different temperatures, due to current flow.

Thermoelectric devices can be used for heating/cooling purposes, or for power generation. The applications for thermoelectric devices can span different areas and industries. Waste-heat energy harvesting in automotives is one promising application for such technology. Remote space missions use thermoelectric devices for power generation.

The main challenge for thermoelectric devices' design is the device performance, and the compromise between material parameters to increase it. The attempts to increase the performance of thermoelectric materials and devices are many, from material parameters optimization, to using going to new material structures, and low-dimensional thermoelement.

Thermoelectric Effects

Thermoelectricity is described by three thermoelectric effects, the "Seebeck Effect" is the conversion of temperature gradient across the junctions of two dissimilar metals to electrical voltage in the range of millivolts per Kelvin difference. The effect is non-linear with temperature, and depends on absolute temperature, type and structure of materials. The Seebeck coefficient S is the amount of voltage difference ΔV generated for an applied temperature difference ΔT.

$$S = \Delta V / \Delta T$$

The voltage difference V, then can be calculated using the following equation,

$$V = \int_{T_c}^{T_h} \left(S_B(T) - S_A(T) \right) dT$$

The "Peltier Effect" is that a current flow causes a temperature gradient across the junctions of two dissimilar metals. The heat transfer is in the same direction of charge carriers. The thermal current density q is given by the following equation, where $\prod$ and j are the Peltier coefficient, and electrical current density.

$$q = \prod j$$

The "Thomson Effect" is the heat flow across a conductor, with terminals at different temperatures, due to current flow. The heat flow is given by:

$$\frac{dQ}{dx} = \mu I \frac{dT}{dx}$$

The Thomson effect is the only measurable effect among the three thermoelectric effects, for a certain material, since the two other effects are related to pairs of materials.

Thomson/Kelvin relationship described in the following equations relates the three thermoelectric effects.

$$\Pi = ST$$

$$\mu = T\frac{dS}{dT}$$

Π, S, T, μ are the Peltier coefficient, the Seebeck coefficient, the absolute temperature, and the Thomson coefficient.

Thermoelectric Figure of Merit

The performance of a thermoelectric device is indicated by the figure of merit (FOM) *ZT*, it is a dimensionless quantity, which is given by:

$$ZT = \frac{GS^2T}{K_l + K_e}$$

G, k_l, K_e and T are electrical conductivity, lattice thermal conductivity, electronic thermal conductivity, and absolute average temperature.

For higher *ZT*, a large Seebeck coefficient and electrical conductivity, and small thermal conductivity are required. Thermal conductivity decreases the FOM, since it leads to undesired thermal exchange between hot and cold sides of a thermoelectric device.

The compromise to increase *ZT* is shown in figure below, Increasing the carrier concentration in order to increase electrical conductivity, also gives rise to thermal conductivity, and the Seebeck coefficient starts to decrease with higher thermal conductivity, due to lack of the ability of the material to maintain higher temperature gradient.

Bi_2Te_3 Is the most commonly used semiconductor material because of its relatively high FOM ZT=1 at 300° K.

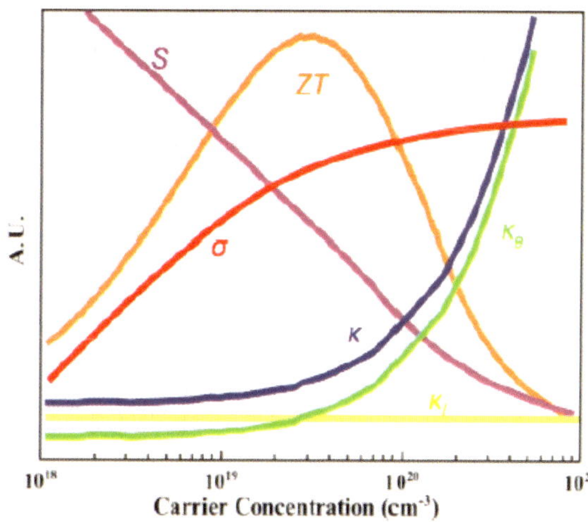

Shows the FOM for some p-type semiconductor materials

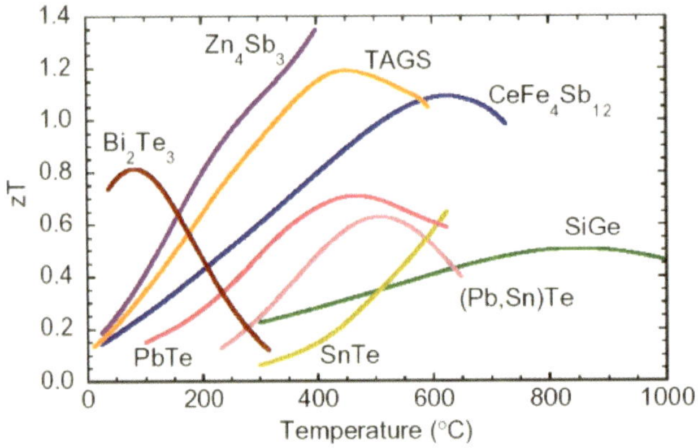

FOM for p-type materials

Solar Cells

Solar cell or photovoltaic cell is a device which converts the energy of light into electrical energy through the photovoltaic effect. The overwhelming majority of solar cells are fabricated from silicon—with increasing efficiency and lowering cost as the materials range from amorphous (noncrystalline) to polycrystalline to crystalline (single crystal) silicon forms. Unlike batteries or fuel cells, solar cells do not utilize chemical reactions or require fuel to produce electric power, and, unlike electric generators, they do not have any moving parts.

Solar cells can be arranged into large groupings called arrays. These arrays, composed of many thousands of individual cells, can function as central electric power stations, converting sunlight into electrical energy for distribution to industrial, commercial, and residential users. Solar cells in much smaller configurations, commonly referred to as solar cell panels or simply solar panels, have been installed by homeowners on their rooftops to replace or augment their conventional electric supply. Solar cell panels also are used to provide electric power in many remote terrestrial locations where conventional electric power sources are either unavailable or prohibitively expensive to install. Because they have no moving parts that could need maintenance or fuels that would require replenishment, solar cells provide power for most space installations, from communications and weather satellites to space stations. (Solar power is insufficient for space probes sent to the outer planets of the solar system or into interstellar space, however, because of the diffusion of radiant energy with distance from the Sun.) Solar cells have also been used in consumer products, such as electronic toys, handheld calculators, and portable radios. Solar cells used in devices of this kind may utilize artificial light (e.g., from incandescent and fluorescent lamps) as well as sunlight.

Working of Solar Cell

A solar cell is a sandwich of n-type silicon (blue) and p-type silicon (red). It generates electricity by using sunlight to make electrons hop across the junction between the different flavors of silicon:

1. When sunlight shines on the cell, photons (light particles) bombard the upper surface.
2. The photons (yellow blobs) carry their energy down through the cell.
3. The photons give up their energy to electrons (green blobs) in the lower, p-type layer.

4. The electrons use this energy to jump across the barrier into the upper, n-type layer and escape out into the circuit.

5. Flowing around the circuit, the electrons make the lamp light up.

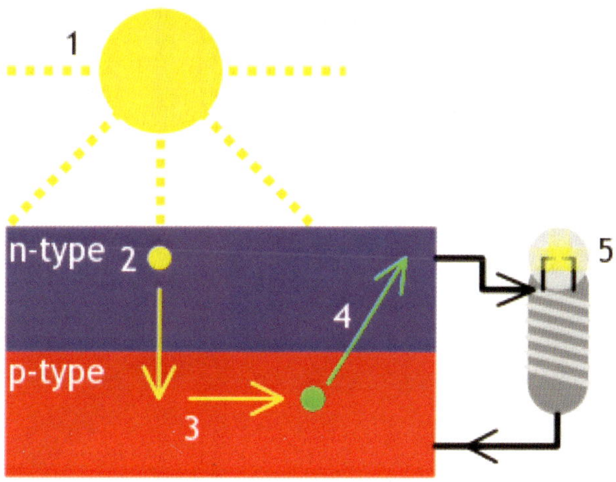

Advantages

- Renewable energy: The energy can be used both to generate electricity and heat in the house. Renewable energy is recovered from the sun, the wind and waves - which in this case is the sun. Solar cells harness the energy from the sun and transform this into usable electricity.

- Economy-friendly energy: Solar cells provide a great opportunity to create savings on your electric bill since you do not pay for the energy that you generate. At the same time you have the opportunity to monetize your photovoltaic system, if you have a solar system with a grid connected installation, you can buy and sell electricity to the collective electricity network. At the same time, you can obtain several grants for solar cells and there will be more economic benefits to be gained in the future.

- Environmentally friendly energy: With solar cells occurs almost no pollution. The discharge of waste and pollution is unavoidable in relation to the production of solar cells, the transport of these and when you install them. However, this is a minimal fraction, compared to if one draws its energy from elsewhere.

- Innovative energy: Photovoltaics is a popular topic in green energy and is considered to be a good solution to prevent climate change. Therefore, this is an innovative market under continuous research and development.

- Infinite Energy: When you have the opportunity to extract energy from the sun's rays, this is a source of energy that will never be exhausted, therefore there will always be a source for electricity production.

- Long term energy: PV systems often have a long life and a good durability. At the same time, there is often a guarantee of minimum 20 years on your solar panels, guaranteeing you, should there be any complications.

- Selling energy: If your home has solar cells, it is often easier to sell the property at a higher price.

Disadvantages

- Interior needs: Not all households that can satisfy their requirements and get the optimum out of their solar cells yet. Solar cells are very sensitive in terms of their location, which means that if there is shade on your lot, it is difficult to exploit solar installation optimally. The solution to this is that you can be connected to the grid and hence can buy energy from others.
- High investment: One-time cost of acquiring a photovoltaic system and have it installed are relatively high. However, one must bear in mind that producing energy then, is free. The solution to this drawback is that by most banks provide the opportunity to take an energy lending, which gives a low interest to you as a customer who invests in green energy.
- Seasonal energy: Compared to other types of renewable energy, the solar power plant is highly seasonal, since we can have periods of limited sun in the UK. The solution to this is to grid connect solar installations and purchase energy from the public electricity network during periods where there is less energy to collect. Investing in a solar battery storage system is also a good choice, since it can store the energy generated during peak hours and make it readily available for cloudy and rainy days.
- Solar cells on your accommodation: It might be harder to install solar panels on older households, as they often have different designs that can provide shade. At the same time flat roofs where drifting snow may fall below the racks, becomes too heavy for a roof with solar cells. Therefore it is important that you inquire about these things when you obtain offers.

References

- Electric-field, electromagnetics, electrostatics: physics-and-radio-electronics.com, Retrieved 15 March 2018
- Electric-current-and-theory-of-electricity: electrical4u.com, Retrieved 20 May 2018
- Electric-generator, technology: britannica.com, Retrieved 12 July 2018
- How-Generators-Work: dieselserviceandsupply.com, Retrieved 19 May 2018
- Working-of-generators: elprocus.com, Retrieved 27 March 2018
- Solar-cell, technology: britannica.com, Retrieved 20 July 2018
- Advantages-and-disadvantages-of-solar-cells: greenmatch.co.uk, Retrieved 14 April 2018

2

Electrical Circuits: Analysis Techniques

Electrical circuits are electrical networks that consist of electrical components in a closed loop thus giving a return path for the passing of the electric current. Finding voltages and currents through each component in the network is achieved through certain network analysis techniques. The aim of this chapter is to provide a detailed examination of the major concepts of network analysis and significant analysis techniques, such as circuit elements, circuit diagram, phasor diagram, Kirchhoff's circuit laws, Thévenin's theorem, Norton's theorem, etc.

Electrical Circuit

An electrical circuit is a path or line through which an electrical current flows. The path may be closed (joined at both ends), making it a loop. A closed circuit makes electrical current flow possible. It may also be an open circuit where the electron flow is cut short because the path is broken. An open circuit does not allow electrical current to flow.

Below is a basic set of symbols that you may find on circuit diagrams,

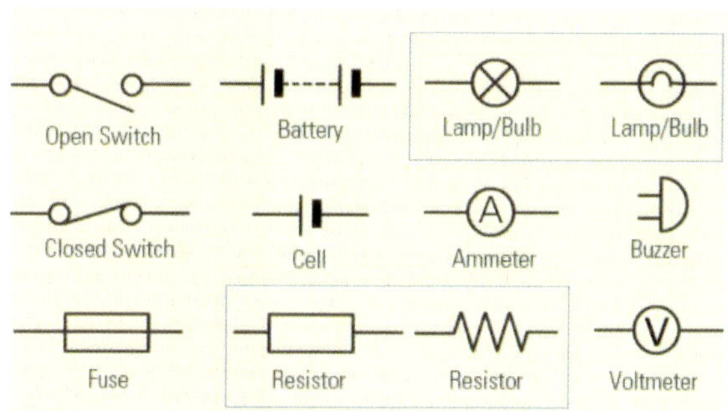

It is very important to know the basic parts of a simple circuit and the symbols that relate to them. A simple circuit has conductors, a switch, a load and a power source. Here are the functions of each part:

Conductors

These are usually copper wires with no insulation. They make the path through which the electricity flows. One piece of the wire connects the current from the power source (cell) to the load. The other piece connects the load back to the power source.

Switch

The switch is simply a small gap in the conductor where you can close or open the circuit. When the switch is closed, the circuit is closed and electricity flows.

The Load

The load is a small light bulb or buzzer that lights when the circuit is turned on. The load is also known as a resistor.

Cell

The power source is a cell. (Note that more than one cell put together is known as a battery).

The diagram below shows how a basic circuit looks like.

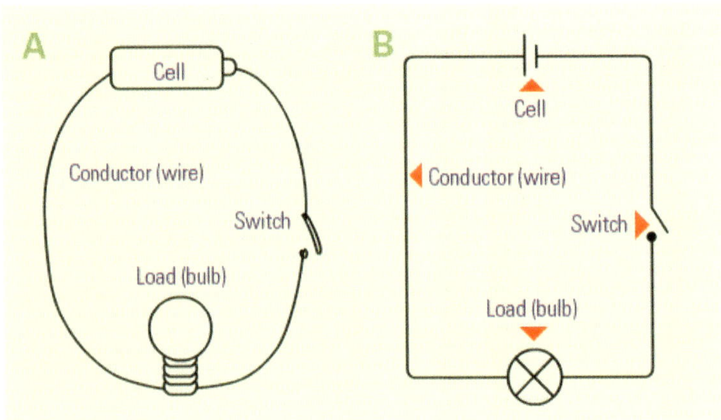

It is important to draw circuits with clean straight lines, as shown in diagram B. Avoid realistic sketches. It is important to know that a circuit can have more than the basic components in the diagram. It can have two or more batteries or two or more bulbs.

Basic Properties of Electric Circuits

- A circuit is always a closed path.
- A circuit always contains at least an energy source which acts as a source of electrons.
- The electric elements include uncontrolled and controlled source of energy, resistors, capacitors, inductors, etc.
- In an electric circuit flow of electrons takes place from negative terminal to positive terminal.
- Direction of flow of conventional current is from positive to negative terminal.
- Flow of current leads to potential drop across the various elements.

Types of Electric Circuit

The electric circuit can be categorized in three different ways:

- Open circuit
- Closed circuit
- Short circuit

Open Circuit

If due to disconnection of any part of an electric circuit if there is no flow of current through the circuit, is said to be an open circuited.

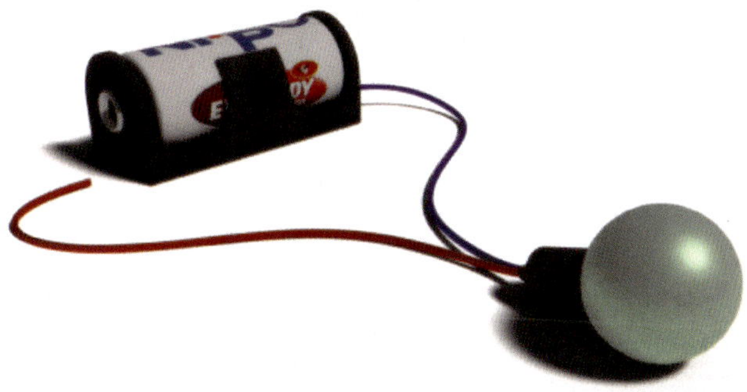

Closed Circuit

If there is no discontinuity in the circuit and current can flow from one part to another part of the circuit, the circuit is said to be closed circuit.

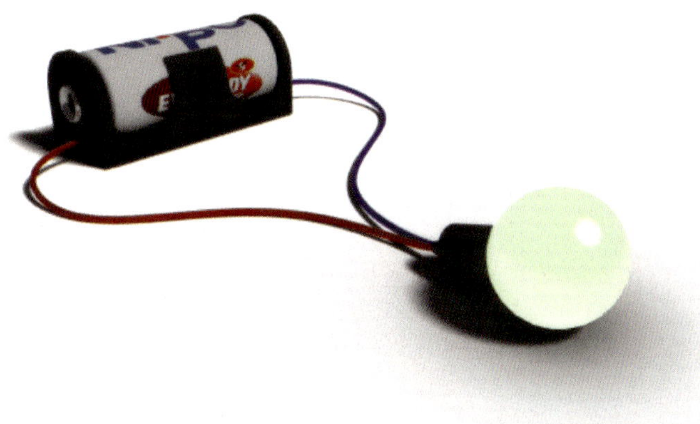

Short Circuit

If two or more phases, one or more phases and earth or neutral of AC system or positive and negative wires or positive or negative wires and earth of DC system touch together directly or connected together by a zero impedance path then the circuit is said to be short circuited.

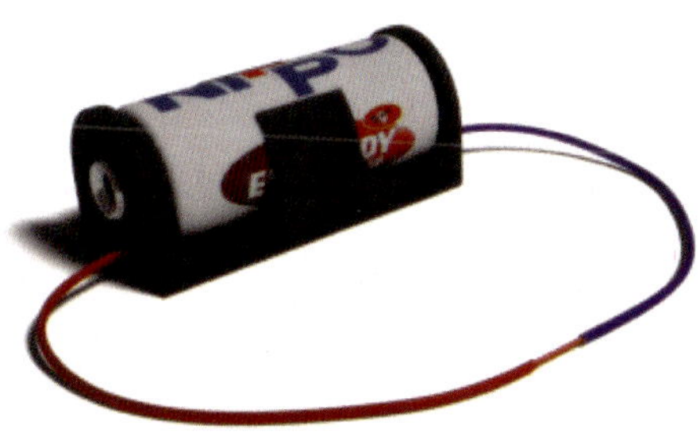

Series Circuit

A series circuit is one that has more than one resistor, but only one path through which the electricity (electrons) flows. From one end of the cell (battery), the electrons move along one path with NO branches, through the resistors, to the other end of the cell. All the components in a series circuit are connected end-to-end.

A resistor in a circuit is anything that uses some of the power from the cell. In the example below, the resistors are the bulbs. In a series circuit, the components are arranged in a line, one after the other.

Take a look at the diagram below:

The schematic drawing is a better way to draw a series circuit.

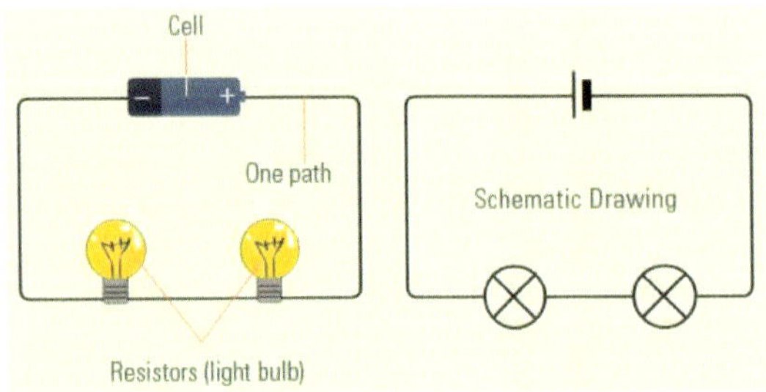

Each time there is damage (break) in any one of the resistors the entire circuit will not function. For example, if one light bulb goes out, all the other lights will go off because the electricity path in the broken bulb is cut off.

Do you put Christmas lights on the trees at home during Christmas? If the lights are in a series circuit, one burned out bulb will keep all the lights off. That is one disadvantage of series circuits. One advantage though is that you will always know if there is a break in a series circuit.

If there are many bulbs in a circuit with a battery (cell), it is very likely that the light will be dimmer because many resistors are acting on the same voltage of power from the battery.

RL Circuit

A circuit that contains a pure resistance R ohms connected in series with a coil having pure inductance of L (Henry) is known as R L Series Circuit. When an AC supply voltage V is applied the current, I flows in the circuit. IR and IL will be the current flowing in the resistor and inductor respectively, but the amount of current flowing through both the elements will be same as they are connected in series with each other. The circuit diagram of RL Series Circuit is shown below:

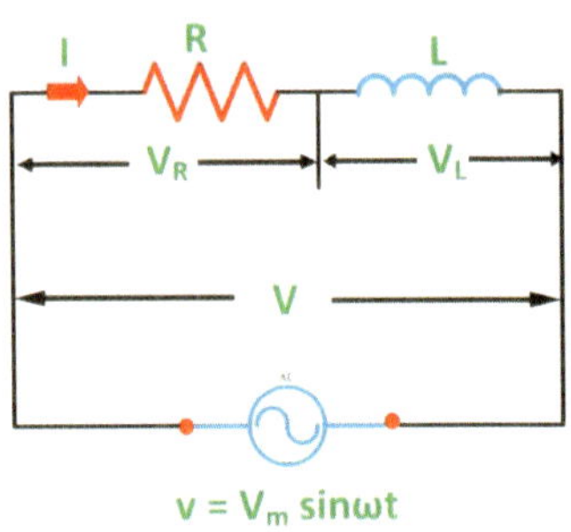

Where,

- V_R – voltage across the resistor R
- V_L – voltage across the inductor L
- V – Total voltage of the circuit

Phasor Diagram of the RL Series Circuit

The phasor diagram of the RL Series circuit is shown below:

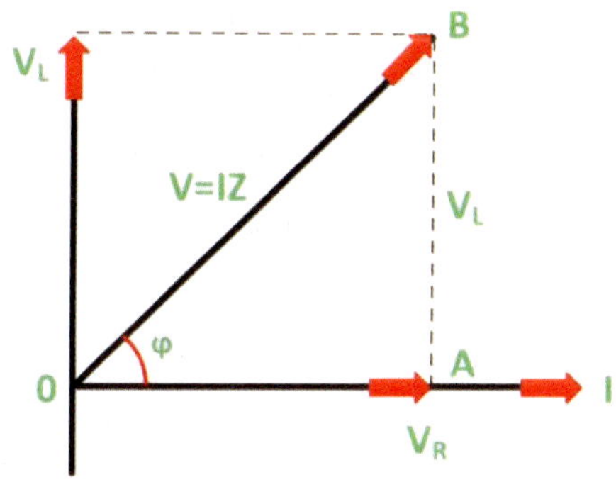

Steps to draw the Phasor Diagram of RL Series Circuit. The following steps are given below which are followed to draw the phasor diagram step by step:

- Current I is taken as a reference.
- The Voltage drop across the resistance $V_R = I_R$ is drawn in phase with the current I.
- The voltage drop across the inductive reactance $V_L = IX_L$ is drawn ahead of the current I. As the current lags voltage by an angle of 90 degrees in the pure Inductive circuit.
- The vector sum of the two voltages drops V_R and V_L is equal to the applied voltage V.

Now,

In right angle triangle OAB

$$V_R = I_R \text{ and } V_L = IX_L \text{ where } X_L = 2\pi fL$$

$$V = \sqrt{(V_R)^2 + (V_L)^2} = \sqrt{(IR)^2 + (IX_L)^2}$$

$$V = \sqrt{R^2\ X_L^2} \quad \text{or}$$

$$I = L = \frac{V}{Z}$$

Where,

$$Z = \sqrt{R^2\ X_L^2}$$

Z is the total opposition offered to the flow of alternating current by an *RL* Series circuit and is called impedance of the circuit. It is measured in ohms (Ω).

Phase Angle

In *RL* Series Circuit the current lags the voltage by 90-degree angle known as phase angle. It is given by the equation,

$$\tan\varphi = \frac{V_L}{V_R} = \frac{IX_L}{IR} = \frac{X_L}{R} \text{ or}$$

$$\varphi = \tan^{-1}\frac{X_L}{R}$$

Power in R L Series Circuit

If the alternating voltage applied across the circuit is given by the equation,

$$v = V_m \mathrm{Sin}\,\omega t$$

The equation of current I is given as,

$$i = I_m \sin(\omega t - \varphi)$$

Than the instantaneous power is given by the equation,

$$p = vi$$

Putting the value of v and i from the $v = V_m \mathrm{Sin}\,\omega t$ and $i = I\ \sin(\omega t - \varphi)$ equations in the $p = vi$ equation we will get,

$$P = (V_m \mathrm{Sin}\,\omega t) x\, I_m \sin(\omega t - \varphi)$$

$$p = \frac{V_m I_m}{2} 2\sin(\omega t - \varphi)\sin\omega t$$

$$P = \frac{V_m}{\sqrt{2}}\frac{I_m}{\sqrt{2}}\left[\cos\varphi - \cos(2\omega t - \varphi)\right]$$

$$P = \frac{V_m}{\sqrt{2}}\frac{I_m}{\sqrt{2}}\cos\varphi - \frac{V_m}{\sqrt{2}}\frac{I_m}{\sqrt{2}}\cos(2\omega t - \varphi)$$

The average power consumed in the circuit over one complete cycle is given by the equation shown below:

$$P = \text{average of } \frac{V_m}{\sqrt{2}}\frac{I_m}{\sqrt{2}}\cos\varphi - \text{average of } \frac{V_m}{\sqrt{2}}\frac{I_m}{\sqrt{2}}\cos(2\omega t - \varphi) \quad or$$

$$P = \frac{V_m}{\sqrt{2}}\frac{I_m}{\sqrt{2}}\cos\varphi - \text{Zero} \qquad or$$

$$P = V_{r.m.s} I_{r.m.s} \cos\varphi = VI\cos\varphi$$

Where, *cosϕ* is called the power factor of the circuit.

$$\cos\varphi = \frac{V_R}{V} = \frac{IR}{IZ} = \frac{R}{Z}$$

The power factor is defined as the ratio of resistance to the impedance of an AC Circuit.

Putting the value of *V* and *cosϕ* from the above equation the value of power will be,

$$P = (IZ)(I)(R/Z) = I^2R$$

From the equation above, it can be concluded that the inductor does not consume any power in the circuit.

Waveform and Power Curve of the RL Series Circuit

The waveform and power curve of the RL Series Circuit is shown below:

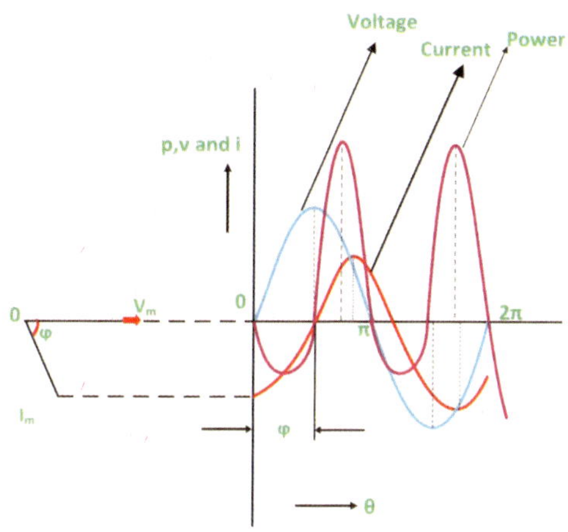

The various points on the power curve are obtained by the product of voltage and current. If you analyze the curve carefully, it is seen that the power is negative between angle 0 and ϕ and between 180 degrees and (180 + ϕ) and during the rest of the cycle the power is positive. The current lags the voltage and thus they are not in phase with each other.

Example

Calculating Characteristic Time and Current in an RL Circuit

(a) What is the characteristic time constant for a 7.50 mH inductor in series with a 3.00 Ω resistor?

(b) Find the current 5.00 ms after the switch is moved to position 2 to disconnect the battery, if it is initially 10.0 A.

Strategy for (a)

The time constant for an *RL* circuit is defined by $\tau = L/R$.

Solution for (a)

Entering known values into the expression for τ given in $\tau = L/R$ yields

$$\tau = \frac{L}{R} = \frac{7.50mH}{3.00\Omega} = 2.50ms$$

Discussion for (a)

This is a small but definitely finite time. The coil will be very close to its full current in about ten time constants, or about 25 ms.

Strategy for (b)

We can find the current by using $I = I_0 e^{-t/T}$, or by considering the decline in steps. Since the time is twice the characteristic time, we consider the process in steps.

Solution for (b)

In the first 2.50 ms, the current declines to 0.368 of its initial value, which is

$$\begin{aligned} I &= 0.368 I_0 = (0.368)(10.0\text{ A}) \\ &= 3.68 A\ at\ t = 2.50ms \end{aligned}$$

After another 2.50 ms, or a total of 5.00 ms, the current declines to 0.368 of the value just found. That is,

$$\begin{aligned} I' &= 0.368 I = (0.368)(3.68\text{ A}) \\ &= 1.35 A\ at\ t = 5.00ms \end{aligned}$$

Discussion for (b)

After another 5.00 ms has passed, the current will be 0.183 A; so, although it does die out, the current certainly does not go to zero instantaneously.

RC Circuit

The combination of a pure resistance R in ohms and pure capacitance C in Farads is called RC circuit. The capacitor store energy and the resistor connect in series with the capacitor control the charging and discharging of a capacitor. The RC circuit is used in camera flashes, pacemaker, timing circuit etc.

The RC signal filters the signals by blocking some frequencies and allowing others to pass through it. It is also called first order RC circuit and is used to filter the signals by passing some frequencies and blocking others. The RC fitlers are mostly use for selecting signals and for rejecting noise.

The high-pass filter and the low-pass filter are the most common type of RC filters. The high pass filter passes the frequency greater than the fixed cutoff frequency and blocks the frequency lower than the fixed cutoff frequency. Similarly, the low-pass filter allows the frequency lower than the fixed cutoff frequency and attenuates the frequency higher than the fixed cutoff frequency.

RC Series Circuit

A circuit containing resistance and capacitance connected in series together is called an RC series circuit.

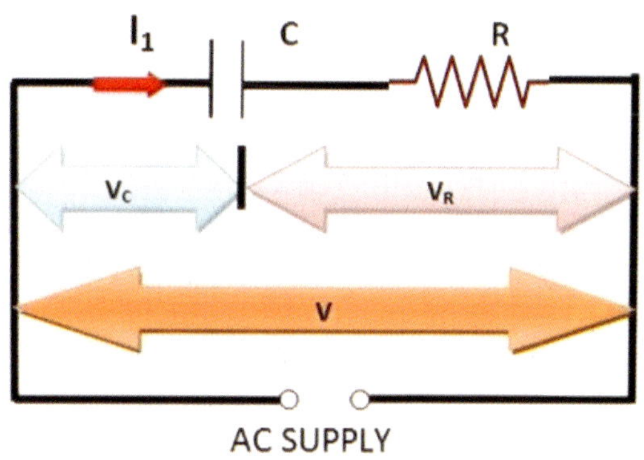

Steps to Draw a Phasor Diagram for an RC Circuit

1. Current I is taken as reference
2. Voltage drop in resistance is (V_R)

   V_R = IR is drawn in phase with the current I
3. Voltage drop in capacitive reactance is (V_C)

   V_C = IXC and is drawn 90° behind the current (as current leads the voltage by 90° in pure capacitive load circuit).
4. The vector sum of the two voltage drops is equal to the applied voltage (V).

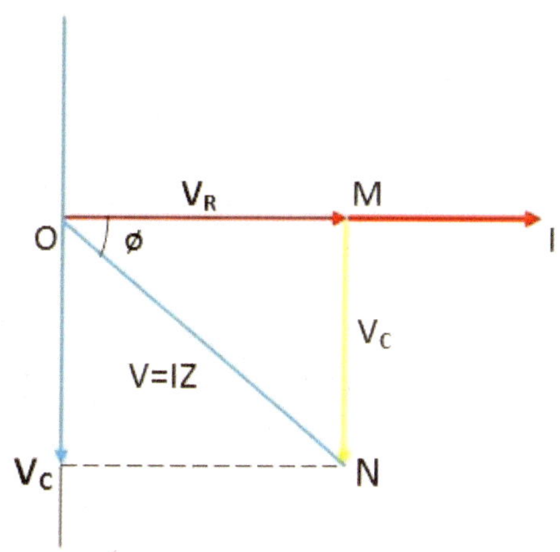

Charging

Figure below shows a simple RC circuit that employs a DC voltage source. The capacitor is initially uncharged. As soon as the switch is closed, current flows to and from the initially uncharged capacitor. As charge increases on the capacitor plates, there is increasing opposition to the flow of charge by the repulsion of like charges on each plate.

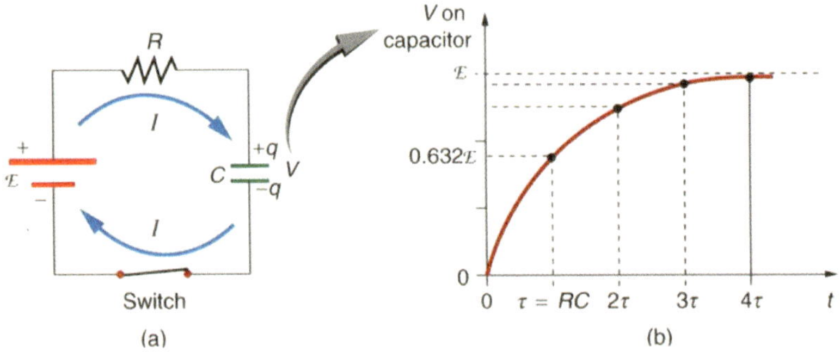

Charging an RC Circuit:

(a) An *RC* circuit with an initially uncharged capacitor. Current flows in the direction shown as soon as the switch is closed. Mutual repulsion of like charges in the capacitor progressively slows the flow as the capacitor is charged, stopping the current when the capacitor is fully charged and *Q=C·emf.*

(b) A graph of voltage across the capacitor versus time, with the switch closing at time *t=o*. (Note that in the two parts of the figure, the capital script *E* stands for emf, *q* stands for the charge stored on the capacitor, and τ is the *RC* time constant).

In terms of voltage, across the capacitor voltage is given by $V_c=Q/C$, where *Q* is the amount of charge stored on each plate and *C* is the capacitance. This voltage opposes the battery, growing from zero to the maximum emf when fully charged. Thus, the current decreases from its initial value of $I_o=emf/R$ to zero as the voltage on the capacitor reaches the same value as the emf. When there is no current, there is no IR drop, so the voltage on the capacitor must then equal the emf of the voltage source.

Initially, voltage on the capacitor is zero and rises rapidly at first since the initial current is a maximum. Figure (b) above shows a graph of capacitor voltage versus time (*t*) starting when the switch is closed at *t=o*. The voltage approaches emf asymptotically since the closer it gets to emf the less current flows. The equation for voltage versus time when charging a capacitor *C* through a resistor *R*, is:

$$\mathrm{V}(\mathrm{t}) = \mathrm{emf}\left(1 - e^{\mathrm{t/RC}}\right),$$

where, *V(t)* is the voltage across the capacitor and emf is equal to the emf of the DC voltage source. (The exact form can be derived by solving a linear differential equation describing the RC circuit, but this is slightly beyond the scope of this Atom.) Note that the unit of RC is second. We define the time constant τ for an RC circuit as τ=RC. τ shows how quickly the circuit charges or discharges.

Discharging

Discharging a capacitor through a resistor proceeds in a similar fashion, as illustrates. Initially, the current is $I_0=V_0/R$, driven by the initial voltage Vo on the capacitor. As the voltage decreases, the current and hence the rate of discharge decreases, implying another exponential formula for *V*. Using calculus, the voltage *V* on a capacitor *C* being discharged through a resistor *R* is found to be,

$$V(t)=V_0e^{-t/RC}.$$

Impedance

Impedance is the measure of the opposition that a circuit presents to the passage of a current when a voltage is applied.

Rather than solving the differential equation relating to circuits that contain resistors and capacitors, we can imagine all sources in the circuit are complex exponentials having the same frequency. This technique is useful in solving problems in which phase relationship is important. The phase of the complex impedance is the phase shift by which the current is ahead of the voltage.

Complex Analysis

For an RC circuit in, the AC source driving the circuit is given as:

$$V_{in}(t)=Ve^{j\omega t},$$

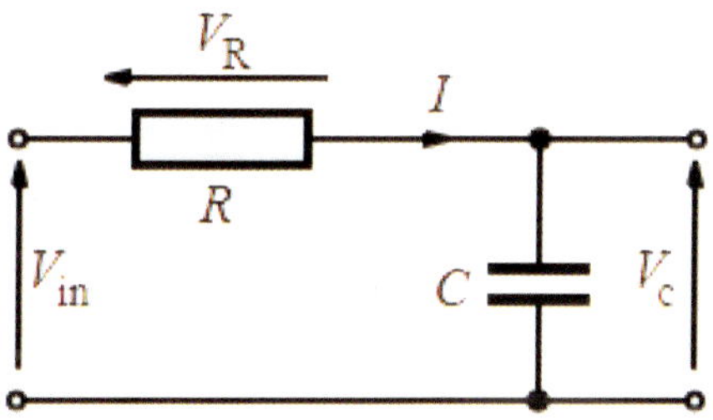

Series RC circuit

where V is the amplitude of the AC voltage, j is the imaginary unit ($j^2=-1$), and ω is the angular frequency of the AC source. Two things to note:

1. We use lower case alphabets for voltages and sources to represent that they are alternating (i.e., we use $v_{in}(t)$ instead of $V_{in}(t)$).
2. The imaginary unit is given the symbol "j", not the usual "i". "i" is reserved for alternating currents.

Complex Impedance

The advantage of assuming sources take this form is that all voltages and currents in the circuit are also complex exponentials (having the same frequency as the source). To appreciate the reason for this, we can investigate how each circuit element behaves when either the voltage or current is a complex exponential. For the resistor, *v=Ri*. From our voltage given above, $i=\frac{V}{R}e^{j\omega t}$. Thus the resistor's voltage is a complex, as is the current with an amplitude $i = \frac{V}{R}$. For a capacitor, $i = C\frac{dv}{dt}$. Letting the voltage be a complex exponential we have $i = j\omega CVe^{j\omega t}$. The amplitude of this complex exponential is $I = j\omega CV$.

The major consequence of assuming complex exponential voltage and currents is that the ratio $Z = \frac{V}{I}$ for rather than depending on time each element depends on source frequency. This quantity is known as the element's (complex) impedance. The magnitude of the complex impedance is the ratio of the voltage amplitude to the current amplitude. Just like resistance in DC cases, impedance is the measure of the opposition that a circuit presents to the passage of a current when a voltage is applied. The impedance of a resistor is R, while that of a capacitor (C) is $\frac{1}{j\omega C}$. In the case of the circuit in, to find the complex impedance of the RC circuit, we add the impedance of the two components, just as with two resistors in series: $Z = R + \frac{1}{j\omega C}$.

Finding Real Currents and Voltages

Since $e^{j\omega t} = \cos(\omega t) + j\sin(\omega t)$, to find the real currents and voltages we simply need to take the real part of the *i(t)* and *v(t)*. The (real value) impedance is the real part of the complex impedance Z. For a series RC circuit, we get $Z = \sqrt{R^2 + \left(\frac{1}{\omega C}\right)^2}$. We see that the amplitude of the current will be $V/Z = \sqrt{R^2 + \left(\frac{1}{\omega C}\right)^2}$.

Phase Angle and Power Factor

In a series RC circuit connected to an AC voltage source, voltage and current maintain a phase difference.

Impedance is an AC (alternating current) analogue to resistance in a DC circuit. As we studied in a previously Atom ("Impedance"), current, voltage and impedance in an RC circuit are related by an AC version of Ohm 's law: $I = \frac{V}{Z}$, where I and V are peak current and peak voltage respectively, and Z is the impedance of the circuit.

In a series RC circuit connected to an AC voltage source as shown in figure, conservation of charge requires current be the same in each part of the circuit at all times. Therefore we can say: the currents in the resistor and capacitor are equal and in phase. (We will represent instantaneous current as *i(t)*.)

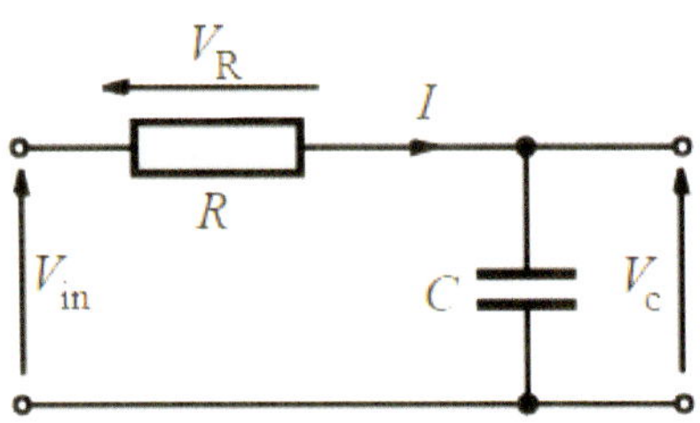

Series RC circuit

On the other hand, because the total voltage should be equal to the sum of voltages on the resistor and capacitor, so we have:

$$v(t) = v_R(t) + v_C(t)$$
$$= i(t)R + i(t)\frac{1}{j\omega C},$$
$$= i(t)\left(R + \frac{1}{j\omega C}\right)$$

where, ω is the angular frequency of the AC voltage source and j is the imaginary unit; $j^2=-1$.

Since the complex number $Z = R + \frac{1}{j\omega C} = \sqrt{R^2 + \left(\frac{1}{\omega C}\right)^2}\, e^{j\phi}$ has a phase angle ϕ that satisfies $\cos\phi = \frac{R}{\sqrt{R^2 + \left(\frac{1}{\omega C}\right)^2}}$,

we notice that voltage v(t) and current i(t) has a phase difference of ϕ.

For R=0, $\phi=90°$. As learned from the preceding series of Atoms—the voltage across the capacitor V_C follows the current by one-fourth of a cycle (or 90°).

Power Factor

Because voltage and current are out of phase, power dissipated by the circuit is not equal to: (*peak voltage*) times (*peak current*). The fact that source voltage and current are out of phase affects the power delivered to the circuit. It can be shown that the average power is $I_{rms}V_{rms}\cos\phi$, where I_{rms} and V_{rms} are the root mean square (rms) averages of the current and voltage, respectively. For this reason, $\cos\phi$ is called the power factor, which can range from 0 to 1.

Parallel Circuit

In a parallel circuit, there is more than one resistor (bulb) and they are arranged on many paths. This means electricity (electrons) can travel from one end of the cell through many branches to the other end of the cell.

Look at the figure below involving two resistors in a parallel circuit:

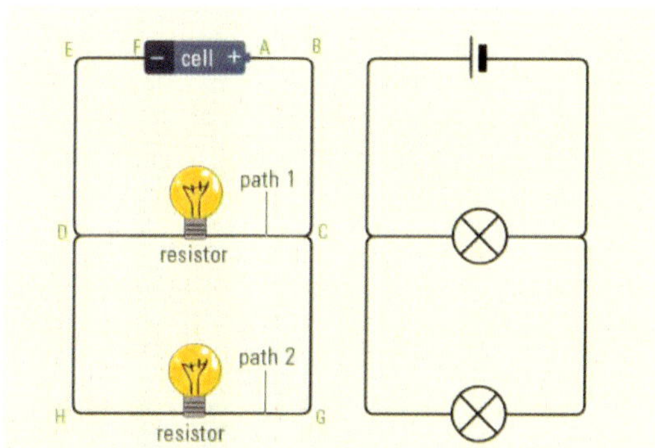

You will notice from the above that there is more than one path:

PATH 1: A-B-C-D-E-F and back to A.

PATH 2: A-B-C-G-H-D-E-F and back to A.

From the above, it is clear that electricity from the cell can take either path A or Path B to return to the cell. The great thing about parallel circuits is that, even when one resistor (bulb) burns out, the other bulbs will work because the electricity is not flowing through one path.

Think of all the light bulbs in your home. If one bulb burns out, the other bulbs in the rooms still work.

Another great thing is that the bulbs in a parallel circuit do not dim out like the case in series circuits. This is because the voltage across one branch is the same as the voltage across all other branches.

Circuit Elements

Circuit elements are those components of an electric circuit or electronic circuit which either can supply energy to the circuit or can take energy from the circuit. All the components of an circuit are connected together in required manner by conductors to form a desired and complete circuit. If the connecting conductor are taken as ideal, all the parts of the circuit, except the connecting conductors are circuit elements.

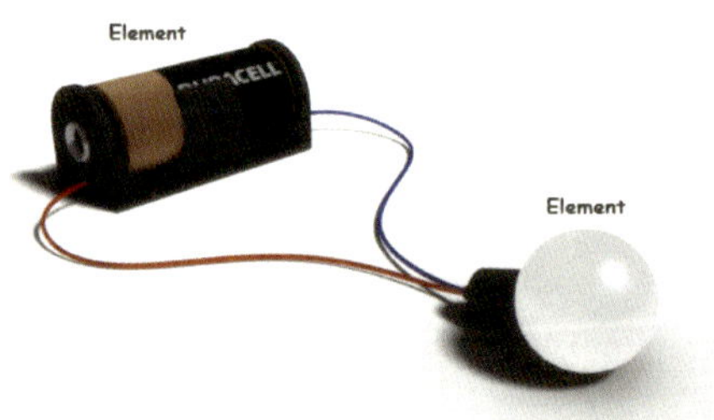

Types of Electric Circuit Elements

The electrical circuit elements either absorb energy from circuit or deliver energy to the circuit. Based on that there are two types of electric circuit elements, namely passive circuit elements and active circuit elements.

Passive Element

The circuit elements that receive energy (or absorb energy) and either convert it into heat or store it in an electric field or a magnetic field are called passive circuit elements.

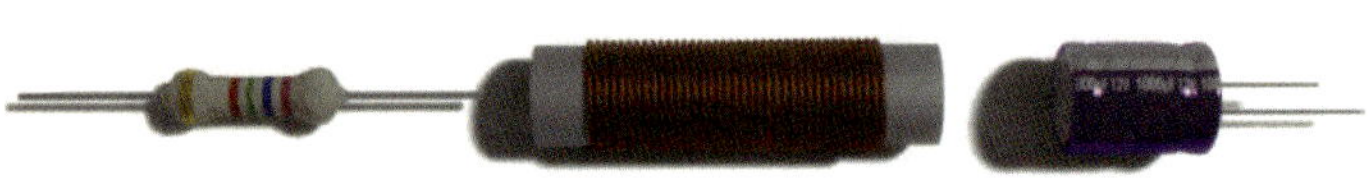

Resistor

Resistor is taken as passive element since it can dissipate energy as heat as long as current flows through it but in any situation a resistor cannot deliver energy to the circuit.

Inductor

An inductor is also considered as passive element of circuit, because it can store energy in it as magnetic field, and can deliver it to the circuit, but not in continuous basis. The energy absorbing and delivering capacity of an inductor are limited and transient in nature. That is why, an inductor is taken as passive element of a circuit.

Capacitor

For same reason, a capacitor is considered as passive element, because it can store energy in it as electric field and deliver it to the circuit, but not in continuous basis. The energy dealing capacity of a capacitor is limited and transient too.

Active Circuit Element

The circuit elements which supply energy to the circuit are called active circuit element.

Examples: Include voltage sources, current sources, and generators such as alternators, DC generators etc.

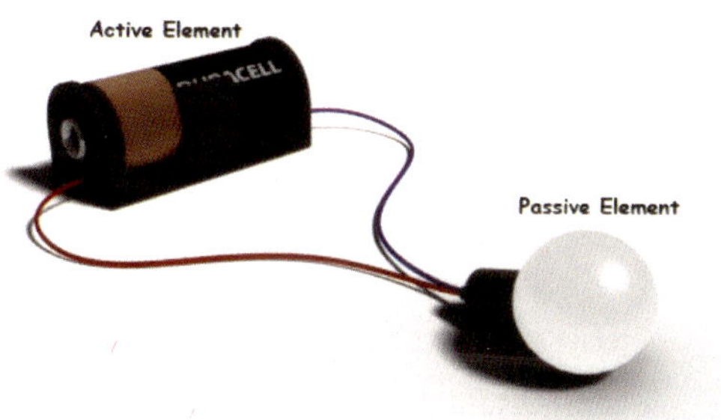

Voltage Source

A voltage source can be considered as an active element. If current leaves from the positive terminal of the voltage source, the energy is delivered from the source to the circuit.

As per definition of active element of electrical circuit, a battery can be considered as active element, as it continuously delivers energy to the circuit during discharging.

Current Source

A current source is also considered as active element. The current supplied to the circuit by an ideal current source is independent of circuit voltage.

Note: A transistor is an active circuit element, because, it can amplify the power of a signal. On the other hand,transformer is not an active circuit element because it does not amplify power and the power remains same both in primary and secondary side. Although, a transistor is an example of active circuit elements, but the transformer is an example of passive circuit elements.

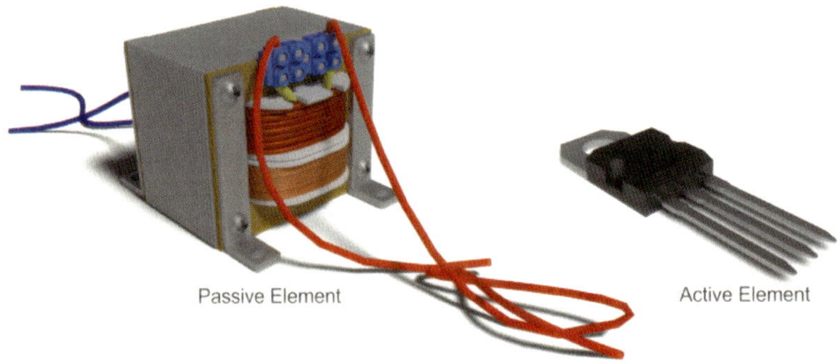

The circuit elements can also be categorised as bilateral and unilateral.

Bilateral Circuit Element

Conduction of current in both directions in a circuit element with same magnitude is termed as bilateral circuit element. It offers same resistance to the current of either directions.

Example: resistor, inductor, capacitor etc.

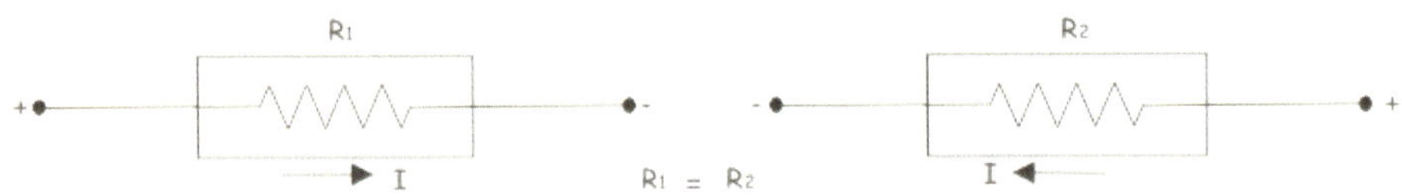

This figure shows a bilateral circuit element can conduct from both sides and offers same resistance for current from either side.

Unilateral Circuit Element

Conduction of current in one direction is termed as unilateral circuit element. It does not offer same resistance to the current of either direction. The resistance of the unilateral circuit element is different for forward current than that of reverse current.

Example: diode, transistor etc.

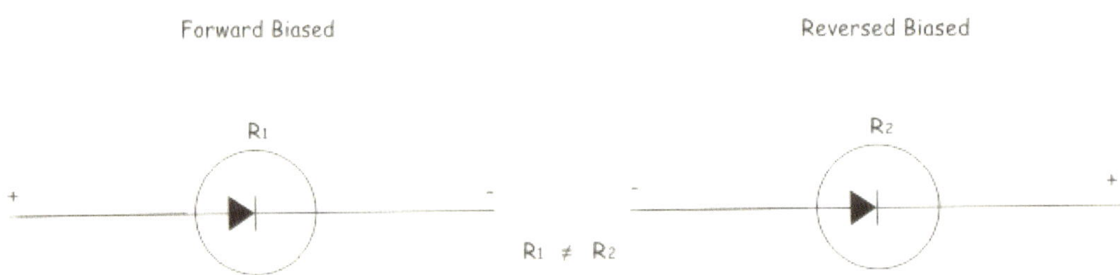

This figure shows, a diode as unilateral circuit element. When diode is forward biased it offers very small resistance and conducts. While it is reverse biased, it offers very high resistance and doesn't conduct. The circuit element can be categorized in another manner, such as lumped and distributed circuit elements.

Lumped Circuit Elements

When the voltage across and current through the element don't vary with dimension of the element, it is called lumped circuit elements.

Example: Resistor connected in any electrical circuit.

Distributed Circuit Element

When the voltage across and current through the element change with dimensions of the element, it is called distributed circuit element.

Example: Resistance of the transmission line. It varies with the length of the line.

Circuit Diagram

Circuit diagram is a simplified conventional graphical representation of an electrical circuit, showing how electrical components are connected together. It's usually used by engineers and electricians to explain symbolically parts and paths of an electrical circuit, which is important in design, construction, and maintenance of electrical and electronic equipment.

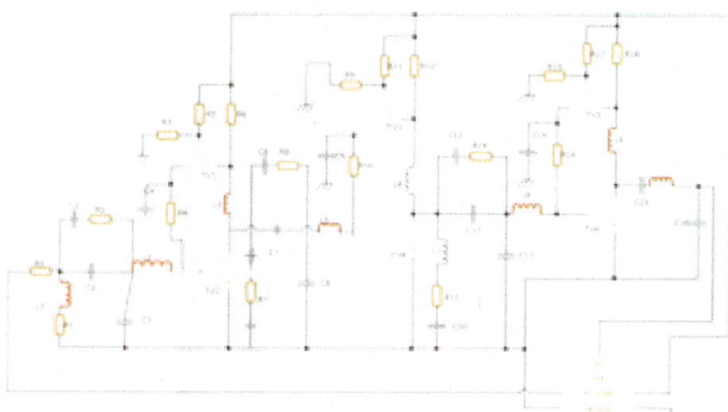

Standard Circuit Symbols for Designing

Circuit diagrams use international standard symbols to provide schematic diagrams of the circuit and its components. Each symbol is intended to represent some feature of the physical construction of the device. First you need to identify the symbols used to represent the components you wish to draw. Edraw provides you with lots of circuits symbols including cell, battery, source, resister, switch, lamp, motor, bell, buzzer, relay, capacitor, etc.

Below are some frequently used circuits symbols:

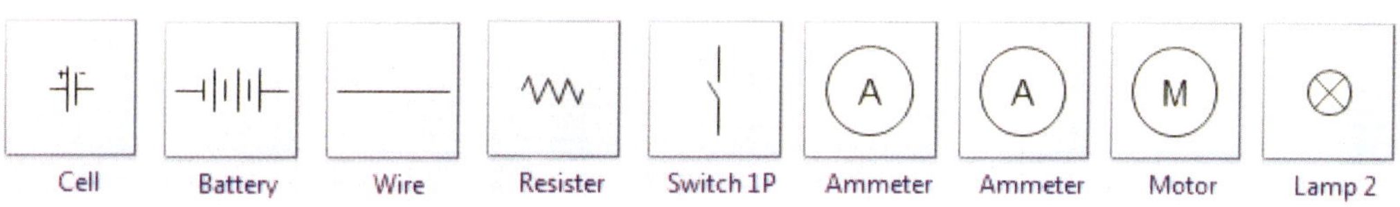

- Cell: A cell stands for electrical energy supply. A single cell is often called a battery, but strictly a battery is two or more cells joined together. It's represented by a long and a short parallel line.
- Battery: A battery is more than one cell. The larger terminal (on the left) is positive (+).It is represented by a collection of long and short parallel lines.
- Wire: Pass current from one part of the circuit to another. It's a connecting wire between two components.
- Resister: A resister is used to restrict the flow of a current. It's represented by a zigzag line.
- Switch: An on-off switch allows current to flow only when it is in the closed (on) position. It's generally represented by providing a break in a straight line by lifting a portion of the line upward at a diagonal.
- Ammeter: An ammeter is used to measure current. It's represented by a letter A in a circle.
- Voltmeter: A voltmeter is used to measure voltage. It's represented by a letter V in a circle.
- Motor: A transducer converts electrical energy to kinetic energy. It's represented by a letter M in a circle.
- Lamp: A transducer which converts electrical energy to light.

Types of Circuit Diagram

Circuit diagrams can be divided into two categories - pictorial circuit diagram and schematic circuit diagram.

Pictorial Circuit Diagram

Pictorial diagrams are much easier to understand than schematic circuit diagrams. Connecting realistic electrical components with the wiring, a pictorial diagram makes it easy for viewers to identify the electrical components of a system immediately without professional knowledge required.

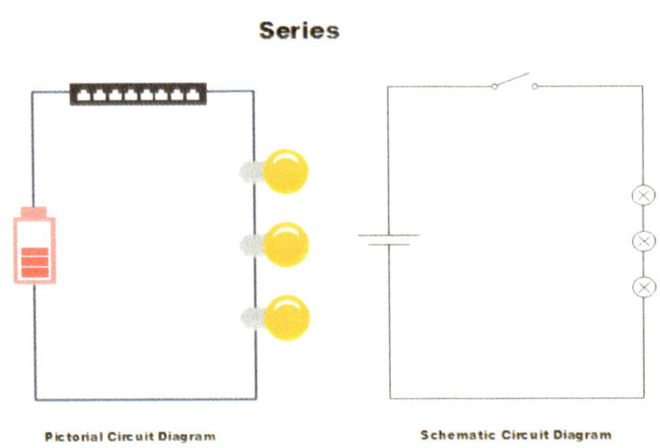

Schematic Circuit Diagram

A schematic circuit diagram represents the electrical system in the form of a diagram that shows the main features or relationships but not the details. In a schematic circuit diagram, the presentation of electrical components and wiring does not completely correspond to the physical arrangements in the real device. If you want to understand a schematic diagram, you are required to master basic knowledge of electricity and physics as well as internationally standardized symbols. Look at the parallel circuits below, you may find that the battery is represented as two short lines, the lights are represented as a circle with a cross inside, and the wiring is represented as a line. This kind of circuit diagram is mainly used by electrical engineers. The following is a schematic circuit diagram of a semiconductor electronic. Schematics and circuit diagrams are both important engineering diagrams.

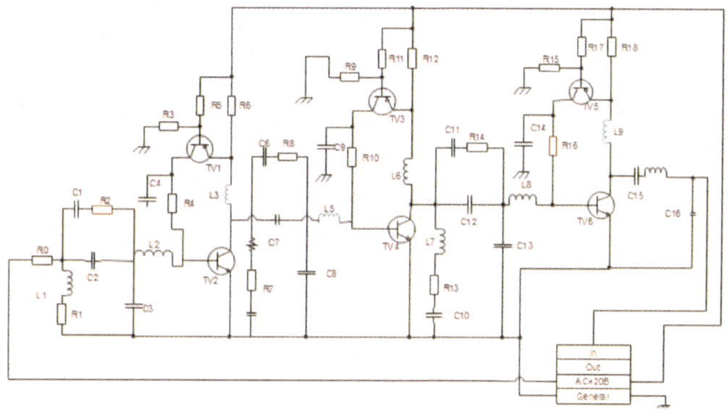

Phasor Diagrams

Phasor diagrams are a graphical way of representing the magnitude and directional relationship between two or more alternating quantities sinusoidal waveforms of the same frequency can have a Phase Difference between themselves which represents the angular difference of the two sinusoidal waveforms. Also the terms "lead" and "lag" as well as "in-phase" and "out-of-phase" are commonly used to indicate the relationship of one waveform to the other with the generalized sinusoidal expression given as: $A_{(t)} = A_m \, sin(\omega t \pm \Phi)$ representing the sinusoid in the time-domain form.

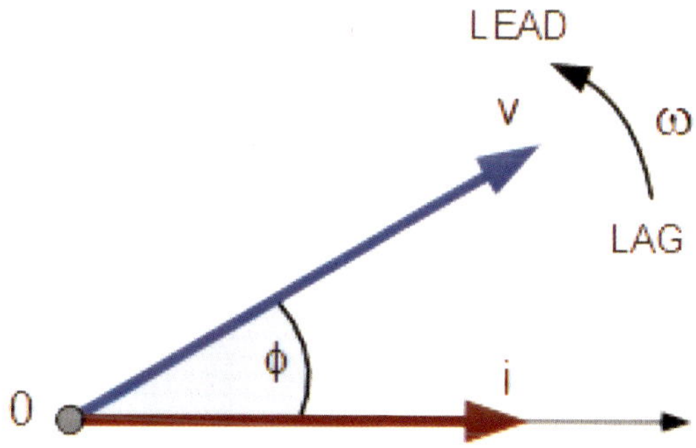

But when presented mathematically in this way it is sometimes difficult to visualise this angular or phasor difference between two or more sinusoidal waveforms. One way to overcome this problem is to represent the sinusoids graphically within the spacial or phasor-domain form by using Phasor diagrams, and this is achieved by the rotating vector method.

Basically a rotating vector, simply called a "Phasor" is a scaled line whose length represents an AC quantity that has both magnitude ("peak amplitude") and direction ("phase") which is "frozen" at some point in time.

A phasor is a vector that has an arrow head at one end which signifies partly the maximum value of the vector quantity (V or I) and partly the end of the vector that rotates.

Generally, vectors are assumed to pivot at one end around a fixed zero point known as the "point of origin" while the arrowed end representing the quantity, freely rotates in an anti-clockwise direction at an angular velocity, (ω) of one full revolution for every cycle. This anti-clockwise rotation of the vector is considered to be a positive rotation. Likewise, a clockwise rotation is considered to be a negative rotation.

Although the both the terms vectors and phasors are used to describe a rotating line that itself has both magnitude and direction, the main difference between the two is that a vectors magnitude is the "peak value" of the sinusoid while a phasors magnitude is the "rms value" of the sinusoid. In both cases the phase angle and direction remains the same.

The phase of an alternating quantity at any instant in time can be represented by a phasor diagram, so phasor diagrams can be thought of as "functions of time". A complete sine wave can be constructed by a single vector rotating at an angular velocity of $\omega = 2\pi f$, where f is the frequency of the waveform. Then a Phasor is a quantity that has both "Magnitude" and "Direction".

Generally, when constructing a phasor diagram, angular velocity of a sine wave is always assumed to be: ω in rad/sec.

Phasor Diagram of a Sinusoidal Waveform

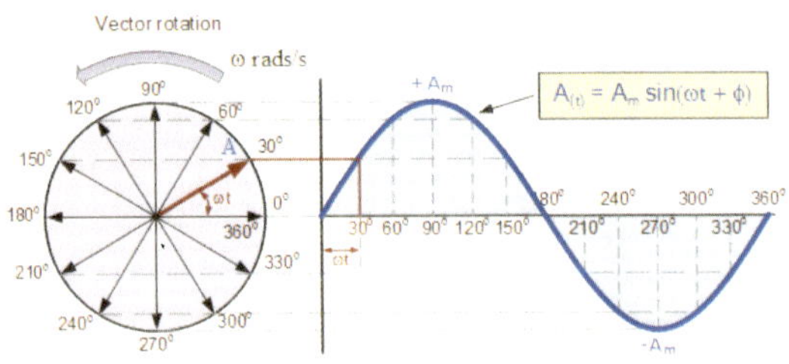

As the single vector rotates in an anti-clockwise direction, its tip at point A will rotate one complete revolution of 360° or 2π representing one complete cycle. If the length of its moving tip is transferred at different angular intervals in time to a graph as shown above, a sinusoidal waveform would be drawn starting at the left with zero time. Each position along the horizontal axis indicates the time that has elapsed since zero time, $t = 0$. When the vector is horizontal the tip of the vector represents the angles at 0°, 180° and at 360°.

Likewise, when the tip of the vector is vertical it represents the positive peak value, (+Am) at 90° or π/2 and the negative peak value, (-Am) at 270° or 3π/2. Then the time axis of the waveform represents the angle either in degrees or radians through which the phasor has moved. So we can say that a phasor represent a scaled voltage or current value of a rotating vector which is "frozen" at some point in time, (t) and in our example above, this is at an angle of 30°.

Sometimes when we are analysing alternating waveforms we may need to know the position of the phasor, representing the Alternating Quantity at some particular instant in time especially when we want to compare two different waveforms on the same axis. For example, voltage and current. We have assumed in the waveform above that the waveform starts at time t = 0 with a corresponding phase angle in either degrees or radians.

But if a second waveform starts to the left or to the right of this zero point or we want to represent in phasor notation the relationship between the two waveforms then we will need to take into account this phase difference, Φ of the waveform.

Phase Difference of a Sinusoidal Waveform

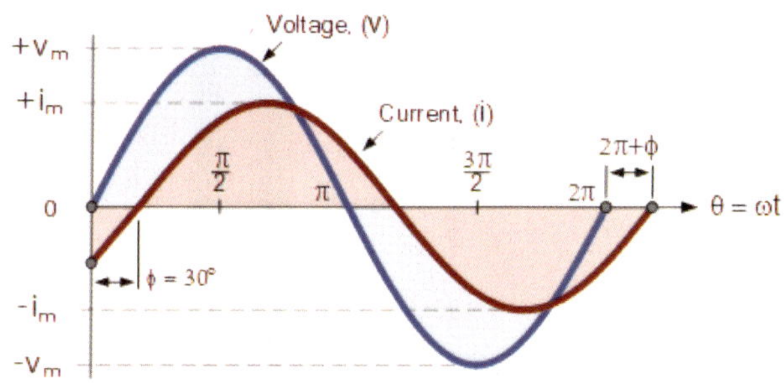

The generalised mathematical expression to define these two sinusoidal quantities will be written as:

$$v_{(t)} = V_m \sin(\omega t)$$
$$i_{(t)} = I_m \sin(\omega t - \phi)$$

The current, i is lagging the voltage, v by angle Φ and in our example above this is 30°. So the difference between the two phasors representing the two sinusoidal quantities is angle Φ and the resulting phasor diagram will be.

Phasor Diagram of a Sinusoidal Waveform

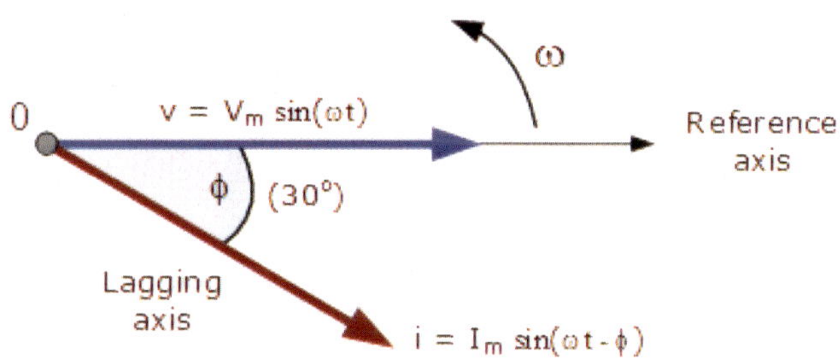

The phasor diagram is drawn corresponding to time zero (t=0) on the horizontal axis. The lengths of the phasors are proportional to the values of the voltage, (V) and the current, (I) at the instant in time that the phasor diagram is drawn. The current phasor lags the voltage phasor by the angle, Φ, as the two phasors rotate in an *anticlockwise*direction as stated earlier, therefore the angle, Φ is also measured in the same anticlockwise direction.

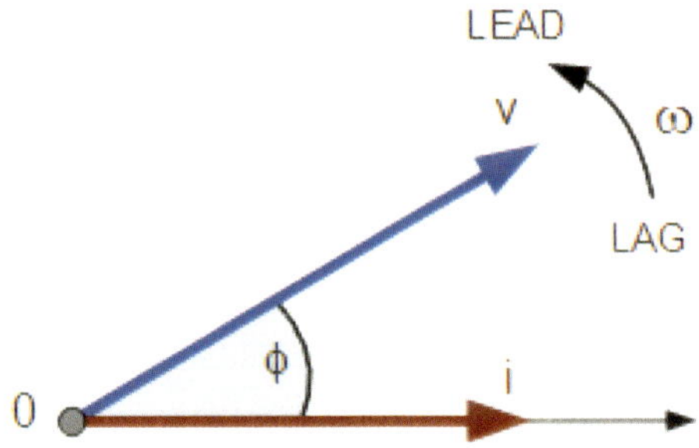

If however, the waveforms are frozen at time, t = 30°, the corresponding phasor diagram would look like the one shown on the right. Once again the current phasor lags behind the voltage phasor as the two waveforms are of the same frequency.

However, as the current waveform is now crossing the horizontal zero axis line at this instant in time we can use the current phasor as our new reference and correctly say that the voltage phasor is "leading" the current phasor by angle, Φ. Either way, one phasor is designated as the *reference* phasor and all the other phasors will be either leading or lagging with respect to this reference.

Phasor Addition

Sometimes it is necessary when studying sinusoids to add together two alternating waveforms, for example in an AC series circuit, that are not in-phase with each other. If they are in-phase that is, there is no phase shift then they can be added together in the same way as DC values to find the algebraic sum of the two vectors. For example, if two voltages of say 50 volts and 25 volts respectively are together "in-phase", they will add or sum together to form one voltage of 75 volts (50 + 25).

If however, they are not in-phase that is, they do not have identical directions or starting point then the phase angle between them needs to be taken into account so they are added together using phasor diagrams to determine their Resultant Phasor or Vector Sum by using the *parallelogram law*.

Consider two AC voltages, V_1 having a peak voltage of 20 volts, and V_2 having a peak voltage of 30 volts where V_1 leads V_2 by 60°. The total voltage, V_T of the two voltages can be found by firstly drawing a phasor diagram representing the two vectors and then constructing a parallelogram in which two of the sides are the voltages, V_1 and V_2 as shown below.

Phasor Addition of Two Phasors

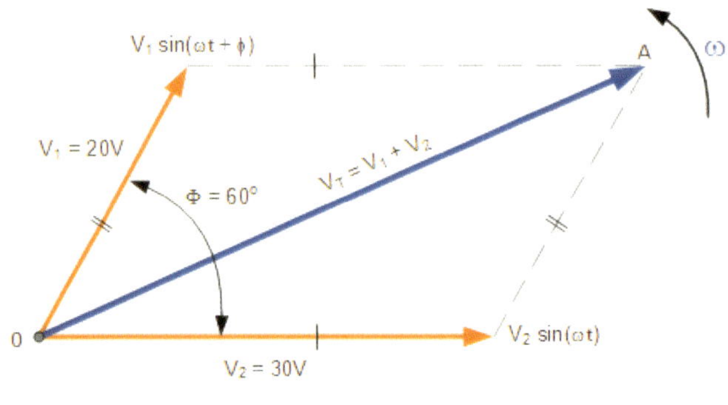

By drawing out the two phasors to scale onto graph paper, their phasor sum $V_1 + V_2$ can be easily found by measuring the length of the diagonal line, known as the "resultant r-vector", from the zero point to the intersection of the construction lines 0-A. The downside of this graphical method is that it is time consuming when drawing the phasors to scale.

Also, while this graphical method gives an answer which is accurate enough for most purposes, it may produce an error if not drawn accurately or correctly to scale. Then one way to ensure that the correct answer is always obtained is by an analytical method.

Mathematically we can add the two voltages together by firstly finding their "vertical" and "horizontal" directions, and from this we can then calculate both the "vertical" and "horizontal" components for the resultant "r vector", V_T. This analytical method which uses the cosine and sine rule to find this resultant value is commonly called the Rectangular Form.

In the rectangular form, the phasor is divided up into a real part, x and an imaginary part, y forming the generalised expression Z = x ± jy. This then gives us a mathematical expression that represents both the magnitude and the phase of the sinusoidal voltage as:

Complex Sinusoid

$$V_m = \cos(\phi) + jV_m(\sin\phi)$$

So the addition of two vectors, A and B using the previous generalised expression is as follows:

$$A = x + jy \quad \textit{and} \quad B = w + jz$$
$$A + B = (x + w) + j(y + z)$$

Phasor Addition using Rectangular Form

Voltage, V_2 of 30 volts points in the reference direction along the horizontal zero axis, then it has a horizontal component but no vertical component as follows:

- Horizontal Component = 30 cos 0° = 30 volts;
- Vertical Component = 30 sin 0° = 0 volts.

This then gives us the rectangular expression for voltage V_2 of: 30 + j0

Voltage, V_1 of 20 volts leads voltage, V_2 by 60°, then it has both horizontal and vertical components as follows:

- Horizontal Component = 20 cos 60° = 20 x 0.5 = 10 volts;
- Vertical Component = 20 sin 60° = 20 x 0.866 = 17.32 volts.

This then gives us the rectangular expression for voltage V_1 = 10 + j17.32

The resultant voltage, V_T is found by adding together the horizontal and vertical components as follows:

$V_{Horizontal}$ = sum of real parts of V_1 and V_2 = 30 + 10 = 40 volts

$V_{Vertical}$ = sum of imaginary parts of V_1 and V_2 = 0 + 17.32 = 17.32 volts.

Now that both the real and imaginary values have been found the magnitude of voltage, V_T is determined by simply using Pythagoras's Theorem for a 90° triangle as follows:

$$V_T = \sqrt{\begin{pmatrix}\text{Real or Horizontal} \\ \text{Component}\end{pmatrix}^2 + \begin{pmatrix}\text{Imaginary or Vertical} \\ \text{Component}\end{pmatrix}^2}$$

$$V_T = \sqrt{40^2 + 17.32^2}$$

$$\therefore V_T = 43.6 volts$$

Then the resulting phasor diagram will be,

Resultant Value of V_T

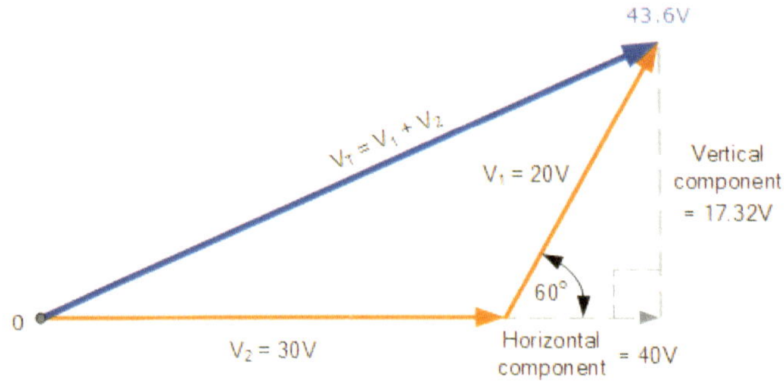

Phasor Subtraction

Phasor subtraction is very similar to the above rectangular method of addition, except this time the vector difference is the other diagonal of the parallelogram between the two voltages of V_1 and V_2 as shown.

Vector Subtraction of Two Phasors

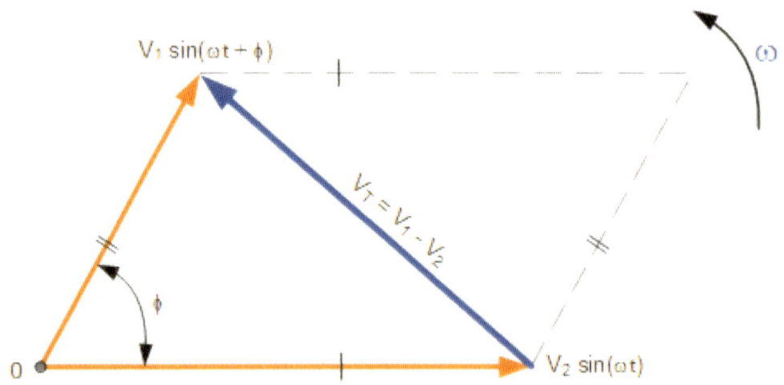

This time instead of "adding" together both the horizontal and vertical components we take them away, subtraction.

$$A = x + jy \quad and \quad B = w + jz$$

$$A - B = (x - w) + j(y - z)$$

The 3-phase Phasor Diagram

Previously we have only looked at single-phase AC waveforms where a single multi-turn coil rotates within a magnetic field. But if three identical coils each with the same number of coil turns are placed at an electrical angle of 120° to each other on the same rotor shaft, a three-phase voltage supply would be generated.

A balanced three-phase voltage supply consists of three individual sinusoidal voltages that are all equal in magnitude and frequency but are out-of-phase with each other by exactly 120° electrical degrees.

Standard practice is to colour code the three phases as Red, Yellow and Blue to identify each individual phase with the red phase as the reference phase. The normal sequence of rotation for a three phase supply is Red followed by Yellow followed by Blue, (R, Y, B).

As with the single-phase phasors above, the phasors representing a three-phase system also rotate in an anti-clockwise direction around a central point as indicated by the arrow marked ω in rad/s. The phasors for a three-phase balanced star or delta connected system are shown below.

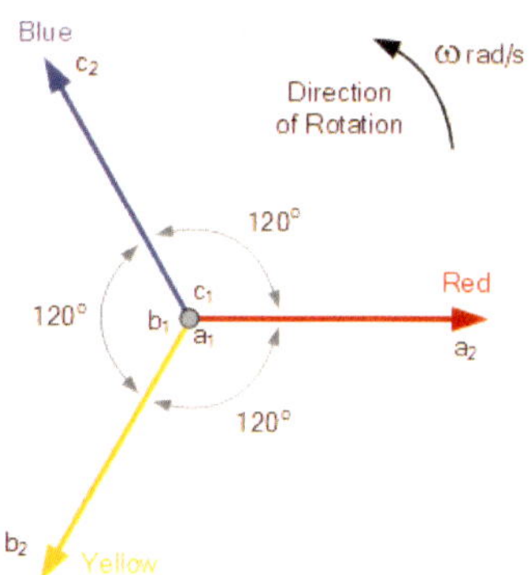

The phase voltages are all equal in magnitude but only differ in their phase angle. The three windings of the coils are connected together at points, a_1, b_1 and c_1 to produce a common neutral connection for the three individual phases. Then if the red phase is taken as the reference phase each individual phase voltage can be defined with respect to the common neutral as.

Three-phase Voltage Equations

$$\text{Red Phase: } V_{RN} = V_m \sin\theta$$

$$\text{Yellow Phase: } V_{YN} = V_m \sin\left(\theta - 120^0\right)$$

$$\text{Blue Phase } V_{BN} = V_m \sin\left(\theta - 240^0\right)$$

$$or$$

$$V_{RN} = V_m \sin\left(\theta + 120^0\right)$$

If the red phase voltage, V_{RN} is taken as the reference voltage as stated earlier then the phase sequence will be R – Y – B so the voltage in the yellow phase lags V_{RN} by 120°, and the voltage in the blue phase lags V_{YN} also by 120°. But we can also say the blue phase voltage, V_{BN} leads the red phase voltage, V_{RN} by 120°.

One final point about a three-phase system. As the three individual sinusoidal voltages have a fixed relationship between each other of 120° they are said to be "balanced" therefore, in a set of balanced three phase voltages their phasor sum will always be zero as: $V_a + V_b + V_c = 0$

Electrical Impedance

Electrical impedance (also known as "impedance" in short) is an addition of the definition of resistance to analternating current(AC). Which means that impedance includes both resistance (opposition of the electric current that causes heat) and reactance (a measure of such an oppositional current alternates) – in detail, the opposition adjacent to the electric currents. In the direct current (DC), electrical impedance is same as a resistance, except that it does not hold true in AC circuits.

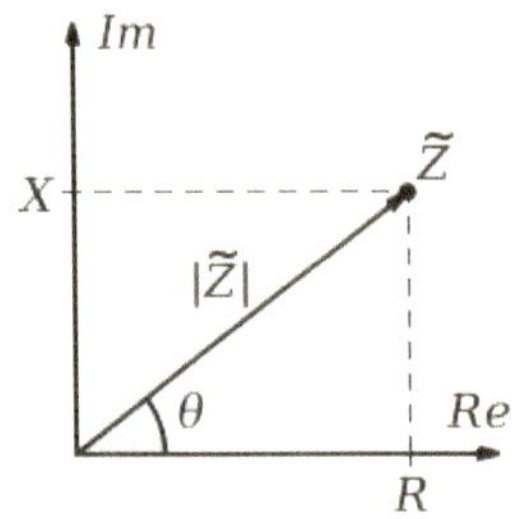

Electrical Impedance

Impedance can also be dissimilar from resistance when a DC circuit changes flow in one way or other- similar to the opening and closing of an electrical switch, as is observed in the computers when they open and close switches to represent ones and zeros (binary language). The opposite of an impedance is admittance, which is the measure of the allowance of current. The figure to the left is a complex impedance plane, in which impedance is represented by a Z, resistance is depicted as R, and the reactance is depicted with X.

Electrical Impedance Tomography (EIT)

The fundamental principle of electrical impedance tomography (EIT) is akin to electrical resistance tomography (ERT) such that several measurements at the periphery of a process vessel or tube are taken and combined to give information on the electrical properties of the process volume.

Electrical Impedance Tomography

Electrical Impedance Tomography (EIT) is a non-invasive medical imaging method in which a figure of the conductivity or permitivity of a part of the body is incidental from the surface electrode measurements. Electrical conductivity depends on the free ion content and differs significantly between different biological tissues (absolute EIT) or dissimilar practical states of one and other similar tissues or organs (relative or functional EIT). Majority of EIT systems apply little irregular currents at a single frequency; however, some EIT systems use various frequencies to better discriminate between usual and suspected abnormal tissue within the same organ (multifrequency-EIT or electrical impedance spectroscopy).

Complex Impedance

A resistor with a value of R has an impedance of R ohms, a real number. An ideal inductor has a complex impedance of,

$$Z = j2\pi L$$

Where, 'f' is the frequency in Hertz and L is the inductance in Henries. It is imaginary because an a ideal inductor can simply store and release electrical energy. It can't dissipate it as heat like a resistor. Similarly, an ideal capacitor has a complex impedance of,

$$Z = \frac{-j}{2\pi fc}$$

Where 'C' is the capacitance in farads.

Use of Complex Impedance

The behavior of the impedance of an AC circuit with various components quickly becomes unmanageable if sines and cosines are used to present the voltages and current. A mathematical build which eases the complexity use of complex exponential functions. The necessary parts of the strategy are as follows.

Math relationship underlaying the technique,

$$ej\omega t = cos\omega t + sin\omega t$$

The real part of a complex exponential function can be used to represent an AC voltage or current.

$$V = Vm\ COS\omega t$$
$$I = Im\ COS\ (\omega t - \varphi)$$

The impedance can then be expressed as a complex exponential,

$$Z = Vm /\ Im\ e - jØ = R + jX$$

The impedance of the individual circuit elements can then be expressed as pure real or imaginary numbers.

$$R\ - j / \omega c\ j\omega L$$

Complex Impedance for RL and RC

Using complex impedance is a significant technique for handling multi-component AC circuits. If a complex plane is used with resistance along the real axis, then the reactance of a capacitor and inductor are treated as imaginary numbers. For series combinations of the components such as RL and RC combinations, the component values are added as if they were components of a vector. Shown now is the Cartesian form of the complex impedance. They can also be written in polar form. Impedances in combination circuits like the RLC parallel circuit.

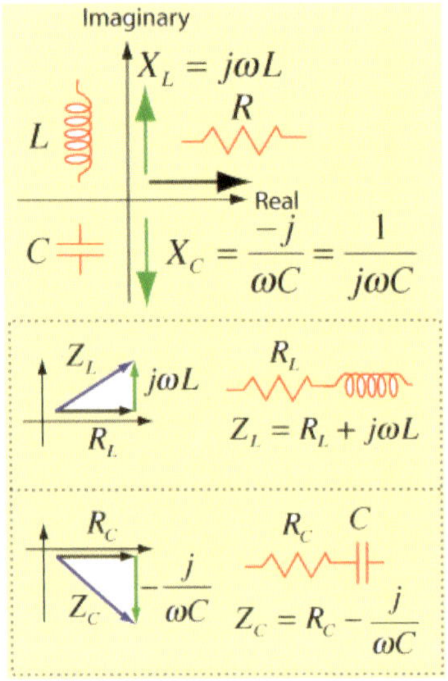

Complex Impedance for RL and RC

Resistance and Reactance

Resistance is fundamentally a friction against the motion of electrons. It is there in all conductors to some extent (except superconductors!), and most notably in resistors. When alternating current goes through a resistance, a voltage drop is formed that is in-phase with the current. Resistance is mathematically symbolized by the letter "R" and is measured in the unit of ohms (Ω).

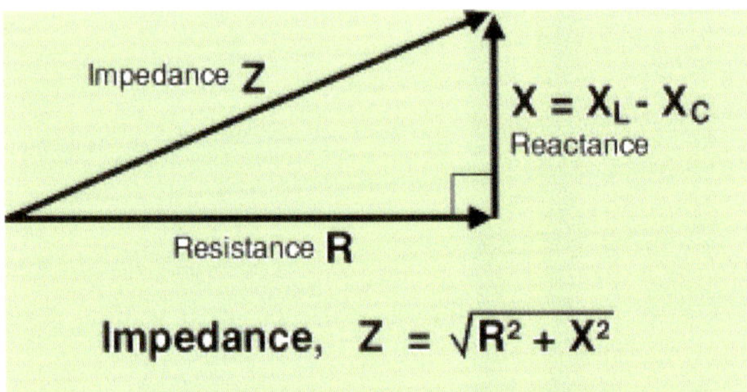

Resistance and Reactance Circuit

Reactance is essentially inactive against the motion of electrons. It is present anywhere electric or magnetic fields are developed in proportion to an applied voltage or current, correspondingly; but most notably in capacitors and inductors. When alternating current goes through a pure reactance, a voltage drop is produced, which is 900 out of phase with the current. Reactance is a mathematically symbolized by the letter "X" and is measured in the unit of Ohms (Ω).

Network Analysis

Network Analysis for electric circuits are the different useful techniques related to several currents, emfs, and resistance voltages in such circuit. This is somewhat the collection of techniques of finding the voltages and currents in every component of the network.

Nodal Analysis

Nodal analysis is a method that provides a general procedure for analyzing circuits using node voltages as the circuit variables. Nodal Analysis is also called the Node-Voltage Method.

Some features of nodal analysis are as follows:

- Nodal Analysis is based on the application of the Kirchhoff's Current Law (KCL).
- Having 'n' nodes there will be 'n-1' simultaneous equations to solve.
- Solving 'n-1' equations all the nodes voltages can be obtained.
- The number of non reference nodes is equal to the number of Nodal equations that can be obtained.

Types of Nodes in Nodal Analysis

- Non Reference Node: It is a node which has a definite Node Voltage. e.g. Here Node 1 and Node 2 are the Non Reference nodes.
- Reference Node: It is a node which acts a reference point to all the other node. It is also called the Datum Node.

Types of Reference Nodes

1. Chassis Ground: This type of reference node acts a common node for more than one circuits.

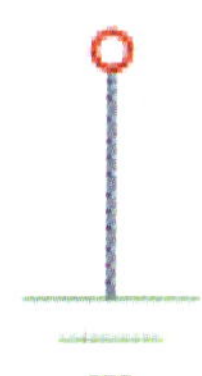

2. Earth Ground: When earth potential is used as a reference in any circuit then this type of reference node is called Earth Ground.

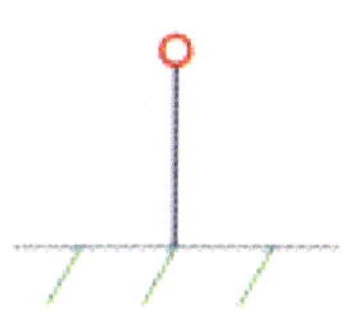

Solving of Circuit using Nodal Analysis

Basic Steps used in Nodal Analysis

- Select a node as the reference node. Assign voltages V_1, V_2... V_{n-1} to the remaining nodes. The voltages are referenced with respect to the reference node.
- Apply KCL to each of the non reference nodes.
- Use Ohm's law to express the branch currents in terms of node voltages.

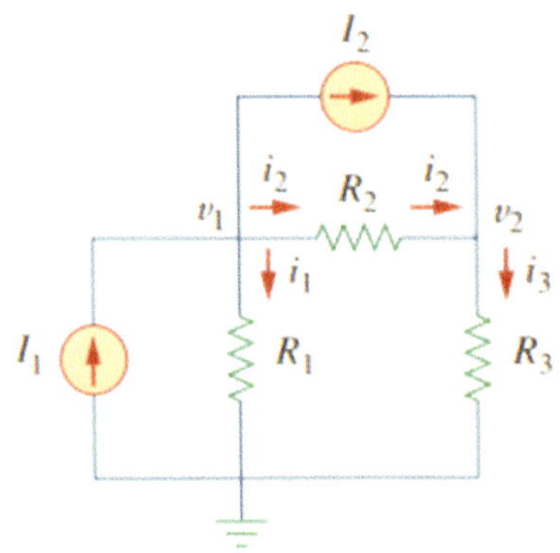

Node always assumes that current flows from a higher potential to a lower potential in resistor. Hence, current is expressed as follows,

$$I = \frac{V_{high} - V_{low}}{R}$$

- After the application of Ohm's Law get the 'n-1' node equations in terms of node voltages and resistances.
- Solve 'n-1' node equations for the values of node voltages and get the required node Voltages as result.

Nodal Analysis with Current Sources

Nodal analysis with current sources is very easy and it is discussed with a example below.

Example: Calculate Node Voltages in following circuit,

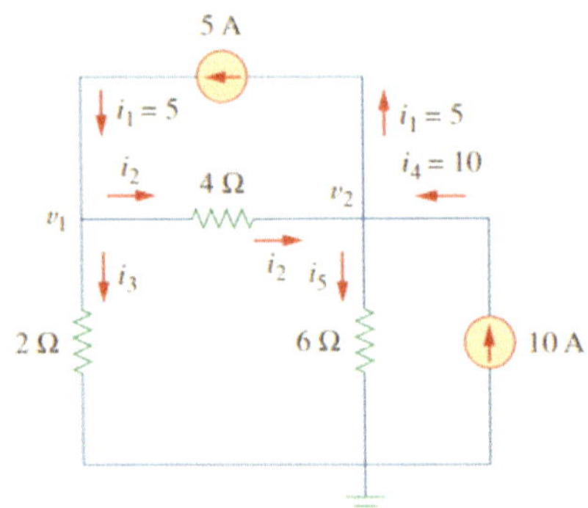

In the following circuit we have 3 nodes from which one is reference node and other two are non reference nodes - Node 1 and Node 2.

Step I. Assign the nodes voltages as v_1 and v_2 and also mark the directions of branch currents with respect to the reference nodes.

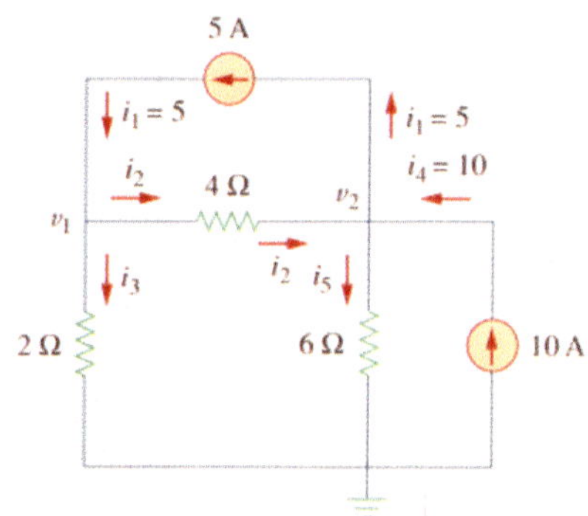

Step II. Apply KCL to Nodes 1 and 2.

KCL at Node 1,

$$i_1 = i_2 + i_3$$

KCL at Node 2,

$$i_2 + i_4 = i_1 + i_5$$

Step III. Apply Ohm's Law to KCL equations,

- Ohm's law to KCL equation at Node 1,

$$i_1 = i_2 + i_3 \Rightarrow 5 = \frac{v_1 - v_2}{4} + \frac{v_2 - 0}{4}$$

Simplifying the above equation we get,

$$3v_1 - v_2 = 20$$

- Now, Ohm's Law to KCL equation at Node 2,

$$i_2 + i_4 = i_1 + i_5 \Rightarrow \frac{v_1 - v_2}{4} + 10 = 5 + \frac{v_2 - 0}{6}$$

Simplifying the above equation we get,

$$-3v_1 + 5v_2 = 60$$

Step IV. Now solve the above two equations to get the values of v_1 and v_2 as,

Using elimination method,

$$3v_1 - v_2 = 20$$
$$-3v_1 + 5v_2 = 60$$
$$\Rightarrow 4v_2 = 80 \Rightarrow v_2 = 20\,Volts$$

And substituting value v_2 = 20 Volts in equation mentioned before the above one, we get,

$$3v_1 - 20 = 20 \Rightarrow v_1 = \frac{40}{3} = 13.333\,Volts$$

Hence node voltages are as v1 = 13.33 Volts and v2 = 20 Volts.

Nodal Analysis with Voltage Sources

Case I. If a voltage source is connected between the reference node and a non reference node, we simply set the voltage at the non-reference node equal to the voltage of the voltage source and its analysis can be done as we done with current sources. v_1 = 10 Volts.

Case II. If the voltage source is between the two non reference nodes then it forms a supernode whose analysis is done as following:

Supernode Analysis

Whenever a voltage source (Independent or Dependent) is connected between the two non reference nodes then these two nodes form a generalized node called the Super node. So, Super node can be regarded as a surface enclosing the voltage source and its two nodes.

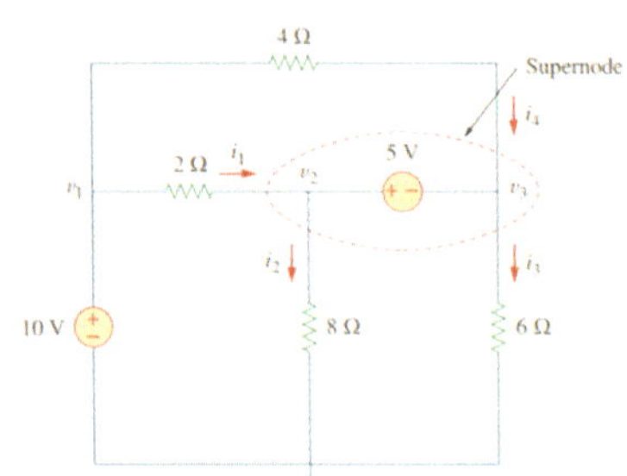

In the above figure source is connected between two non reference nodes Node - 2 and Node - 3. So here Node - 2 and Node - 3 form the Super node.

Properties of Supernode

- Always the difference between the voltage of two non reference nodes is known at Supernode.
- A supernode has no voltage of its own.
- A supernode requires application of both KCL and KVL to solve it.
- Any element can be connected in parallel with the voltage source forming the supernode.
- A Supernode satisfies the KCL as like a simple node.

Solving Any Circuit Containing Supernode

Let's take a example to understand how to solve circuit containing Supernode.

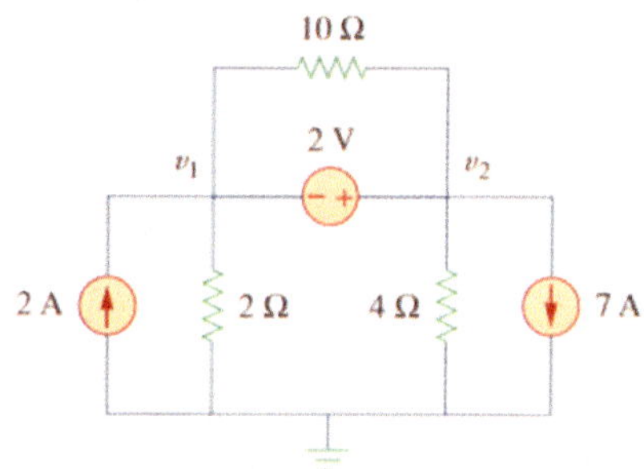

Here 2V voltage source is connected between Node-1 and Node-2 and it forms a Supernode with a 10Ω resistor in parallel.

Note - Any element connected in parallel with the voltage source forming Super node doesn't make any difference because v_2- v_1 = 2V always whatever may be the value of resistor. Thus 10 Ω can be removed and circuit is redrawn and applying KCL to the supernode as shown in figure gives,

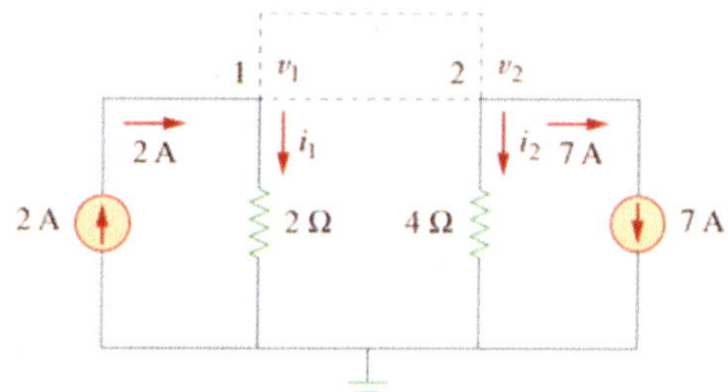

Expressing and in terms of the node voltages.

$$2 = \frac{v_1 - 0}{2} + \frac{v_2 - 0}{4} + 7 \Rightarrow 8 = 2v_1 + v_2 + 28 \Rightarrow v_2 = -2v_1 - 20$$

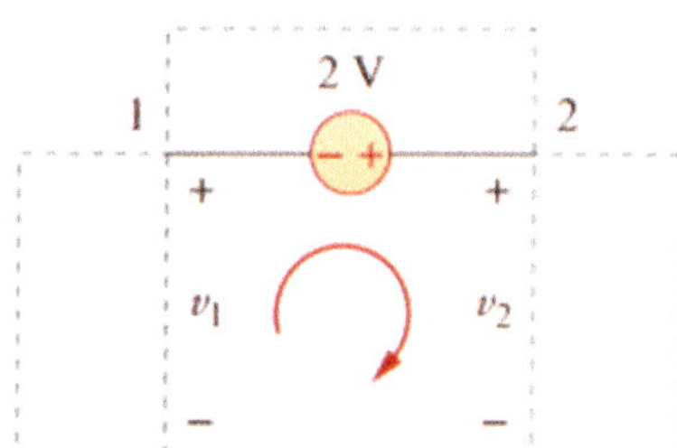

$$v_1 + 2 - v_2 = 0 \Rightarrow v_2 = v_1 + 2$$

From above two equations we can write as,

$$v_2 = v_1 + 2 = -2v_1 - 20 \Rightarrow 3v_1 = -22$$
$$\Rightarrow v_1 = -7.333V \;\&\; v_2 = v_1 + 2 = -7.333 + 2 = -5.333V$$

Hence, $v_1 = -7.333V$ and $v_2 = -5.333V$ which is required answer.

Mesh Analysis

Mesh Current Analysis is a technique used to find the currents circulating around a loop or mesh with in any closed path of a circuit.

Mesh Current Analysis Circuit

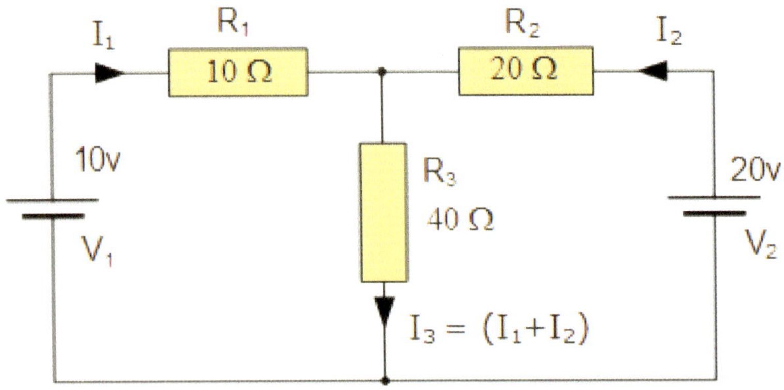

One simple method of reducing the amount of math's involved is to analyse the circuit using Kirchhoff's Current Law equations to determine the currents, I_1 and I_2 flowing in the two resistors. Then there is no need to calculate the current I_3 as its just the sum of I_1 and I_2. So Kirchhoff's second voltage law simply becomes:

$$10 = 50I_1 + 40I_2$$

$$20 = 40I_1 + 60I_2$$

therefore, one line of math's calculation have been saved.

Mesh Current Analysis

An easier method of solving the above circuit is by using Mesh Current Analysis or Loop Analysis which is also sometimes called Maxwell´s Circulating Currents method. Instead of labelling the branch currents we need to label each "closed loop" with a circulating current.

As a general rule of thumb, only label inside loops in a clockwise direction with circulating currents as the aim is to cover all the elements of the circuit at least once. Any required branch current may be found from the appropriate loop or mesh currents as before using Kirchhoff´s method.

For example: $i_1 = I_1$, $i_2 = -I_2$ and $I_3 = I_1 - I_2$

We now write Kirchhoff's voltage law equation in the same way as before to solve them but the advantage of this method is that it ensures that the information obtained from the circuit equations is the minimum required to solve the circuit as the information is more general and can easily be put into a matrix form.

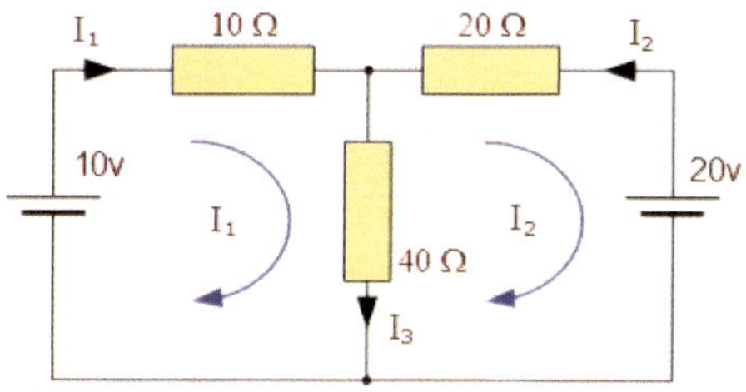

$10 = 50I_1 - 40I_2$

$-20 = .40I_1 + 60I_2$

or $[V]=[R][I]$

These equations can be solved quite quickly by using a single mesh impedance matrix Z. Each element ON the principal diagonal will be "positive" and is the total impedance of each mesh. Where as, each element OFF the principal diagonal will either be "zero" or "negative" and represents the circuit element connecting all the appropriate meshes.

First we need to understand that when dealing with matrices, for the division of two matrices it is the same as multiplying one matrix by the inverse of the other as shown.

$$[V]=[I]\times[R] \text{ or } [R]\times[I]=[V]$$

$$\begin{bmatrix} 50 & -40 \\ -40 & 60 \end{bmatrix} \times \begin{bmatrix} I_1 \\ I_2 \end{bmatrix} = \begin{bmatrix} 10 \\ -20 \end{bmatrix}$$

$$I = \frac{V}{R} = R^{-1} \times V$$

$$\text{Inverse of } R = \begin{bmatrix} 60 & 40 \\ 40 & 50 \end{bmatrix}$$

$$|R| = (60 \times 50) - (40 \times 40) = 1400$$

$$\therefore R^{-1} = \frac{1}{1400}\begin{bmatrix} 60 & 40 \\ 40 & 50 \end{bmatrix}$$

Having found the inverse of R, as V/R is the same as V x R^{-1}, we can now use it to find the two circulating currents.

$$[I] = \left[R^{-1}\right] \times [V]$$

$$\begin{bmatrix} I_1 \\ I_2 \end{bmatrix} = \frac{1}{1400}\begin{bmatrix} 60 & 40 \\ 40 & 50 \end{bmatrix} \times \begin{bmatrix} 10 \\ -20 \end{bmatrix}$$

$$I_1 = \frac{(60\times 10)+(40\times -20)}{1400} = \frac{-200}{1400} = -0.143A$$

$$I_2 = \frac{(40\times 10)+(50\times -20)}{1400} = \frac{-600}{1400} = -0.429A$$

Where,

[V] gives the total battery voltage for loop 1 and then loop 2

[I] states the names of the loop currents which we are trying to find

[R] is the resistance matrix

[R^{-1}] is the inverse of the [R] matrix.

and this gives I_1 as -0.143 Amps and I_2 as -0.429 Amps

$I_3 = I_1 - I_2$

The combined current of I_3 is therefore given as : -0.143 – (-0.429) = 0.286 Amps

which is the same value of 0.286 amps.

Ohm's Law

Ohm's Law states that the current flowing in a circuit is directly proportional to the applied potential difference and inversely proportional to the resistance in the circuit.

In other words by doubling the voltage across a circuit the current will also double. However if the resistance is doubled the current will fall by half.

In this mathematical relationship the unit of resistance is measured in Ohms.

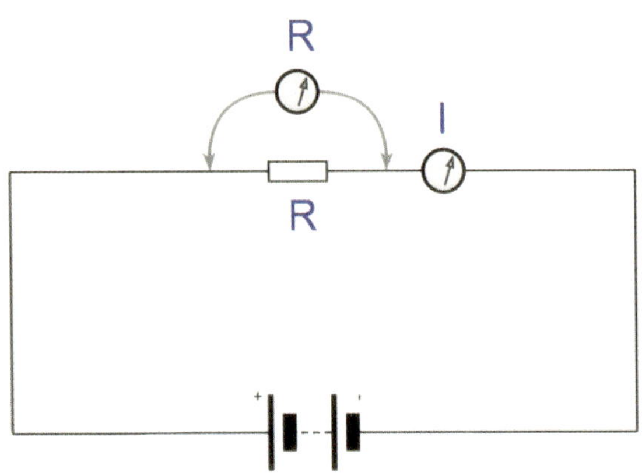

Ohm's Law Formula

The Ohm's Law formula or equation is very straightforward.

Ohm's law can be expressed in a mathematical form:

$$V = IR$$

Where,

V = voltage expressed in Volts

I = current expressed in Amps

R = resistance expressed in Ohms

The formula can be manipulated so that if any two quantities are known the third can be calculated.

$$I = \frac{V}{R}$$

$$R = \frac{V}{I}$$

Ohm's Law Triangle

To help remember the formula it is possible to use a triangle with one side horizontal and the peak at the top like a pyramid. This is sometimes known as the Ohm's law triangle.

In the top corner of the Ohm's law triangle is the letter V, in the left hand corner, the letter I, and in the right hand bottom corner, R.

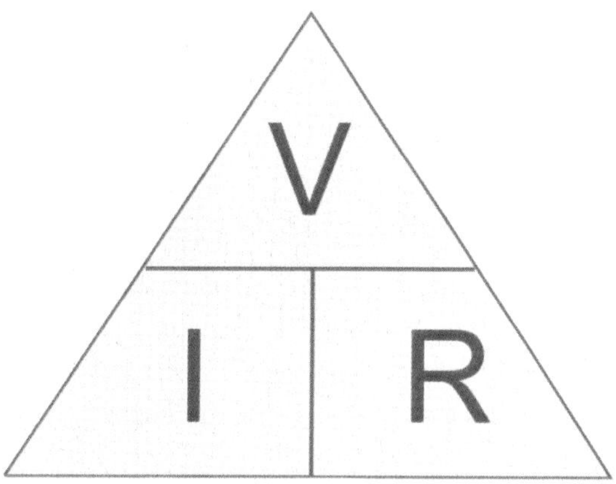

To use the triangle cover up the unknown quantity and then and then calculate it from the other two. If they are in line they are multiplied, but if one is on top of the other then they should be divided. In other words if current has to be calculated the voltage is divided by the resistance i.e. V/R and so forth.

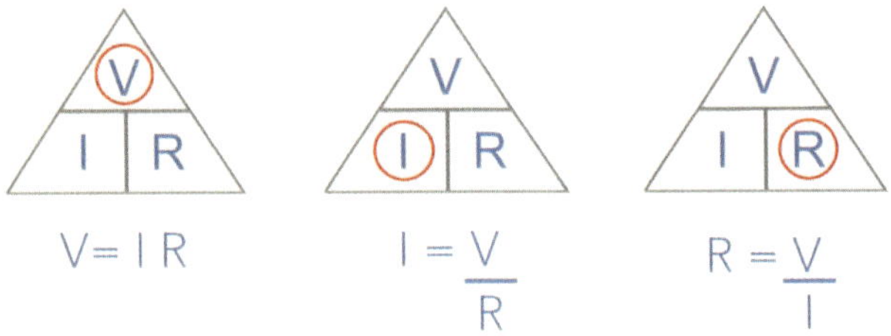

If the voltage has to be calculated then it is found by multiplying the current by the resistance i.e. I x R.

Ohm's Law Calculation Example

If a voltage of 10 volts is placed across a 500 ohm resistor determine the amount of current that will flow.

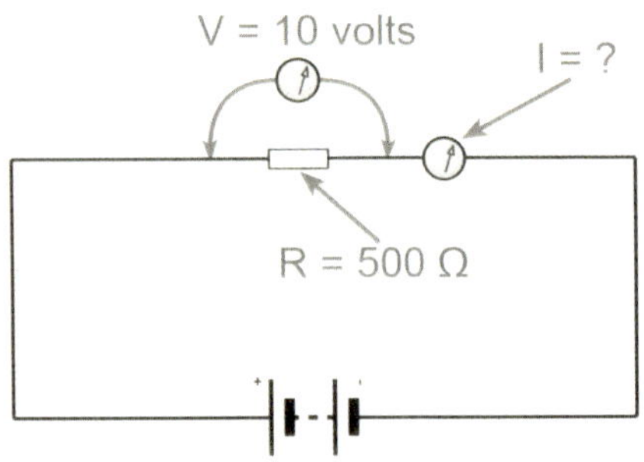

Looking at the Ohm's Law triangle the current is the unknown leaving the voltage and resistance as the known values.

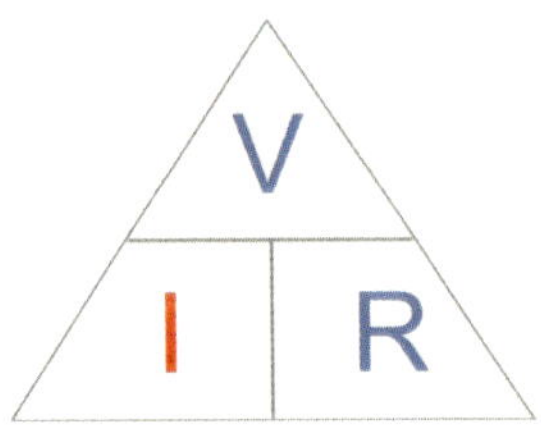

$$I = \frac{V}{R}$$

In this way the current is found by dividing the voltage by the resistance.

$$I = \frac{V}{R} = \frac{10}{500} = 0.02A = 20mA$$

Linear Graph

It can be seen that if the voltage and current were plotted for a fixed resistor or a length of wire, etc, there would be a linear curve.

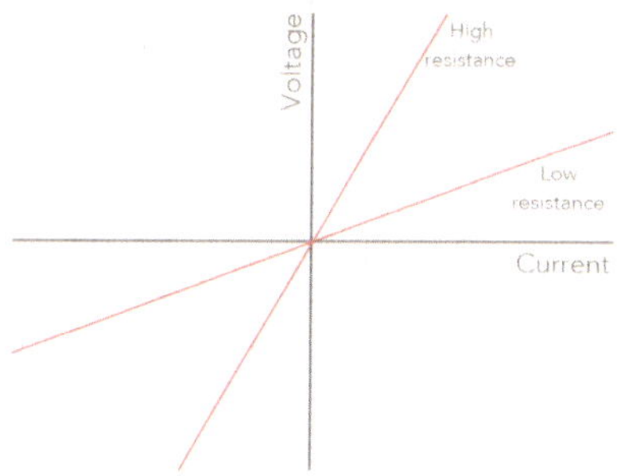

Plot of voltage and current for a linear resistance

It can be seen that doubling the voltage doubles the current that passes through the particular circuit element.

On the graph, there are two lines, one for a higher resistance - this one requires more voltage to be applied for a given current flowing. Accordingly this must have a higher resistance. Conversely the curve for the lower resistance shows a component that requires a lower voltage to be applied for a given current.

Non-linear Components

Ohm's law in its basic form where a doubling in voltage results in a doubling in current applies to linear components like ordinary resistors. Some components like diodes have non-linear curves where the resistance is affected by the applied voltage.

Ohm's Law is one of the most basic concepts within electrical and electronic engineering. The concept of an item having a certain resistance which determines the amount of current that flows through it for a certain voltage is key to the operation virtually all circuits.

Ohm's Law Pie Chart

To help us understand the the relationship between the various values a little further, we can take all of the Ohm's Law equations from above for finding Voltage, Current, Resistance and of course Power and condense them into a simple Ohm's Law pie chart for use in AC and DC circuits and calculations as shown,

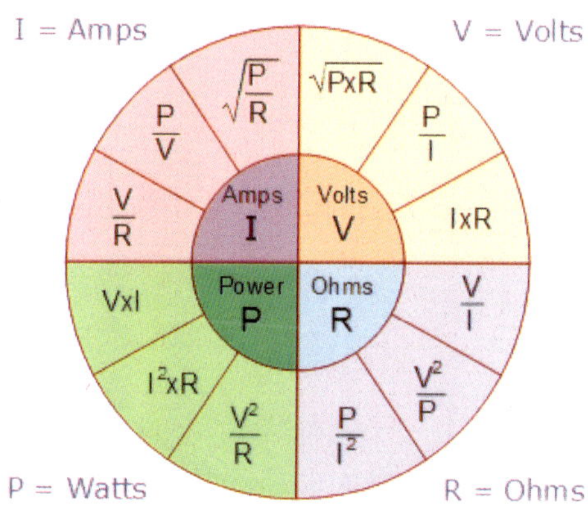

As well as using the Ohm's Law Pie Chart shown above, we can also put the individual Ohm's Law

equations into a simple matrix table as shown for easy reference when calculating an unknown value.

Ohm's Law Matrix Table

Ohms Law Formulas				
Known Values	Resistance (R)	Current (I)	Voltage (V)	Power (P)
Current & Resistance	---	---	$V = I \times R$	$P = I^2 \times R$
Voltage & Current	$R = \frac{V}{I}$	---	---	$P = V \times I$
Power & Current	$R = \frac{P}{I^2}$	---	$V = \frac{P}{I}$	---
Voltage & Resistance	---	$I = \frac{V}{R}$	---	$P = \frac{V^2}{R}$
Power & Resistance	---	$I = \sqrt{\frac{P}{R}}$	$V = \sqrt{P \times R}$	---
Voltage & Power	$R = \frac{V^2}{P}$	$I = \frac{P}{V}$	---	---

Ohm's Law Example

For the circuit shown below find the Voltage (V), the Current (I), the Resistance (R) and the Power (P).

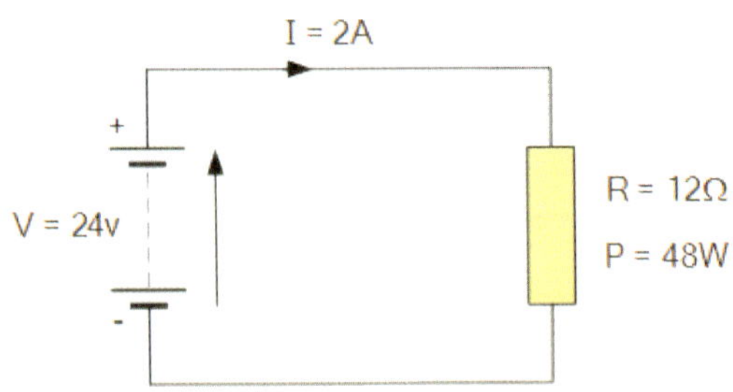

Voltage [V = I x R] = 2 x 12Ω = 24V

Current [I = V ÷ R] = 24 ÷ 12Ω = 2A

Resistance [R = V ÷ I] = 24 ÷ 2 = 12 Ω

Power [P = V x I] = 24 x 2 = 48W

Power within an electrical circuit is only present when BOTH voltage and current are present. For example, in an open-circuit condition, voltage is present but there is no current flow I = 0 (zero), therefore V*0 is 0 so the power dissipated within the circuit must also be 0. Likewise, if we have a short-circuit condition, current flow is present but there is no voltage V = 0, therefore 0*I = 0 so again the power dissipated within the circuit is 0.

As electrical power is the product of V*I, the power dissipated in a circuit is the same whether the circuit contains high voltage and low current or low voltage and high current flow. Generally, electrical power is dissipated in the form of Heat (heaters), Mechanical Work such as motors, Energy in the form of radiated (Lamps) or as stored energy (Batteries).

Kirchhoff's Circuit Laws

Kirchhoff's circuit laws are two equations first published by Gustav Kirchhoff in 1845. Fundamentally, they address conservation of energy and charge in the context of electrical circuits.

Although Kirchhoff's Laws can be derived from the equations of James Clerk Maxwell, Maxwell did not publish his set of differential equations (which form the foundation of classical electrodynamics, optics, and electric circuits) until 1861 and 1862. Kirchhoff, rather, used Georg Ohm's work as a foundation for *Kirchhoff's current law (KCL)* and *Kirchhoff's voltage law (KVL).*

Kirchhoff's laws are extremely important to the analysis of closed circuits. Consider, for example, the circuit illustrated in the figure below, consisting of five resistors in a combination of in series and parallel arrangements. Simplification of this circuit to a combination of series and parallel connections is impossible. However, using Kirchhoff's rules, one can analyze the circuit to determine the parameters of this circuit using the values of the resistors (R_1, R_2, R_3, r_1 and r_2). Also of importance in this example is that the values E_1 and E_2 represent sources of voltage (e.g., batteries).

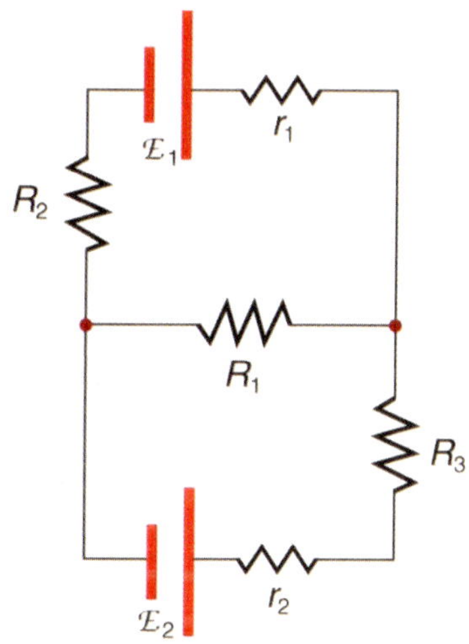

Closed Circuit: To determine all variables (i.e., current and voltage drops across the different resistors) in this circuit, Kirchhoff's rules must be applied

As a final note, Kirchhoff's laws depend on certain conditions. The voltage law is a simplification of Faraday's law of induction, and is based on the assumption that there is no fluctuating magnetic field within the closed loop. Thus, although this law can be applied to circuits containing resistors and capacitors (as well as other circuit elements), it can only be used as an approximation to the behavior of the circuit when a changing current and therefore magnetic field are involved.

The Junction Rule

Kirchhoff's junction rule states that at any circuit junction, the sum of the currents flowing into and out of that junction are equal.

Kirchhoff's junction rule, also known as Kirchhoff's current law (KCL), Kirchoff's first law, Kirchhoff's point rule, and Kirchhoff's nodal rule, is an application of the principle of conservation of electric charge.

Kirchhoff's junction rule states that at any junction (node) in an electrical circuit, the sum of the currents flowing into that junction is equal to the sum of the currents flowing out of that junction. In other words, given that a current will be positive or negative depending on whether it is flowing towards or away from a junction, the algebraic sum of currents in a network of conductors meeting at a point is equal to zero.

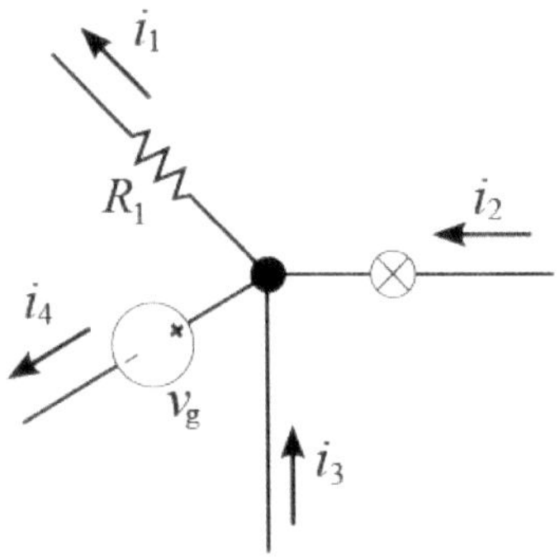

Kirchhoff's Junction Law illustrated as currents flowing into and out of a junction.

Thus, Kirchoff's junction rule can be stated mathematically as a sum of currents (I):

$$\sum_{k=1}^{n} I_k = 0$$

where n is the total number of branches carrying current towards or away from the node.

This law is founded on the conservation of charge (measured in coulombs), which is the product of current (amperes) and time (seconds).

Limitation

Kirchhoff's junction law is limited in its applicability. It holds for all cases in which total electric charge (Q) is constant in the region in consideration. Practically, this is always true so long as the law is applied for a specific point. Over a region, however, charge density may not be constant. Because charge is conserved, the only way this is possible is if there is a flow of charge across the boundary of the region. This flow would be a current, thus violating Kirchhoff's junction law.

The Loop Rule

Kirchhoff's loop rule states that the sum of the emf values in any closed loop is equal to the sum of the potential drops in that loop.

Kirchhoff's loop rule (otherwise known as Kirchhoff's voltage law (KVL), Kirchhoff's mesh rule, Kirchhoff's second law, or Kirchhoff's second rule) is a rule pertaining to circuits, and is based on the principle of conservation of energy.

Conservation of energy—the principle that energy is neither created nor destroyed—is a ubiquitous principle across many studies in physics, including circuits. Applied to circuitry, it is implicit that the directed sum of the electrical potential differences (voltages) around any closed network is equal to zero. In other words, the sum of the electromotive force (emf) values in any closed loop is equal to the sum of the potential drops in that loop (which may come from resistors).

Another equivalent statement is that the algebraic sum of the products of resistances of conductors (and currents in them) in a closed loop is equal to the total electromotive force available in that loop. Mathematically, Kirchhoff's loop rule can be represented as the sum of voltages in a circuit, which is equated with zero:

$$\sum_{k=1}^{n} V_k = 0.$$

Here, V_k is the voltage across element k, and n is the total number of elements in the closed loop circuit. An illustration of such a circuit is shown in figure. In this example, the sum of v_1, v_2, v_3, and v_4 (and v_5 if it is included), is zero.

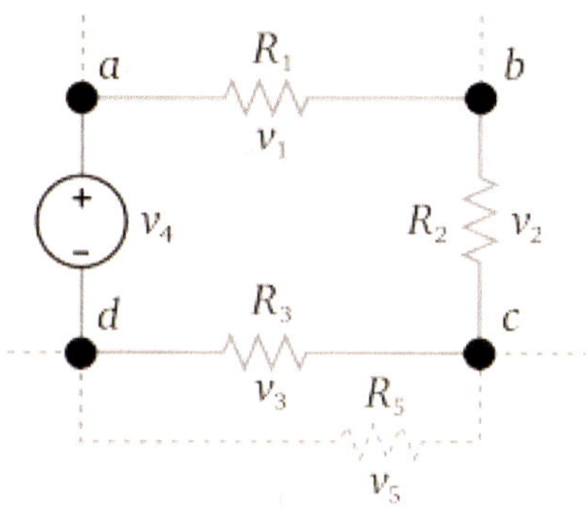

Kirchhoff's Loop Rule: Kirchhoff's loop rule states that the sum of all the voltages around the loop is equal to zero: $v1 + v2 + v3 - v4 = 0$

Given that voltage is a measurement of energy per unit charge, Kirchhoff's loop rule is based on the law of conservation of energy, which states: the total energy gained per unit charge must equal the amount of energy lost per unit of charge.

Example

The following illustrates the changes in potential in a simple series circuit loop. Kirchhoff's second rule requires $emf - Ir - IR_1 - IR_2 = 0$. Rearranged, this is $emf = Ir + IR_1 + IR_2$, meaning that the emf equals the sum of the IR (voltage) drops in the loop. The emf supplies 18 V, which is reduced to zero by the resistances, with 1 V across the internal resistance, and 12 V and 5 V across the two load resistances, for a total of 18 V.

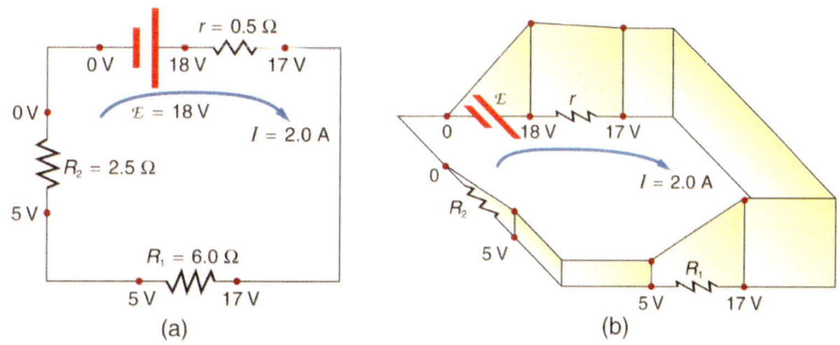

The Loop Rule: An example of Kirchhoff's second rule where the sum of the changes in potential around a closed loop must be zero,

(a) In this standard schematic of a simple series circuit, the emf supplies 18 V, which is reduced to zero by the resistances, with 1 V across the internal resistance, and 12 V and 5 V across the two load resistances, for a total of 18 V.

(b) This perspective view represents the potential as something like a roller coaster, where charge is raised in potential by the emf and lowered by the resistances.

Limitation

Kirchhoff's loop rule is a simplification of Faraday's law of induction, and holds under the assumption that there is no fluctuating magnetic field linking the closed loop. In the presence of a variable magnetic field, electric fields could be induced and emf could be produced, in which case Kirchhoff's loop rule breaks down.

Applications

Kirchhoff's rules can be used to analyze any circuit and modified for those with EMFs, resistors, capacitors and more. Kirchhoff's rules can be used to analyze any circuit by modifying them for those circuits with electromotive forces, resistors, capacitors and more. Practically speaking, however, the rules are only useful for characterizing those circuits that cannot be simplified by combining elements in series and parallel.

Combinations in series and parallel are typically much easier to perform than applying either of Kirchhoff's rules, but Kirchhoff's rules are more broadly applicable and should be used to solve problems involving complex circuits that cannot be simplified by combining circuit elements in series or parallel.

Example of Kirchoff's Rules

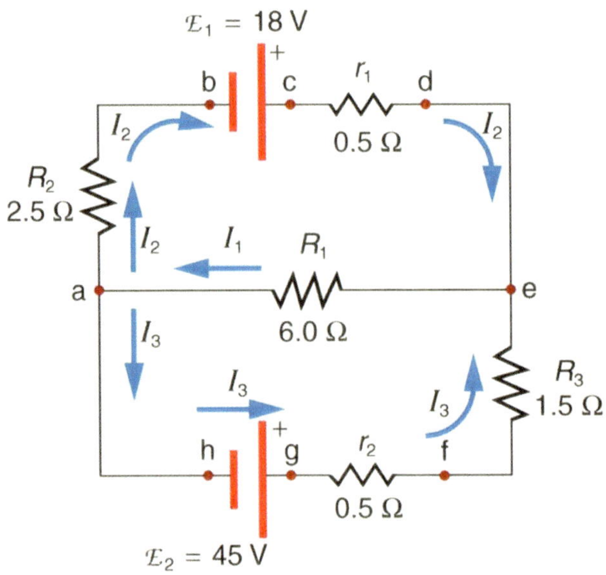

Shows a very complex circuit, but Kirchhoff's loop and junction rules can be applied. To solve the circuit for currents I_1, I_2, and I_3, both rules are necessary:

Kirchhoff's Rules sample problem: This figure shows a very complicated circuit, which can be reduced and solved using Kirchoff's Rules.

Applying Kirchhoff's junction rule at point a, we find:

$$I_1 = I_2 + I_3$$

because I_1 flows into point a, while I_2 and I_3 flow out. The same can be found at point e. We now must solve this equation for each of the three unknown variables, which will require three different equations.

Considering loop abcdea, we can use Kirchhoff's loop rule:

$$-I_2R_2 + emf_1 - I_2r_1 - I_1R_1 = -I_2(R_2) + r_1) + emf_1 - I_1R_1 = 0$$

Substituting values of resistance and emf from the figure diagram and canceling the ampere unit gives:

$$-3I_2 + 18 - 6I_1 = 0$$

This is the second part of a system of three equations that we can use to find all three current values. The last can be found by applying the loop rule to loop aefgha, which gives:

$$I_1R_1 + I_3R_3 + I_3r_2 - emf_2 = I_1R_1 + I_3(R_3 + r_2) - emf_2 = 0$$

Using substitution and simplifying, this becomes:

$$6I_1 + 2I_3 - 45 = 0$$

In this case, the signs were reversed compared with the other loop, because elements are traversed in the opposite direction.

We now have three equations that can be used in a system. The second will be used to define I_2, and can be rearranged to:

$$I_2 = 6 - 2I_1$$

The third equation can be used to define I_3, and can be rearranged to:

$$I_3 = 22.5 - 3I_1$$

Substituting the new definitions of I_2 and I_3 (which are both in common terms of I_1), into the first equation ($I_1=I_2+I_3$), we get:

$$I_1 = (6 - 2I_1) + (22.5 - 3I_1) = 28.5 - 5I_1$$

Simplifying, we find that I_1=4.75 A. Inserting this value into the other two equations, we find that I_2=-3.50 A and I_3=8.25 A.

Source Transformation

Source Transformation simply means replacing one source by an equivalent source. A practical voltage source can be transformed into an equivalent practical current source and similarly a practical current source into voltage source.

Any practical voltage source or simply a voltage source consists of an ideal voltage source in series with an internal resistance or impedance (for any ideal source this impedance will be zero), the output voltage becomes independent of the load current. Cells, batteries and generators are the example of the voltage source.

For any practical current source or simply current source, is an ideal current source in parallel with the internal resistance or impedance, for ideal current source this parallel impedance is infinity. The semiconductor devices like transistors, etc are treated as a current source or an output produce by direct or alternating voltage source is called as direct and alternating current source, respectively.

The voltage and current source are mutually transferable or in other words the source transformation i.e. voltage to current source and current to voltage source can be done. Let us understand this by considering a circuit given below:

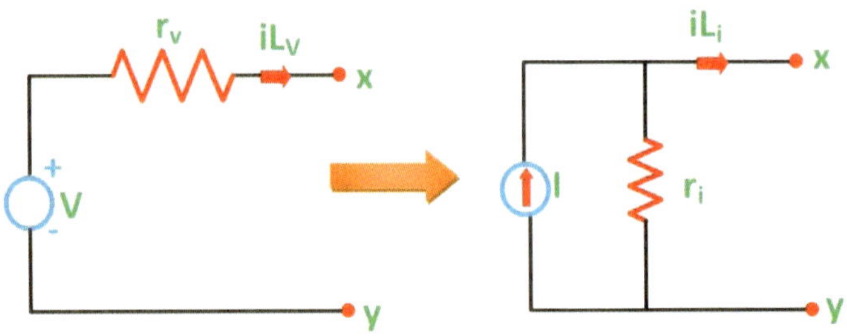

Figure A Figure B

Figure A represents a practical voltage source in series with the internal resistance r_v, while figure B represents a practical current source with parallel internal resistance r_i

For the practical voltage source the load current will be given by the equation,

$$iL_v = \frac{V}{r_V + r_L}$$

Where,

iL_v is the load current for the practical voltage source

V is the voltage

r_v is the internal resistance of the voltage source

r_L is the load resistance

It is assumed that the load resistance r_L is connected at the terminal x-y. Similarly for the practical current source the load current is given as:

$$iL_i = I\frac{r_i}{r_i + r_L}$$

Where,

iL_i is the load current for the practical current source

I is the current

r_i is the internal resistance of the current source

r_L is the load resistance connected across the terminal x-y in the figure B

Two sources become identical, when we will equate the above two equation,

$$\frac{V}{r_V + r_L} = I\frac{r_i}{r_i + r_L}$$

However, for the current source the terminal voltage at x-y would be Iri, x-y terminal are open. i.e.,

$$V = I \times r_i$$

Therefore, we will get,

$$r_V + r_L = r_i + r_L \text{ or}$$
$$r_V = r_L$$

Therefore, for any practical voltage source, if the ideal voltage be V and internal resistance be r_v, the voltage source can be replaced by a current source I with the internal resistance in parallel with the current source.

Source Transformation : Conversion of Voltage Source into Current Source

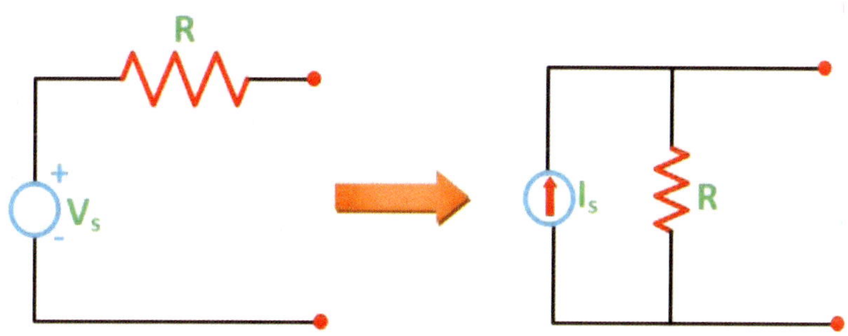

When the voltage source is connected with the resistance in series and it has to be converted into the current source than the resistance is connected in parallel with the current source as shown in the above figure.

Where, $I_s = V_s/R$

Conversion of Current Source into Voltage Source

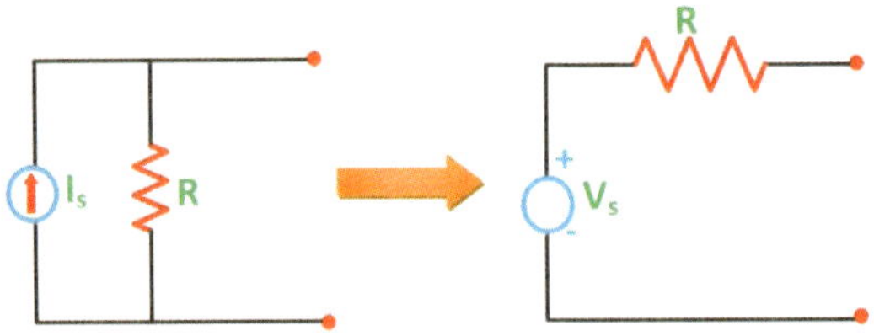

In the above circuit diagram a current source which is connected in parallel with the resistance is transformed into a voltage source by placing the resistance in series with the voltage source.

Where, $V_s = I_s / R$.

Substitution Theorem

Substitution Theorem states that the voltage across any branch or the current through that branch of a network being known, the branch can be replaced by the combination of various elements that will make the same voltage and current through that branch. In other words, the Substitution Theorem says that for branch equivalence, the terminal voltage and current must be same.

The concept of the theorem is based on the substitution of one element from another element. This theorem gives intuition on the behaviour of the circuit. It also helps in proving several other theorems. But the substitution theorem cannot use for solving the theorem which has more than two sources which are neither connected in series nor parallel.

Explanation of Substitution Theorem

Simply one can say that the Substitution Theorem is the replacement of one element with another equivalent element. In a network, if any element is substituted or replaced by a voltage or current source whose voltage and current across or through that element will remain unchanged as the previous network.

Let us understand the theorem with the help of the circuit diagram shown below:

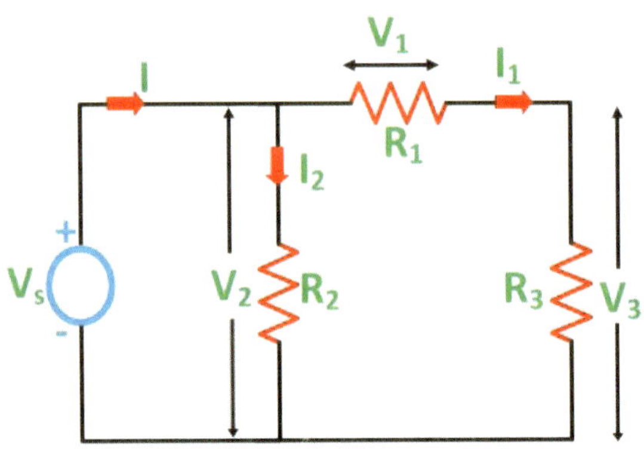

Here various resistances R_1, R_2, R_3 is connected across with the voltage source (V). Current I is flowing through the circuit, which is divided into current I_1 flowing through the resistance R_1 and the current I_2 flowing through the resistance R_2. V_1, V_2 and V_3 are the voltage drop across the resistance R_1, R_2 and R_3 respectively.

Now if the resistance R_3 is substituted by the voltage source V_3 as shown in the circuit diagram below,

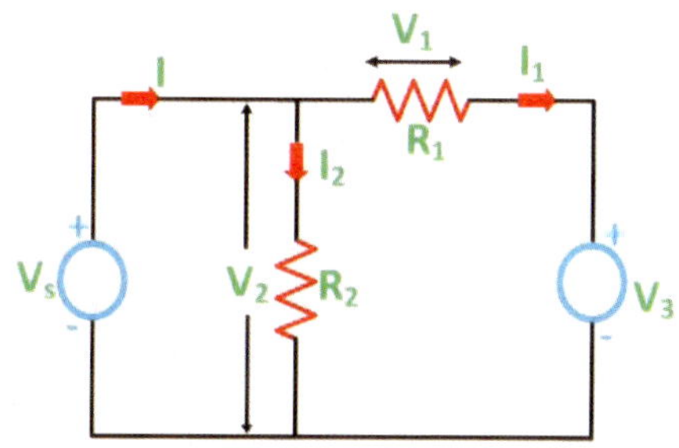

In the circuit diagram shown below the resistance, R_3 is replaced by the current flowing through that element, i.e. I_1.

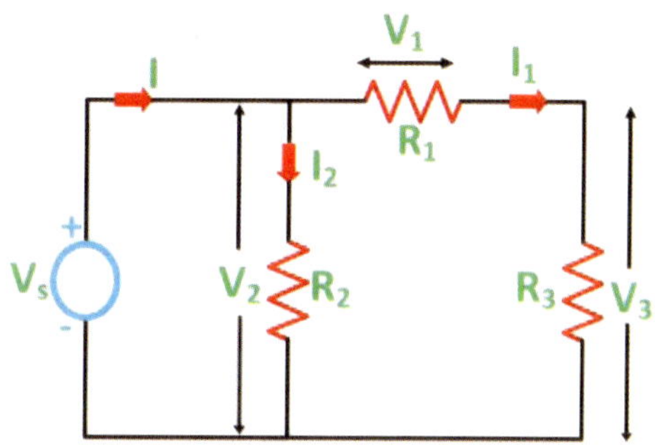

In both the cases shown above if the element is substituted by the voltage source or the current source, then also, the initial conditions of the circuit does not alter. This means that the voltage across the resistance and current flowing through the resistance unaltered even if they are substituted by other sources.

Steps for solving Network using Substitution Theorem

Step 1 – First obtain the concerned branch voltage and current flowing through the branch given by V_{xy} and I_{xy} as shown below in the figure A.

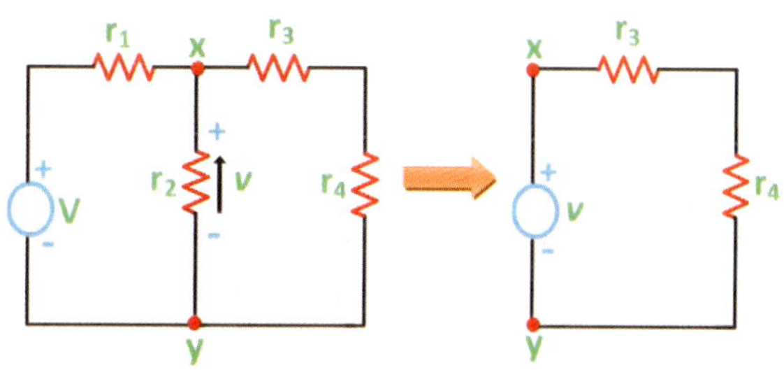

Figure A

Figure B

Step 2 – The branch may be substituted by an independent voltage source as shown above in figure B.

Step 3 – Similarly the branch may be replaced by an independent current source as shown below in the circuit diagram C.

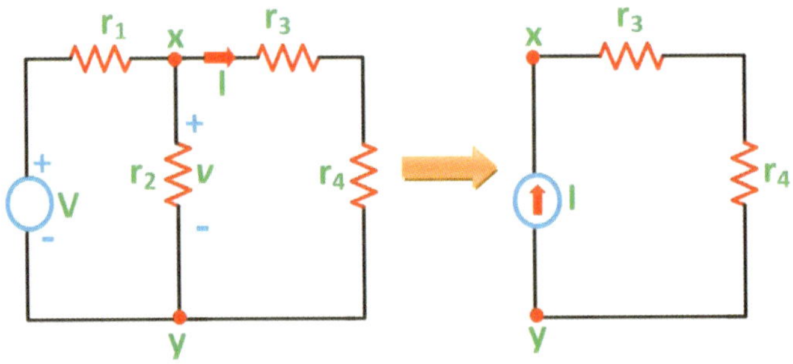

Figure C

Step 4 – Hence it is seen that the voltage drop and the current flowing through the circuit A shown above will be same if it is substituted by any independent voltage or current source shown in the figure B and C.

Norton's Theorem

Norton's theorem states that any linear complex electrical circuit can be reduced into a simple electric circuit with one current and resistance connected in parallel. For understanding in depth regarding norton theory, let us consider Norton's theorem examples as follows:

Examples

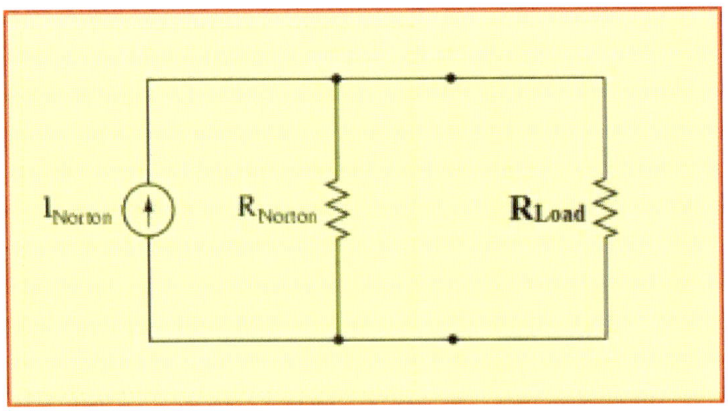

Primarily, let us consider a simple electrical circuit that consists of two voltage sources and three resistors which are connected as shown in the above figure. The above circuit consists of three resistors among which R_2 resistor is considered as load. Then, the circuit can be represented as shown below:

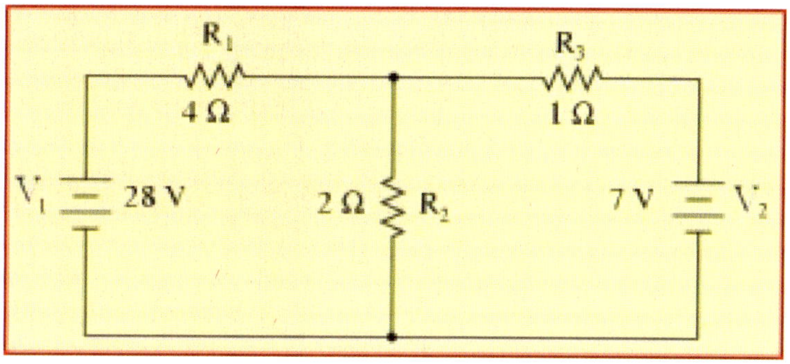

Nortons Theorem Example Circuit with Load Resistor

We know that, if the load changes, then the calculation of various parameters of electric circuits is difficult. So, network theorems are used for calculating the network parameters easily.

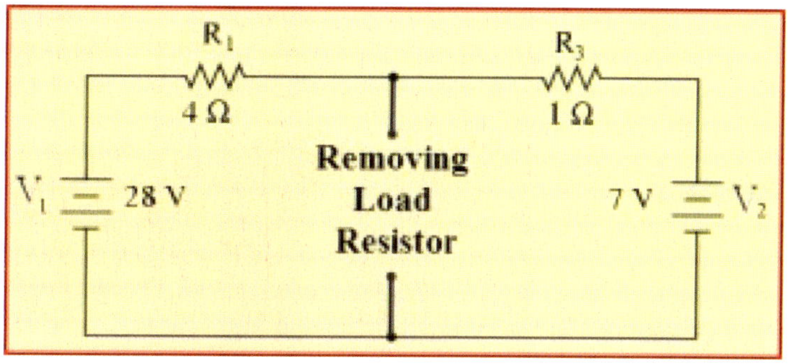

Nortons Theorem Example Circuit After Removing Load Resistor

In this Norton's theorem also we follow the procedure similar to thevenins theorem (up to some extent). Here, primarily remove the load (consider resistor R_2=2 Ohms as load in the circuit) as shown in the above figure. Then, short circuit the load terminals with a wire (exactly opposite to the procedure that we follow in thevenins theorem, i.e., open circuit of load terminals) as shown in the below figure. Now, calculate the resultant current (current through resistors R_1, R_3, and short circuit line after removing R_2) as shown in the figure below:

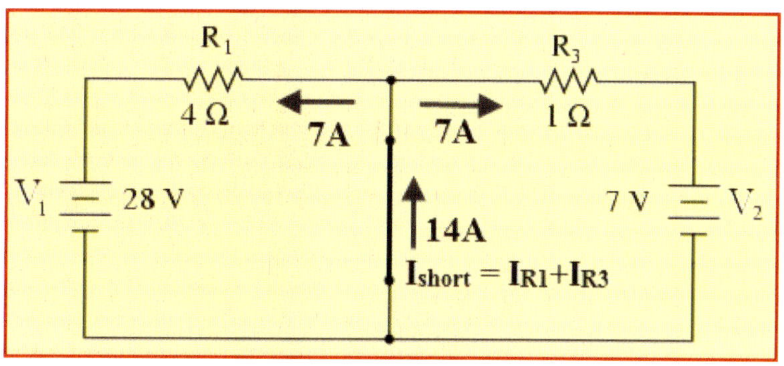

Current Through R1, R3, and Short Circuited Load

From the above figure, the Nortons source current is equal to 14A which is used in the Norton's equivalent circuit as shown in the below figure. Norton's theorem equivalent circuit consists of the Norton current source (INorton) in parallel with Norton's equivalent resistance (RNorton) and load (here R_2=2Ohms).

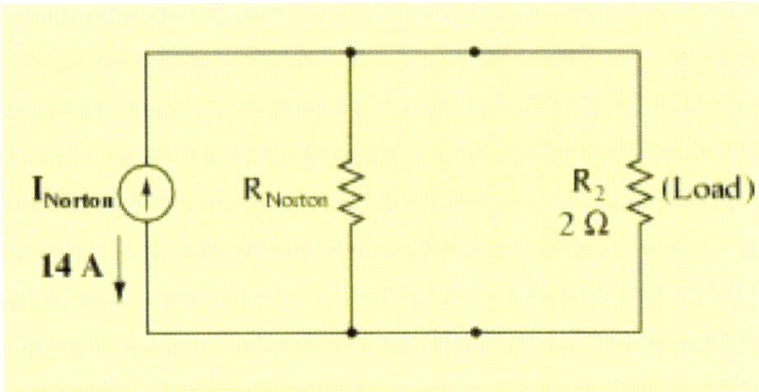

Nortons Equivalent Circuit with INorton, RNorton, RLoad

This Nortorn's theorem equivalent circuit is a simple parallel circuit as shown in the figure. Now, for calculating Norton's equivalent resistance we have to follow two procedures such as Thevenins theorem and Superposition theorem.

Primarily, remove the load resistance (similar to thevenins theorem step of calculating thevenins resistance). Then, replace voltage sources with short circuit (wires in case of ideal voltage sources and in case of practical voltage sources their internal resistances are used). Similarly, current sources with open circuit (breaks in case of ideal current sources and in case of practical current sources their internal resistances are used). Now, the circuit becomes as shown in the figure below and it is a simple parallel circuit with resistors.

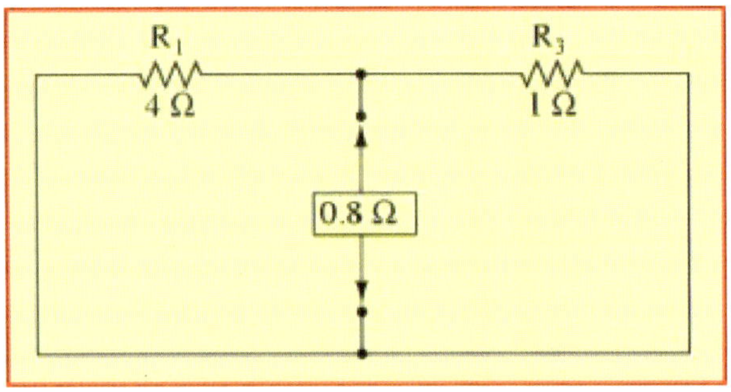

Finding Nortons Resistance

As the resistors R_1 and R_3 are parallel to each other, the value of Norton's resistance is equal to parallel resistance value of R_1 and R_3. Then, the total Norton's theorem equivalent circuit can be represented as shown in the circuit below:

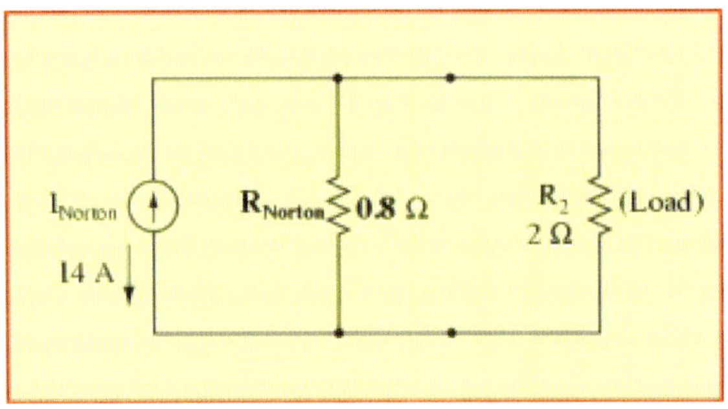

Norton's Theorem Equivalent Circuit

The formula for calculating the load current, Iload can be calculated using various basic laws such as Ohm's law, Krichhoff's voltage law, and Krichhoff's current law.

Thus, the current passing through the load resistor R load (R_2) is given by,

$$I_L = \frac{I_N . R_N}{R_N + R_L}$$

Load Current formula

Where,

I_N = Norton's current (14A)

R_N = Norton's resistance (0.8 Ohms)

R_L = Load resistance (2 Ohms)

Therefore, I load = current passing through load resistance = 4A.

Similarly, the large, complex, linear networks with several numbers of sources (current or voltage sources) and resistors can be reduced to simple parallel circuits with single current source in parallel with Norton's resistance and load.

Thus, the Norton's equivalent circuit with Rn and In can be determined and a simple parallel circuit can be formed (from a complex network circuit). The calculations of the circuit parameters can be easily analyzed. If one resistance in the circuit is changed rapidly (load), then the Norton's theorem can be used to perform calculations easily.

Compensation Theorem

Compensation Theorem states that in a linear time invariant network when the resistance (R) of an uncoupled branch, carrying a current (I), is changed by (ΔR). The currents in all the branches would change and can be obtained by assuming that an ideal voltage source of (V_C) has been connected such that V_C= I (ΔR) in series with (R + ΔR) when all other sources in the network are replaced by their internal resistances.

In Compensation Theorem, the source voltage (V_C) opposes the original current. In simple words compensation theorem can be stated as – the resistance of any network can be replaced by a voltage source, having the same voltage as the voltage drop across the resistance which is replaced.

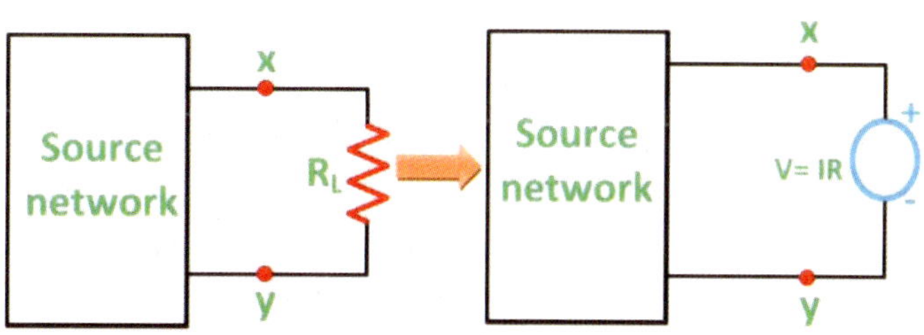

Let us assume a load R_L be connected to a DC source network whose Thevenin's equivalent gives V_o as the Thevenin's voltage and R_{TH} as the Thevenin's resistance as shown in the figure below:

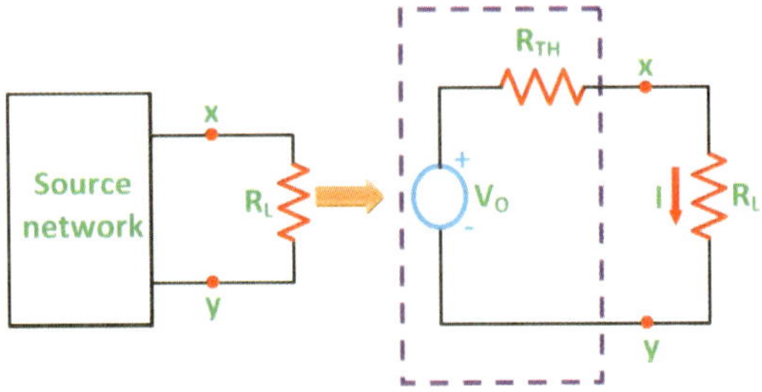

Here,

$$I = \frac{V_0}{R_{TH} + R_L}$$

Let the load resistance R_L be changed to $(R_L + \Delta R_L)$. Since the rest of the circuit remains unchanged, the Thevenin's equivalent network remains the same as shown in the circuit diagram below:

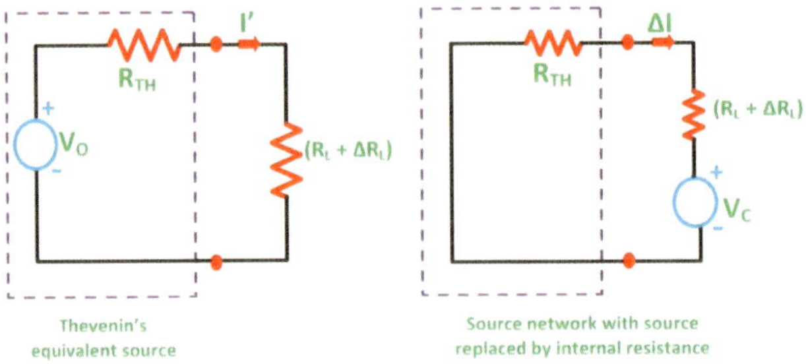

Here,

$$I' = \frac{V_0}{R_{TH} + (R_L + \Delta R_L)}$$

The change of current being termed as ΔI,

Therefore,

$$\Delta I = I' - I$$

Putting the value of I' and I from the above three equation we will get the following equation,

$$\Delta I = \frac{V_0}{R_{TH} + (R_L + \Delta R_L)} - \frac{V_0}{R_{TH} + R_L}$$

$$\Delta I = \frac{V_0\{R_{TH} + R_L - (R_{TH} + R_L + \Delta R_L)\}}{(R_{TH} + R_L + \Delta R_L)(R_{TH} + R_L)}$$

$$\Delta I = -\left[\frac{V_0}{R_{TH} + R_L}\right]\frac{\Delta R_L}{R_{TH} + R_L + \Delta R_L}$$

Now, putting the value of I from the equation mentioned in the starting of this topic in the equation above, we will get the following equation,

$$\Delta I = -\frac{I\Delta R_L}{R_{TH} + R_L + \Delta R_L}$$

As we know, $V_C = I \Delta R_L$ and is known as compensating voltage.

Therefore, the above equation becomes,

$$\Delta I = -\frac{-V_C}{R_{TH} + R_L + \Delta R_L}$$

Hence, Compensation Theorem tells that with the change of branch resistance, branch currents changes and the change is equivalent to an ideal compensating voltage source in series with the branch opposing the original current, all other sources in the network being replaced by their internal resistances.

Superposition Theorem

Superposition theorem is based on the concept of linearity between the response and excitation of an electrical circuit. It states that the response in a particular branch of a linear circuit when multiple independent sources are acting at the same time is equivalent to the sum of the responses due to each independent source acting at a time.

In this method, we will consider only one independent source at a time. So, we have to eliminate the remaining independent sources from the circuit. We can eliminate the voltage sources by shorting their two terminals and similarly, the current sources by opening their two terminals.

Therefore, we need to find the response in a particular branch 'n' times if there are 'n' independent sources. The response in a particular branch could be either current flowing through that branch or voltage across that branch.

Procedure of Superposition Theorem

Follow these steps in order to find the response in a particular branch using superposition theorem.

- Step 1: Find the response in a particular branch by considering one independent source and eliminating the remaining independent sources present in the network.
- Step 2: Repeat Step 1 for all independent sources present in the network.
- Step 3: Add all the responses in order to get the overall response in a particular branch when all independent sources are present in the network.

Example: Find the current flowing through 20 Ω resistor of the following circuit using superposition theorem.

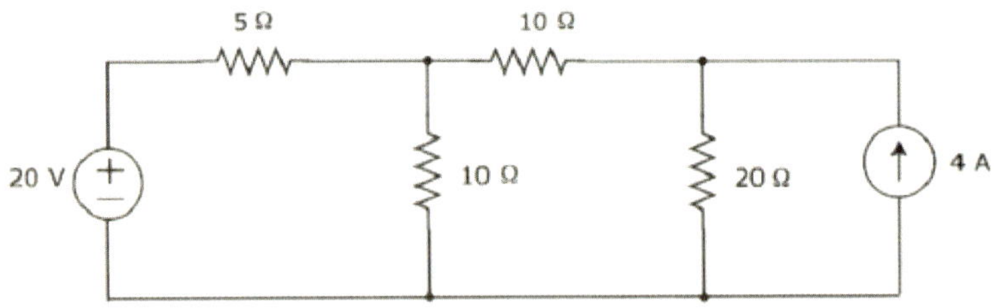

Step 1

Let us find the current flowing through 20 Ω resistor by considering only 20 V voltage source. In this case, we can eliminate the 4 A current source by making open circuit of it. The modified circuit diagram is shown in the following figure.

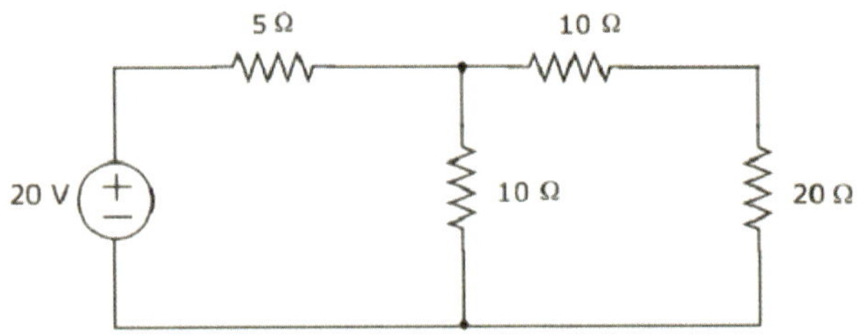

There is only one principal node except Ground in the above circuit. So, we can use nodal analysis method. The node voltage V_1 is labelled in the following figure. Here, V_1 is the voltage from node 1 with respect to ground.

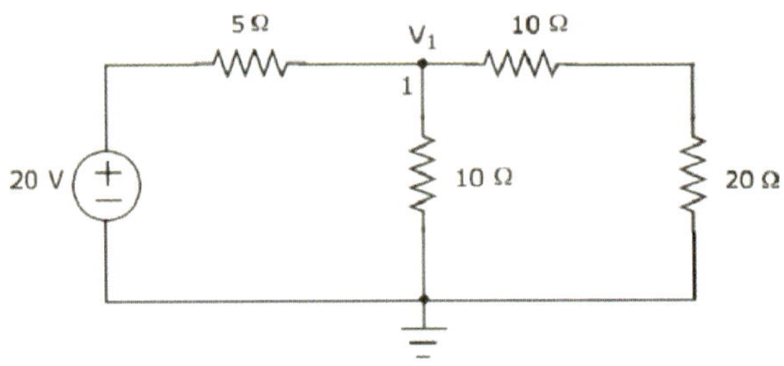

The nodal equation at node 1 is,

$$\frac{V_1 - 20}{5} + \frac{V_1}{10} + \frac{V_1}{10+20} = 0$$

$$\Rightarrow \frac{6V_1 - 120 + 3V_1 + V_1}{30} = 0$$

$$\Rightarrow 10V_1 = 120$$

$$\Rightarrow V_1 = 12\,V$$

The current flowing through 20 Ω resistor can be found by doing the following simplification.

$$I_1 = \frac{V_1}{10+20}$$

Substitute the value of V_1 in the above equation.

$$I_1 = \frac{12}{10+20} = \frac{12}{30} = 0.4\,A$$

Therefore, the current flowing through 20 Ω resistor is 0.4 A, when only 20 V voltage source is considered.

Step 2

Let us find the current flowing through 20 Ω resistor by considering only 4 A current source. In this case, we can eliminate the 20 V voltage source by making short-circuit of it. The modified circuit diagram is shown in the following figure.

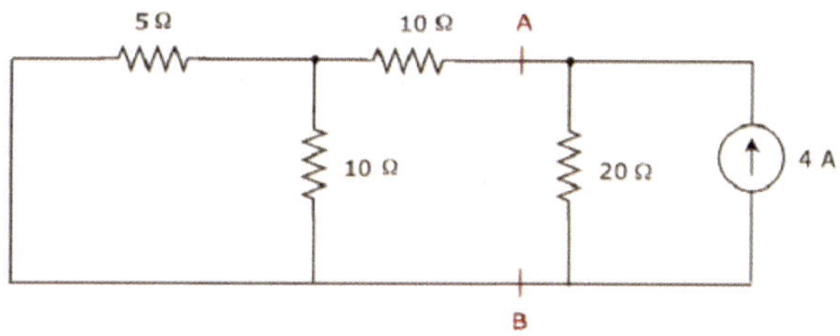

$$R_{AB} = \left(\frac{5\times 10}{5+10}\right) + 10 = \frac{10}{3} + 10 = \frac{40}{3}\Omega$$

The simplified circuit diagram is shown in the following figure.

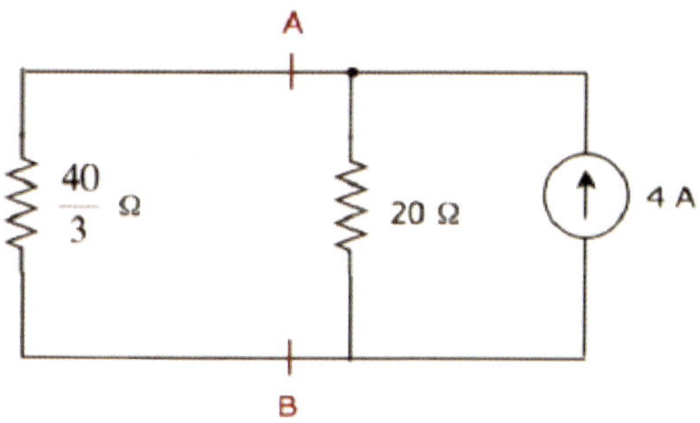

We can find the current flowing through 20 Ω resistor, by using current division principle.

$$I_2 = I_S\left(\frac{R_1}{R_1+R_2}\right)$$

Substitute I_S=4A, $R_1 = \frac{40}{3}$ Ω and R_2=20Ω in the above equation.

$$I_2 = 4\left(\frac{\frac{40}{3}}{\frac{40}{3}+20}\right) = 4\left(\frac{40}{100}\right) = 1.6A$$

Therefore, the current flowing through 20 Ω resistor is 1.6 A, when only 4 A current source is considered.

Step 3

We will get the current flowing through 20 Ω resistor of the given circuit by doing the addition of two currents that we got in step 1 and step 2. Mathematically, it can be written as

$$I = I_1 + I_2$$

Substitute, the values of I_1 and I_2 in the above equation.

$$I = 0.4 + 1.6 = 2A$$

Therefore, the current flowing through 20 Ω resistor of given circuit is 2A.

Thevenin's Theorem

Thevenin's Theorem states that – any complicated network across its load terminals can be substituted by a voltage source with one resistance in series. This theorem helps in the study of the variation of current in a particular branch when the resistance of the branch is varied while the remaining network remains the same for example designing of electronics circuits.

A more general statement of Thevenin's Theorem is that any linear active network consisting of independent or dependent voltage and current source and the network elements can be replaced by an equivalent circuit having a voltage source in series with a resistance, that voltage source being the open circuited voltage across the open circuited load terminals and the resistance being the internal resistance of the source.

In other words, the current flowing through a resistor connected across any two terminals of a network by an equivalent circuit having a voltage source E_{th} in series with a resistor R_{th}. Where Eth is the open circuit voltage between the required two terminals called the Thevenin voltage and the R_{th} is the equivalent resistance of the network as seen from the two terminal with all other sources replaced by their internal resistances called Thevenin resistance.

Explanation of Thevenin's Theorem

The Thevenin's statement is explained with the help of a circuit shown below.

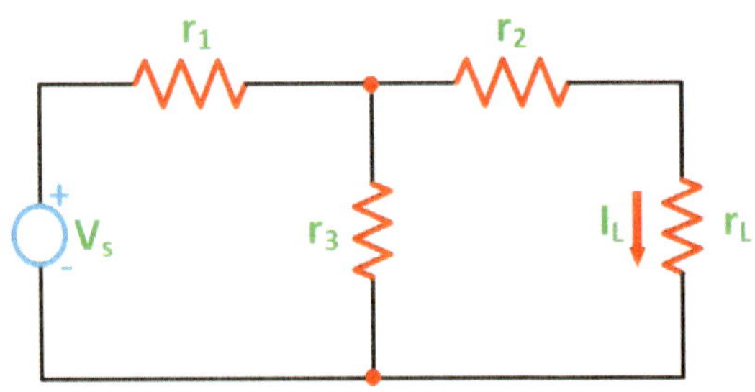

Let us consider a simple DC circuit as shown in the figure above, where we have to find the load current I_L by the Thevenin's theorem. In order to find the equivalent voltage source, r_L is removed from the circuit as shown in the figure below and V_{oc} or V_{TH} is calculated.

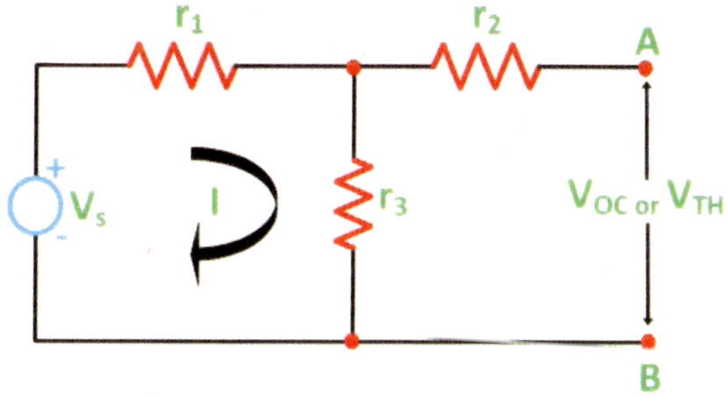

$$V_{OC} = I\, r_3 = \frac{V_S}{r_1 + r_3} r_3$$

Now, to find the internal resistance of the network (Thevenin's resistance or equivalent resistance) in series with the open circuit voltage V_{OC}, also known as Thevenin's voltage V_{TH}, the voltage source is removed or we can say it is deactivated by a short circuit (as the source does not have any internal resistance) as shown in the figure below:

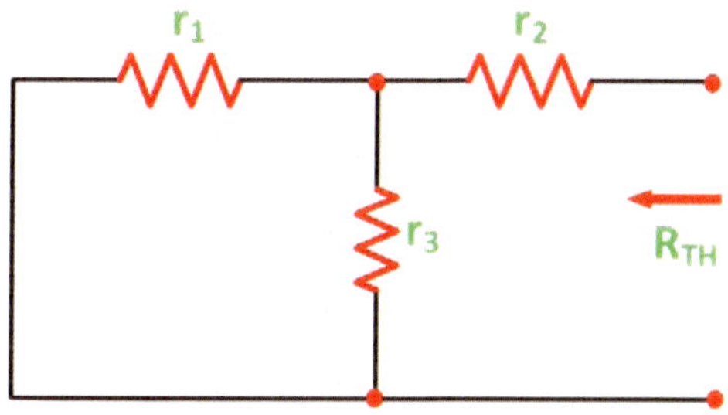

$$R_{TH} = r_2 + \frac{r_1 r_3}{r_1 + r_3}$$

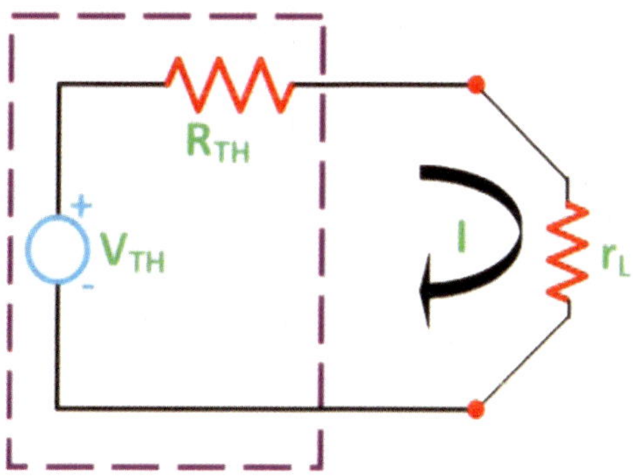

Equivalent Circuit of Thevenin's Theorem

As per Thevenin's Statement, the load current is determined by the circuit shown above and the equivalent Thevenin's circuit is obtained.

The Load current I_L is given as,

$$I_L = \frac{V_{TH}}{R_{TH} + r_L}$$

Where,

V_{TH} is the Thevenin's equivalent voltage. It is an open circuit voltage across the terminal AB known as load terminal R_{TH} is the Thevenin's equivalent resistance, as seen from the load terminals where all the sources are replaced by their internal impedance r_L is the load resistance.

Steps for Solving Thevenin's Theorem

- Step 1: First of all remove the load resistance r_L of the given circuit.
- Step 2: Replace all the impedance source by their internal resistance.
- Step 3: If sources are ideal then short circuit the voltage source and open the current source.
- Step 4: Now find the equivalent resistance at the load terminals know as Thevenin's Resistance (R_{TH}).
- Step 5: Draw the Thevenin's equivalent circuit by connecting the load resistance and after that determine the desired response.

This theorem is possibly the most extensively used networks theorem. It is applicable where it is desired to determine the current through or voltage across any one element in a network. The Thevenin's Theorem is an easy way to solve the complicated network.

Example

For example, consider the following circuit;

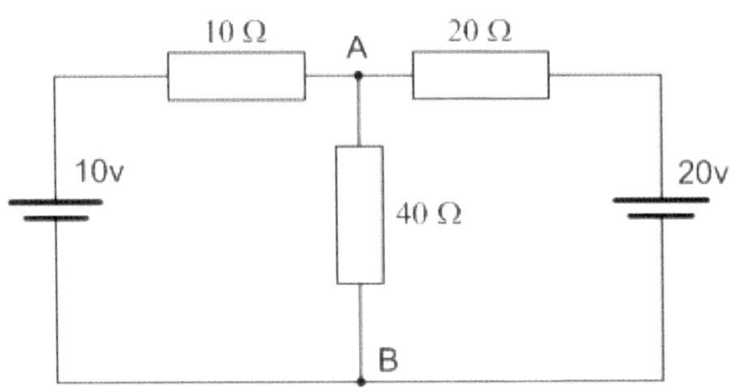

Firstly, we have to remove the centre 40Ω resistor and short out (not physically as this would be dangerous) all the emf´s connected to the circuit, or open circuit any current sources. The value of resistor Rs is found by calculating the total resistance at the terminals A and B with all the emf´s removed, and the value of the voltage required Vs is the total voltage across terminals A and B with an open circuit and no load resistor Rs connected. Then, we get the following circuit;

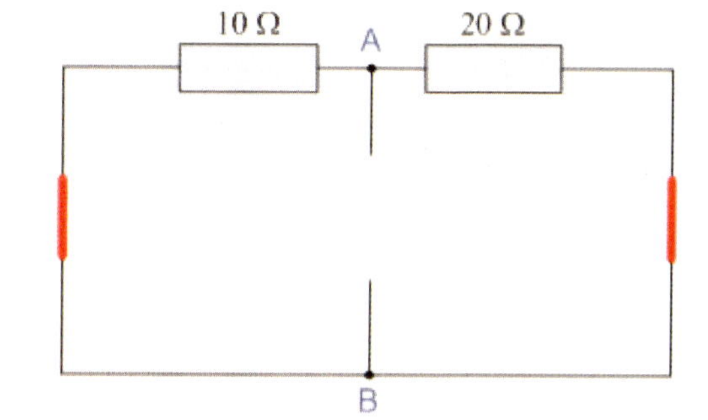

Find the Equivalent Resistance (Rs):

10ΩResistor in parallel with the 20ΩResistor,

$$R_T = \frac{R_1 \times R_2}{R_1 + R_2} = \frac{20 \times 10}{20 + 10} = 6.67\Omega$$

Find the Equivalent voltage (Vs):

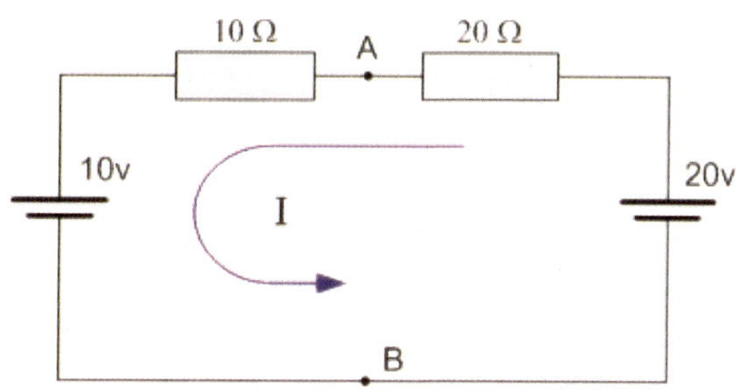

We now need to reconnect the two voltages back into the circuit, and as $V_S = V_{AB}$ the current flowing around the loop is calculated as:

$$I = \frac{20v - 10v}{20\Omega + 10\Omega} = 0.33amps$$

So the voltage drop across the 20Ω resistor can be calculated as:

V_{AB} = 20 - (20Ω x 0.33amps) = 13.33 volts.

Then the Thevenin's Equivalent circuit is shown below with the 40Ω resistor connected.

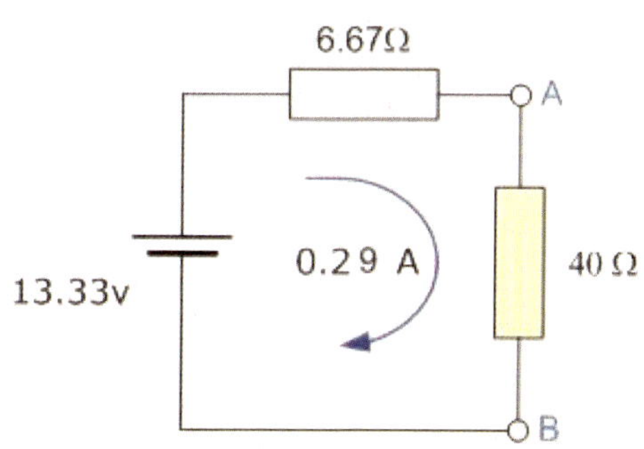

And from this the current flowing in the circuit is given as:

$$I = \frac{13.33v}{6.67\Omega + 40\Omega} = 0.29amps$$

Thevenin's theorem can be used as a circuit analysis method and is particularly useful if the load is to take a series of different values. It is not as powerful as Mesh or Nodal analysis in larger networks because the use of Mesh or Nodal analysis is usually necessary in any Thevenin exercise, so it might well be used from the start. However, Thevenin's equivalent circuits of Transistors, Voltage Sources such as batteries etc, are very useful in circuit design.

Applications of Thevenin's Theorem

- Thevenin's Theorem is especially useful in analyzing power systems and other circuits where one particular resistor in the circuit (called the "load" resistor) is subject to change, and re-calculation of the circuit is necessary with each trial value of load resistance, to determine voltage across it and current through it.
- Source modeling and resistance measurement using the Wheatstone bridge provide applications for Thevenin's theorem.

Practical Limitations

- Many, if not most circuits are only linear over a certain range of values, thus the Thévenin equivalent is valid only within this linear range and may not be valid outside the range.
- The Thévenin equivalent has an equivalent I-V characteristic only from the point of view of the load.
- The power dissipation of the Thévenin equivalent is not necessarily identical to the power dissipation of the real system. However, the power dissipated by an external resistor between the two output terminals is the same however the internal circuit is represented.

References

- What-is-an-electrical-circuit, electricity, science: eschooltoday.com, Retrieved 23 July 2018
- Electric-circuit-or-electrical-network: electrical4u.com, Retrieved 28 May 2018
- What-is-rl-series-circuit: circuitglobe.com, Retrieved 08 March 2018
- What-is-an-rc-resistor-capacitor-circuit: circuitglobe.com, Retrieved 28 May 2018
- Active-and-passive-elements-of-electrical-circuit: electrical4u.com, Retrieved 15 April 2018
- Electrical-impedance-and-its-applications: elprocus.com, Retrieved 17 March 2018
- Network-analysis-for-electric-circuits: electricalengineeringforbeginners.blogspot.com, Retrieved 30 April 2018
- What-is-substitution-theorem: circuitglobe.com, Retrieved 09 July 2018

3

Diodes and Transistors

A diode is an electrical component with two terminals, which conducts current in only one direction. A transistor is any semiconductor device that is used for amplifying electronic signals and electrical power. This chapter has been carefully written to provide an easy understanding of the varied aspects of diodes and transistors. It discusses in detail topics like Zener diode, bridge rectifier, field-effect transistor, bipolar junction transistor, transistor biasing, etc.

Diode

A *diode* is an electrical device allowing current to move through it in one direction with far greater ease than in the other. The most common kind of diode in modern circuit design is the *semiconductor* diode, although other diode technologies exist. Semiconductor diodes are symbolized in schematic diagrams such as figure below. The term "diode" is customarily reserved for small signal devices, I ≤ 1 A. The term *rectifier* is used for power devices, I > 1 A.

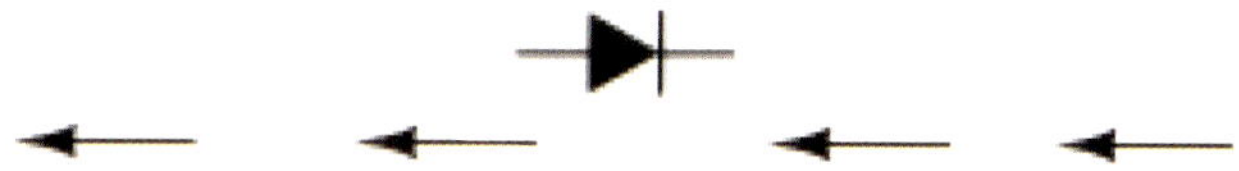

Semiconductor diode schematic symbol: Arrows indicate the direction of electron current flow

When placed in a simple battery-lamp circuit, the diode will either allow or prevent current through the lamp, depending on the polarity of the applied voltage.

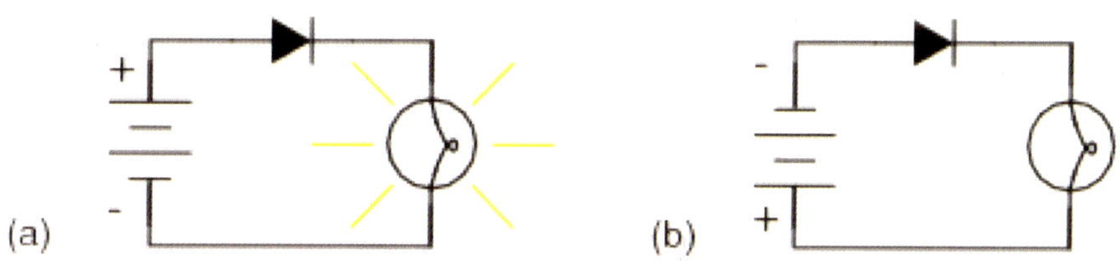

Diode operation: (a) Current flow is permitted; the diode is forward biased.
(b) Current flow is prohibited; the diode is reversed biased

When the polarity of the battery is such that electrons are allowed to flow through the diode, the diode is said to be *forward-biased*. Conversely, when the battery is "backward" and the diode blocks current, the diode is said to be *reverse-biased*. A diode may be thought of as like a switch: "closed" when forward-biased and "open" when reverse-biased.

Oddly enough, the direction of the diode symbol's "arrowhead" points *against* the direction of electron flow. This is because the diode symbol was invented by engineers, who predominantly use *conventional flow* notation in their schematics, showing current as a flow of charge from the positive (+) side of the voltage source to the negative (-). This convention holds true for all semiconductor symbols possessing "arrowheads:" the arrow points in the permitted direction of conventional flow, and against the permitted direction of electron flow.

Hydraulic Check Valve Analogy

Diode behavior is analogous to the behavior of a hydraulic device called a *check valve*. A check valve allows fluid flow through it in only one direction as in figure below:

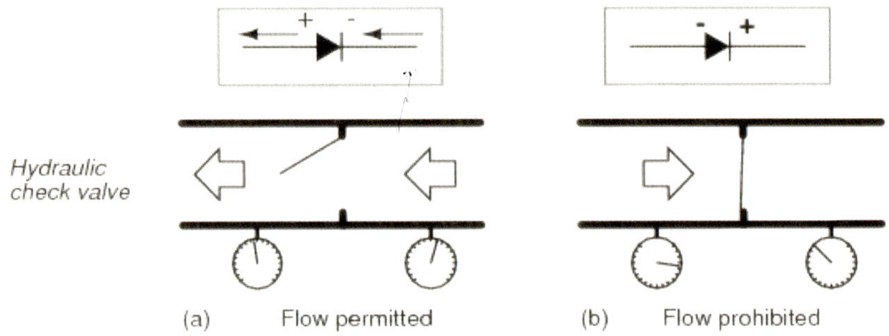

Hydraulic check valve analogy: (a) Electron current flow permitted. (b) Current flow prohibited.

Check valves are essentially pressure-operated devices: they open and allow flow if the pressure across them is of the correct "polarity" to open the gate (in the analogy shown, greater fluid pressure on the right than on the left). If the pressure is of the opposite "polarity," the pressure difference across the check valve will close and hold the gate so that no flow occurs.

Like check valves, diodes are essentially "pressure-" operated (voltage-operated) devices. The essential difference between forward-bias and reverse-bias is the polarity of the voltage dropped across the diode. Let's take a closer look at the simple battery-diode-lamp circuit shown earlier, this time investigating voltage drops across the various components in figure below.

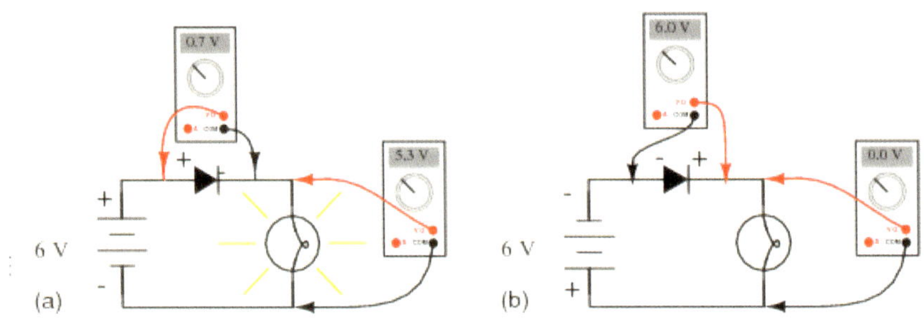

Diode circuit voltage measurements: (a) Forward biased. (b) Reverse biased.

A forward-biased diode conducts current and drops a small voltage across it, leaving most of the battery voltage dropped across the lamp. If the battery's polarity is reversed, the diode becomes reverse-biased, and drops *all* of the battery's voltage leaving none for the lamp. If we consider the diode to be a self-actuating switch (closed in the forward-bias mode and open in the reverse-bias mode), this behavior makes sense. The most substantial difference is that the diode drops a lot more voltage when conducting than the average mechanical switch (0.7 volts versus tens of millivolts).

This forward-bias voltage drop exhibited by the diode is due to the action of the depletion region formed by the P-N junction under the influence of an applied voltage. If no voltage applied is across a semiconductor diode, a thin depletion region exists around the region of the P-N junction, preventing current flow. The depletion region is almost devoid of available charge carriers, and acts as an insulator:

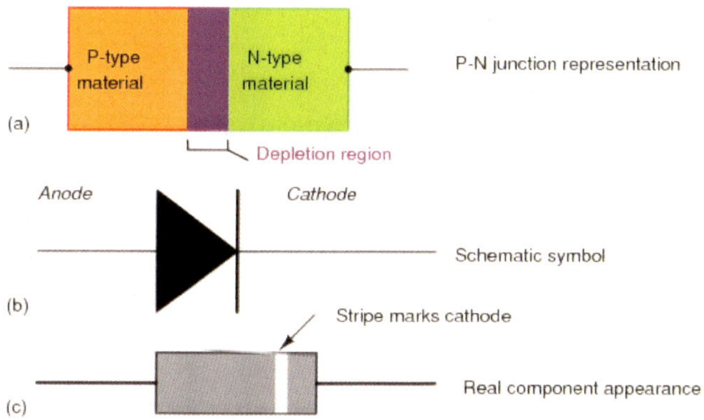

Diode representations: PN-junction model, schematic symbol, physical part

The schematic symbol of the diode is shown in figure above (b) such that the anode (pointing end) corresponds to the P-type semiconductor at (a). The cathode bar, non-pointing end, at (b) corresponds to the N-type material at (a). Also note that the cathode stripe on the physical part (c) corresponds to the cathode on the symbol.

If a reverse-biasing voltage is applied across the P-N junction, this depletion region expands, further resisting any current through it.

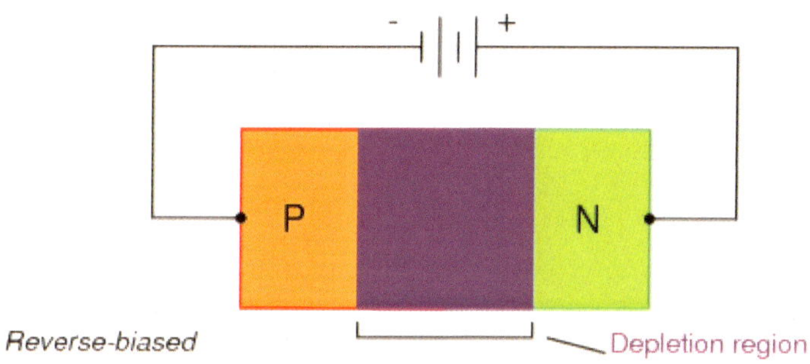

Depletion region expands with reverse bias

Conversely, if a forward-biasing voltage is applied across the P-N junction, the depletion region collapses becoming thinner. The diode becomes less resistive to current through it. In order for a sustained current to go through the diode; though, the depletion region must be fully collapsed by the applied voltage. This takes a certain minimum voltage to accomplish, called the *forward voltage* as illustrated in figure below.

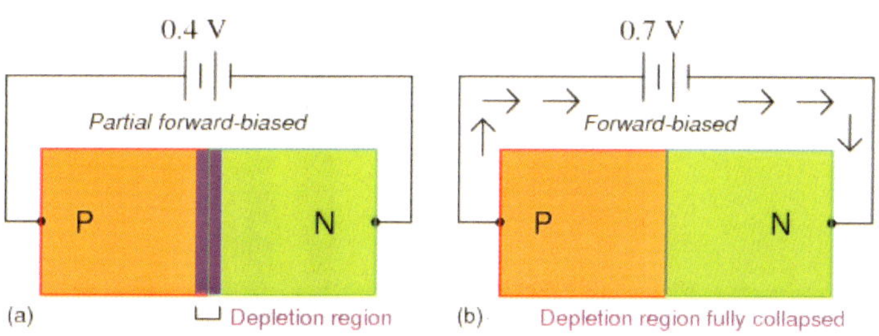

Inceasing forward bias from (a) to (b) decreases depletion region thickness

For silicon diodes, the typical forward voltage is 0.7 volts, nominal. For germanium diodes, the forward voltage is only 0.3 volts. The chemical constituency of the P-N junction comprising the diode accounts for its nominal forward voltage figure, which is why silicon and germanium diodes have such different forward voltages. Forward voltage drop remains approximately constant for a wide range of diode currents, meaning that diode voltage drop is not like that of a resistor or even a normal (closed) switch. For most simplified circuit analysis, the voltage drop across a conducting diode may be considered constant at the nominal figure and not related to the amount of current.

Diode Equation

Actually, forward voltage drop is more complex. An equation describes the exact current through a diode, given the voltage dropped across the junction, the temperature of the junction, and several physical constants. It is commonly known as the *diode equation*:

$$I_D = I_S\left(e^{qV_D/NKT} - 1\right)$$

Where,

I_D = Diode current in amps

I_S = Saturation current in amps

$(\text{typically } 1 \times 10^{-12} \text{ amps})$

e = Euler's constant (~ 2.718281828)

q = charge ofelectron $(1.6 \times 10^{-19} \text{ coulombs})$

V_D = Voltage applied across diode in volts

N = "Nonideality" or "emission" coefficient

(typically between 1and 2)

K = Boltzmann's constant (1.38×10^{-23})

T = Junction temperature in kelvins

The term kT/q describes the voltage produced within the P-N junction due to the action of temperature, and is called the *thermal voltage*, or V_t of the junction. At room temperature, this is about 26 millivolts. Knowing this, and assuming a "nonideality" coefficient of 1, we may simplify the diode equation and re-write it as such:

$$I_D = I_S\left(e^{qV_D/NKT} - 1\right)$$

Where,

I Diode current in amps

I Saturation current in amps

$(\text{typically } 1 \times 10 \ ^{12} \text{ amps})$

e Euler s constant (2.718281828)

V Voltage applied across diode in volts

You need not be familiar with the "diode equation" to analyze simple diode circuits. Just understand that the voltage dropped across a current-conducting diode *does* change with the amount of current going through it, but that this change is fairly small over a wide range of currents. This is why many textbooks simply say the voltage drop across a conducting, semiconductor diode remains constant at 0.7 volts for silicon and 0.3 volts for germanium. However, some circuits intentionally make use of the P-N junction's inherent exponential current/voltage relationship and thus can only be understood in the context of this equation. Also, since temperature is a factor in the diode equation, a forward-biased P-N junction may also be used as a temperature-sensing device, and thus can only be understood if one has a conceptual grasp on this mathematical relationship.

A reverse-biased diode prevents current from going through it, due to the expanded depletion region. In actuality, a very small amount of current can and does go through a reverse-biased diode, called the *leakage current*, but it can be ignored for most purposes. The ability of a diode to withstand reverse-bias voltages is limited, as it is for any insulator. If the applied reverse-bias voltage becomes too great, the diode will experience a condition known as *breakdown*, which is usually destructive. A diode's maximum reverse-bias voltage rating is known as the *Peak Inverse Voltage*, or *PIV*, and may be obtained from the manufacturer. Like forward voltage, the PIV rating of a diode varies with temperature, except that PIV *increases* with increased temperature and *decreases* as the diode becomes cooler—exactly opposite that of forward voltage.

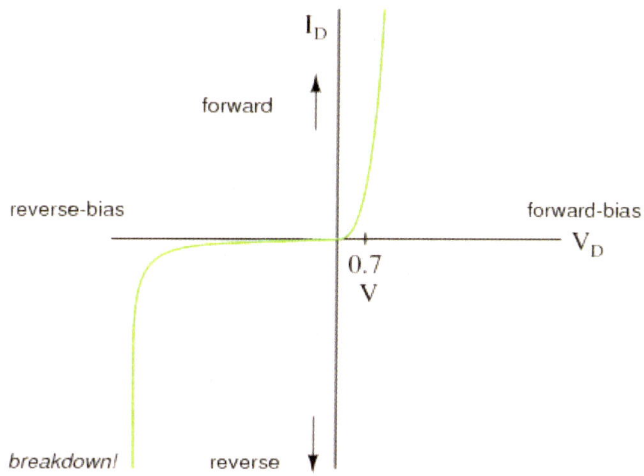

Diode curve: showing knee at 0.7 V forward bias for Si, and reverse breakdown

Typically, the PIV rating of a generic "rectifier" diode is at least 50 volts at room temperature. Diodes with PIV ratings in the many thousands of volts are available for modest prices.

Zener diode

The name zener diode was named after the American physicist Clarance Melvin Zener who discovered the zener effect. Zener diodes are the basic building blocks of electronic circuits. They are widely used in all kinds of electronic equipments. Zener diodes are mainly used to protect electronic circuits from over voltage.

A zener diode is a special type of device designed to operate in the zener breakdown region. Zener diodes acts like normal p-n junction diodes under forward biased condition. When forward biased voltage is applied to the zener diode it allows large amount of electric current and blocks only a small amount of electric current.

Zener diode is heavily doped than the normal p-n junction diode. Hence, it has very thin depletion region. Therefore, zener diodes allow more electric current than the normal p-n junction diodes.

Zener diode allows electric current in forward direction like a normal diode but also allows electric current in the reverse direction if the applied reverse voltage is greater than the zener voltage. Zener diode is always connected in reverse direction because it is specifically designed to work in reverse direction.

Breakdown in Zener Diode

There are two types of reverse breakdown regions in a zener diode: avalanche breakdown and zener breakdown.

Avalanche Breakdown

The avalanche breakdown occurs in both normal diodes and zener diodes at high reverse voltage. When high reverse voltage is applied to the p-n junction diode, the free electrons (minority carriers) gains large amount of energy and accelerated to greater velocities.

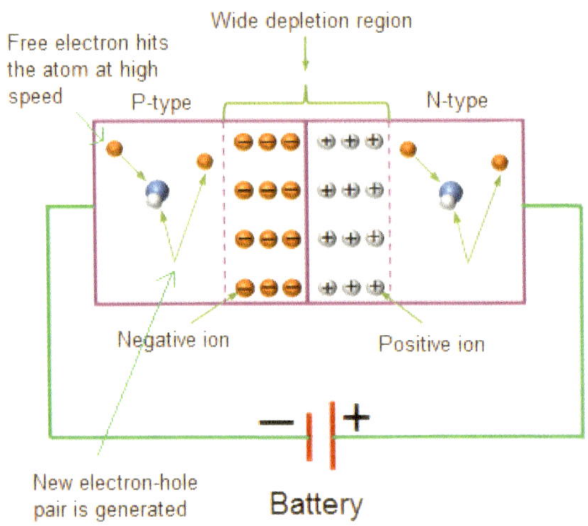

The free electrons moving at high speed will collides with the atoms and knock off more electrons. These electrons are again accelerated and collide with other atoms. Because of this continuous collision with the atoms, a large number of free electrons are generated. As a result, electric current in the diode increases rapidly. This sudden increase in electric current may permanently destroys the normal diode. However, avalanche diodes may not be destroyed because they are carefully designed to operate in avalanche breakdown region. Avalanche breakdown occurs in zener diodes with zener voltage (Vz) greater than 6V.

Zener Breakdown

The zener breakdown occurs in heavily doped p-n junction diodes because of their narrow depletion region. When reverse biased voltage applied to the diode is increased, the narrow depletion region generates strong electric field.

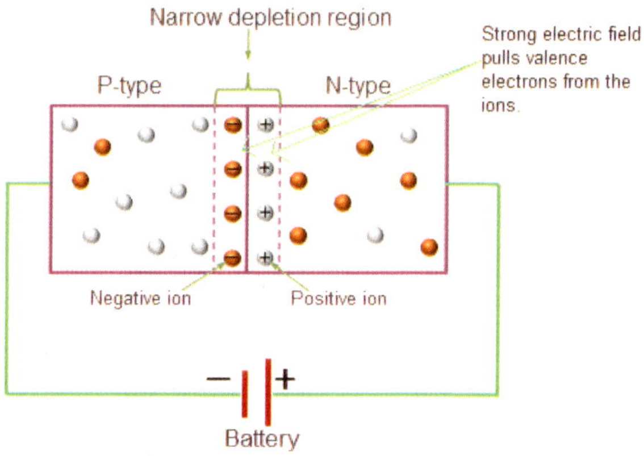

When reverse biased voltage applied to the diode reaches close to zener voltage, the electric field in the depletion region is strong enough to pull electrons from their valence band. The valence electrons which gains sufficient energy from the strong electric field of depletion region will breaks bonding with the parent atom. The valance electrons which break bonding with parent atom will become free electrons. This free electrons carry electric current from one place to another place. At zener breakdown region, a small increase in voltage will rapidly increases the electric current.

- Zener breakdown occurs at low reverse voltage whereas avalanche breakdown occurs at high reverse voltage.
- Zener breakdown occurs in zener diodes because they have very thin depletion region.
- Breakdown region is the normal operating region for a zener diode.
- Zener breakdown occurs in zener diodes with zener voltage (Vz) less than 6V.

Symbol of Zener Diode

The symbol of zener diode is shown in below figure. Zener diode consists of two terminals: cathode and anode.

Zener diode symbol

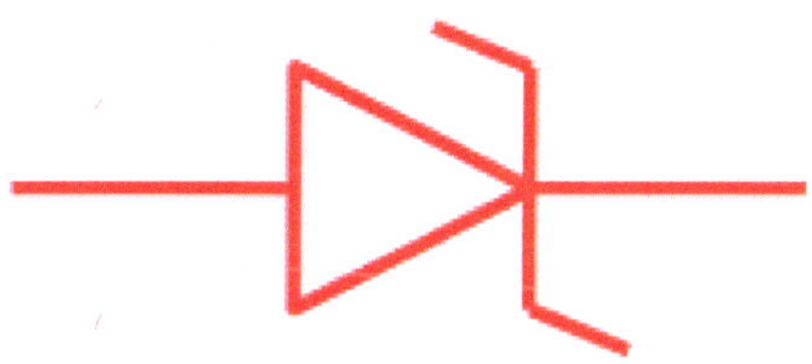

In zener diode, electric current flows from both anode to cathode and cathode to anode.

The symbol of zener diode is similar to the normal p-n junction diode, but with bend edges on the vertical bar.

Zener Diode Circuit Diagram

The circuit diagram of the Zener diode is shown in the figure below. The Zener diode is employed in reverse biasing. The reverse biasing means the n-type material of the diode is connected to the positive terminal of the supply and the P-type material is connected to the negative terminal of the supply. The depletion region of the diode is very thin because it is made of the heavily doped semiconductor material.

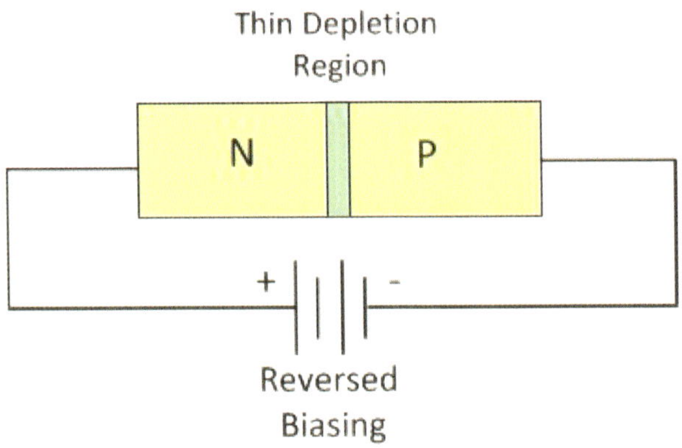

Working of Zener Diode

The Zener diode is made up of heavily doped semiconductor material. The heavily doped means the high-level impurities is added to the material for making it more conductive. The depletion region of the Zener diode is very thin because of the impurities. The heavily doping material increases the intensity of the electric field across the depletion region of the Zener diode even for the small reverse voltage.

When no biasing is applied across the Zener diode, the electrons remains in the valence band of the p-type material and no current flow through the diode. The band in which the valence electrons (outermost orbit electron) place is known as the valence band electron. The electrons of the valence band easily move from one band to another when the external energy is applied across it.

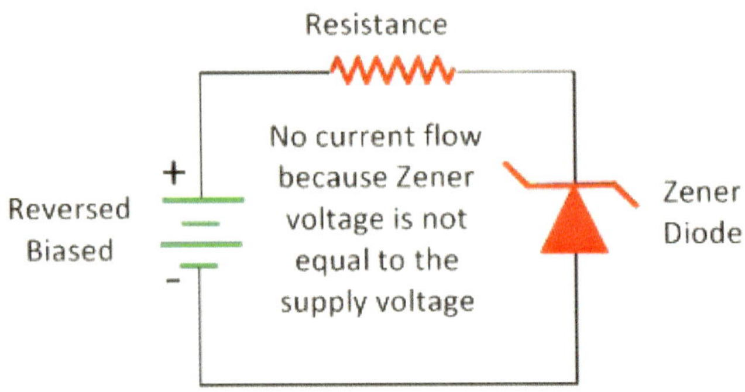

When the reverse bias applies across the diode and the supply voltage is equal to the Zener voltage then it starts conducting in the reverse bias direction. The Zener voltage is the voltage at which the depletion region completely vanish.

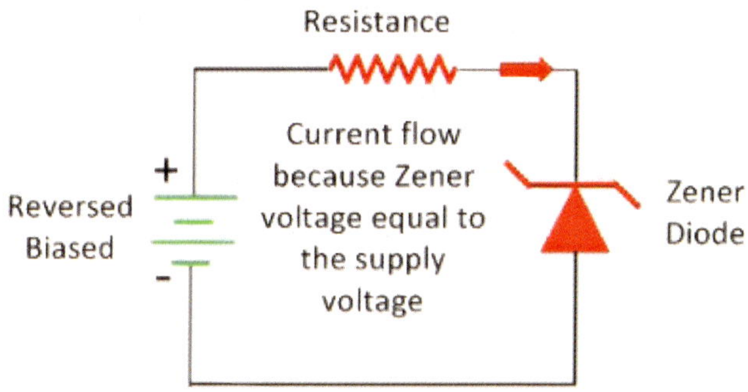

The reverse bias applies across the diode increases the intensity of electric field across the depletion region. Thus, it allows the electrons to move from the valence band of P-type material to the conduction band of N-type material. This transferring of valence band electrons to the conduction band reduces the barrier between the p and n-type material. When the depletion region become completely vanish the diode starts conducting in the reverse biased.

Characteristics of Zener Diode

The six characteristics of a zener diode is shown in the below figure. When forward biased voltage is applied to the zener diode, it works like a normal diode. However, when reverse biased voltage is applied to the zener diode, it works in different manner.

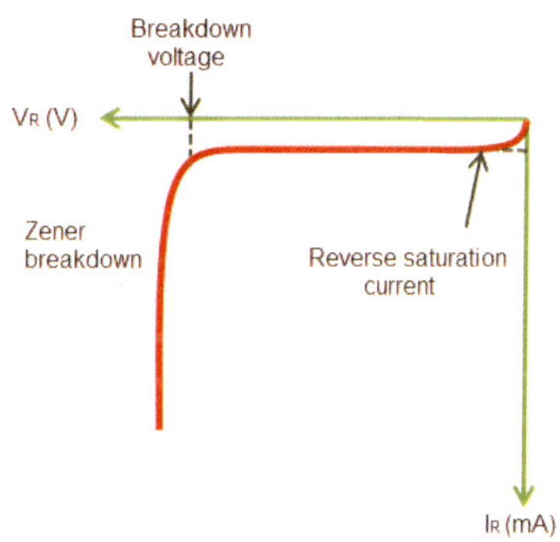

Fig: Zener breakdown

When reverse biased voltage is applied to a zener diode, it allows only a small amount of leakage current until the voltage is less than zener voltage. When reverse biased voltage applied to the zener diode reaches zener voltage, it starts allowing large amount of electric current. At this point, a small increase in reverse voltage will rapidly increases the electric current. Because of this sudden rise in electric current, breakdown occurs called zener breakdown. However, zener diode exhibits a controlled breakdown that does damage the device.

The zener breakdown voltage of the zener diode is depends on the amount of doping applied. If the diode is heavily doped, zener breakdown occurs at low reverse voltages. On the other hand, if the diode is lightly doped, the zener breakdown occurs at high reverse voltages. Zener diodes are available with zener voltages in the range of 1.8V to 400V.

The voltage across the Zener is represented by the equation shown below:

$$V = V_Z + I_Z R_Z$$

Zener Diode as Voltage Regulator

The term regulator means which regulates. The Zener diode can work as a voltage regulator if it is introduced in a circuit. The output across the diode will be constant. It is driven by a current source. As we know if the voltage across the diode exceeds a certain value it would draw excessive current from the supply. The basic diagram of Zener diode as voltage regulator is given below:

To fix the current through the Zener diode series resistance R is introduced whose value can be chosen from the following equation:

Resistor value (ohms) = (V1 – V2) / (Zener current + load current)

The above diagram is of a shunt regulators because the regulating element is parallel to the load element. The Zener diode produces a stable reference voltage across the load which fulfills the criteria of regulator requirement.

The Zener diode allows current to flow in the forward direction in the same manner as an ideal diode. It also permits to flow in the reverse direction when the voltage is above a certain value known as breakdown voltage.

This device is named after Zener. Zener discovered this electrical property. A Zener diode is one in which the reverse breakdown occurs due to electron quantum tunneling under high electric field strength called Zener effect. Many diodes described as Zener diodes rely instead on avalanche breakdown. Both types are used with the Zener effect predominating under 5.6 V and avalanche breakdown above. Regular applications include providing a reference voltage for voltage regulators. This is to protect devices from momentary voltage pulses.

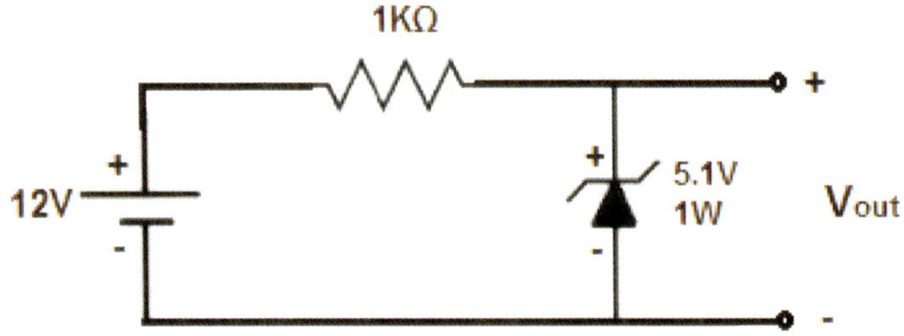

Zener Diode Connectivity

These devices are also encountered in series with a base emitter junction. At transistor stages where selective choice of a device centered around the avalanche or Zener point. It can be used to introduce compensating temperature coefficient balancing of the transistor. DC error amplifier used in a regulated power supply circuit feedback loop system is the on of the example.

Applications of Zener Diode

The Zener diode is mostly used in the commercial and industrial applications. The following are the main application of the Zener diode.

- As Voltage Stabilizer – The Zener diode is used for regulating the voltage. It provides the constant voltage from the fluctuating voltage source to the load. The Zener diode is connected in parallel across the load and maintain the constant voltage V_Z and hence stabilises the voltage.
- For Meter Protection – The Zener diode is generally used in multimeters for controlling the movement of the meter against accidental overloads. It is connected in parallel with the diode. When the overload occurs across the diode most of the current pass through the diode. Thus, protects the meter from damage.
- For Wave Shaping – The Zener diode is used for converting the sine wave into the square wave.This can be done by placing the two Zener Diodes in series with the resistance. The diode is connected back to back and in the opposite direction.

When the voltage applied across the terminal is less than the Zener voltage the diodes offer a high resistive path to the current and the input voltage applied across a diode appeared at the output terminal. When the voltage rises beyond the Zener voltage they offer a low resistance path and large current flow through the diode. Due to which the heavy voltage drop occurs across the resistance and the input wave cutoff at the peak. Thus, the square wave appears across the output terminal.

Bridge Rectifier

A bridge rectifier is a type of full wave rectifier which uses four or more diodes in a bridge circuit configuration to efficiently convert the Alternating Current (AC) into Direct Current (DC).

Bridge Rectifier Construction

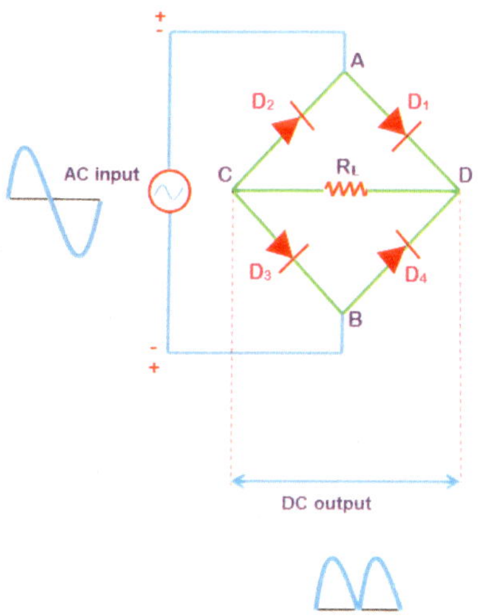

Fig: Bridge Rectifier

The construction diagram of a bridge rectifier is shown in the below figure. The bridge rectifier is made up of four diodes namely D1, D2, D3, D4 and load resistor RL. The four diodes are connected in a closed loop (Bridge) configuration to efficiently convert the Alternating Current (AC) into Direct Current (DC). The main advantage of this bridge circuit configuration is that we do not require an expensive center tapped transformer, thereby reducing its cost and size.

The input AC signal is applied across two terminals A and B and the output DC signal is obtained across the load resistor RL which is connected between the terminals C and D.

The four diodes D1, D2, D3, D4 are arranged in series with only two diodes allowing electric current during each half cycle. For example, diodes D1 and D3 are considered as one pair which allows electric current during the positive half cycle whereas diodes D2 and D4 are considered as another pair which allows electric current during the negative half cycle of the input AC signal.

Working of Bridge Rectifier

When input AC signal is applied across the bridge rectifier, during the positive half cycle diodes D1 and D3 are forward biased and allows electric current while the diodes D2 and D4 are reverse biased and blocks electric current. On the other hand, during the negative half cycle diodes D2 and D4 are forward biased and allows electric current while diodes D1 and D3 are reverse biased and blocks electric current.

During the positive half cycle, the terminal A becomes positive while the terminal B becomes negative. This causes the diodes D1 and D3 forward biased and at the same time, it causes the diodes D2 and D4 reverse biased.

The current flow direction during the positive half cycle is shown in the figure A (I.e. A to D to C to B):

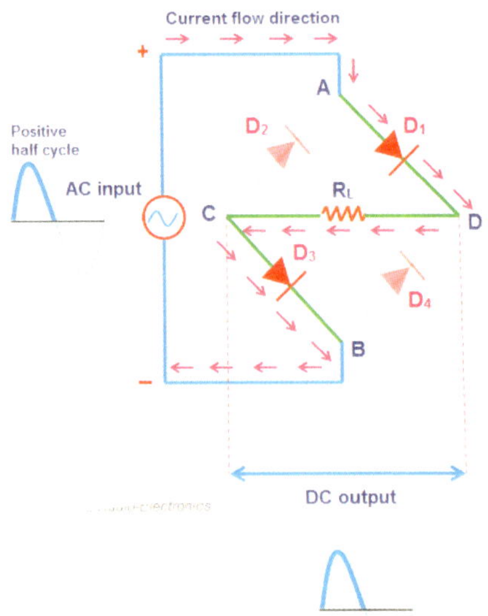

Bridge rectifier during positive half cycle

During the negative half cycle, the terminal B becomes positive while the terminal A becomes negative. This causes the diodes D2 and D4 forward biased and at the same time, it causes the diodes D1 and D3 reverse biased.

The current flow direction during negative half cycle is shown in the figure B (I.e. B to D to C to A).

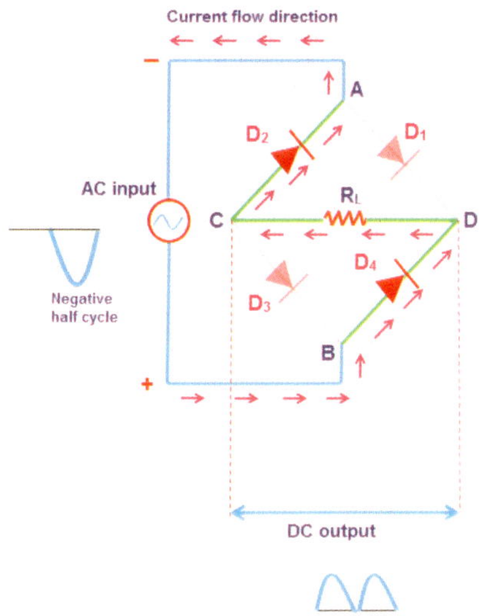

Bridge rectifier during negative half cycle

From the above two figures (A and B), we can observe that the direction of current flow across load resistor R_L is same during the positive half cycle and negative half cycle. Therefore, the polarity of the output DC signal is same for both positive and negative half cycles. The output DC signal polarity may be either completely positive or negative. In our case, it is completely positive. If the direction of diodes is reversed then we get a complete negative DC voltage.

Thus, a bridge rectifier allows electric current during both positive and negative half cycles of the input AC signal.

The output waveforms of the bridge rectifier is shown in the below figure:

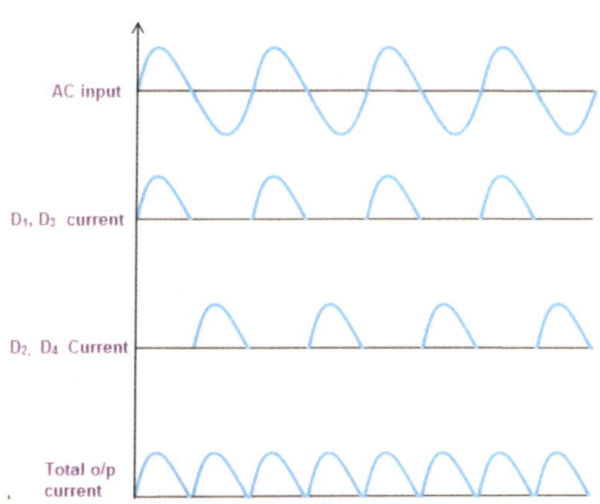

Characteristics of Bridge Rectifier

Peak Inverse Voltage (PIV)

The maximum voltage a diode can withstand in the reverse bias condition is called Peak Inverse Voltage (PIV).

or

The maximum voltage that the non-conducting diode can withstand is called Peak Inverse Voltage (PIV).

During the positive half cycle, the diodes D1 and D3 are in the conducting state while the diodes D2 and D4 are in the non-conducting state. On the other hand, during the negative half cycle, the diodes D2 and D4 are in the conducting state while the diodes D1 and D3 are in the non-conducting state.

The Peak Inverse Voltage (PIV) for a bridge rectifier is given by,

$$PIV = VS_{max}$$

Ripple Factor

The smoothness of the output DC signal is measured by using a factor known as ripple factor. The output DC signal with very fewer ripples is considered as the smooth DC signal while the output DC signal with high ripples is considered as the high pulsating DC signal.

Ripple factor is mathematically defined as the ratio of ripple voltage to the pure DC voltage.

The ripple factor for a bridge rectifier is given by,

$$\gamma = \sqrt{\left(\frac{V_{rms}}{V_{DC}}\right)^2 - 1}$$

The ripple factor of the bridge rectifier is 0.48 which is same as the center tapped full wave rectifier.

Rectifier Efficiency

The rectifier efficiency determines how efficiently the rectifier converts Alternating Current (AC) into Direct Current (DC).

High rectifier efficiency indicates a most reliable rectifier while the low rectifier efficiency indicates a poor rectifier.

Rectifier efficiency is defined as the ratio of the DC output power to the AC input power.

$$\eta = \frac{\text{DC output power}}{\text{AC input power}} = \frac{P_{DC}}{P_{AC}}$$

The maximum rectifier efficiency of a bridge rectifier is 81.2% which is same as the center tapped full wave rectifier.

Advantages of Bridge Rectifier

Low Ripples in the Output DC Signal

The DC output signal of the bridge rectifier is smoother than the half wave rectifier. In other words, the bridge rectifier has fewer ripples as compared to the half wave rectifier. However, the ripple factor of the bridge rectifier is same as the center tapped full wave rectifier.

High Rectifier Efficiency

The rectifier efficiency of the bridge rectifier is very high as compared to the half wave rectifier. However, the rectifier efficiency of bridge rectifier and center tapped full wave rectifier is same.

Low Power Loss

In half wave rectifier only one half cycle of the input AC signal is allowed and the remaining half cycle of the input AC signal is blocked. As a result, nearly half of the applied input power is wasted.

However, in the bridge rectifier, the electric current is allowed during both positive and negative half cycles of the input AC signal. So the output DC power is almost equal to the input AC power.

Disadvantages of Bridge Rectifier

Bridge Rectifier Circuit Looks Very Complex

In a half wave rectifier, only a single diode is used whereas in a center tapped full wave rectifier two diodes are used. But in the bridge rectifier, we use four diodes for the circuit operation. So the bridge rectifier circuit looks more complex than the half wave rectifier and center tapped full wave rectifier.

More Power Loss as Compared to the Center Tapped full Wave Rectifier

In electronic circuits, the more diodes we use the more voltage drop will occur. The power loss in bridge rectifier is almost equal to the center tapped full wave rectifier. However, in a bridge rectifier, the voltage drop is slightly high as compared to the center tapped full wave rectifier. This is due to two additional diodes (total four diodes).

In center tapped full wave rectifier, only one diode conducts during each half cycle. So the voltage drop in the circuit is 0.7 volts. But in the bridge rectifier, two diodes which are connected in series conduct during each half cycle. So the voltage drop occurs due to two diodes which is equal to 1.4 volts (0.7 + 0.7 = 1.4 volts). However, the power loss due to this voltage drop is very small.

Wave-shaping Circuits

Wave shaping circuit is a circuit in which an RC network can change the shape of complex waveforms so that the output barely resembles the input.

Types of Wave Shaping

There are two main types of wave shaping. They are:

- Linear wave shaping
- Non-linear wave shaping

Linear Wave Shaping

Linear elements such as resistors, capacitors and inductors are employed to shape a signal in this linear wave shaping. A Sine wave input has a sine wave output and hence the nonsinusoidal inputs are more prominently used to understand the linear wave shaping.

Filtering is the process of attenuating the unwanted signal or to reproduce the selected portions of the frequency components of a particular signal.

Filters

In the process of shaping a signal, if some portions of the signal are felt unwanted, they can be cut off using a Filter Circuit. A Filter is a circuit that can remove unwanted portions of a signal at its input. The process of reduction in the strength of the signal is also termed as Attenuation.

We have few components which help us in filtering techniques:

- A Capacitor has the property to allow AC and to block DC
- An Inductor has the property to allow DC but blocks AC.

Using these properties, these two components are especially used to block or allow AC or DC. The Filters can be designed depending upon these properties.

We have four main types of filters:

- Low pass filter
- High pass filter
- Band pass filter
- Band stop filter

Let us now discuss these types of filters in detail.

Low Pass Filter

A Filter circuit which allows a set of frequencies that are below a specified value can be termed as a Low pass filter. This filter passes the lower frequencies. The circuit diagram of a low pass filter using RC and RL are as shown below:

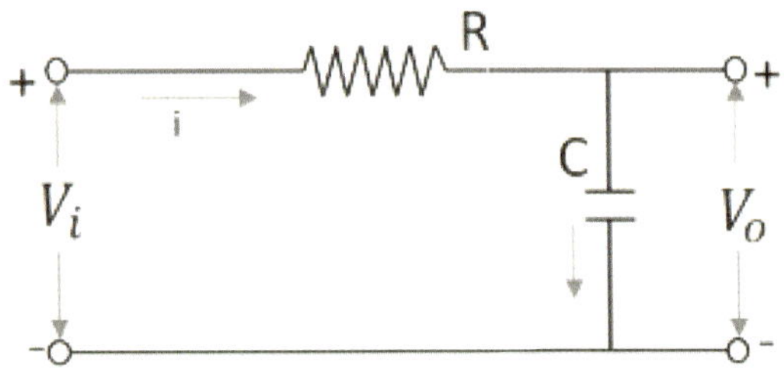

Low Pass RC Circuit

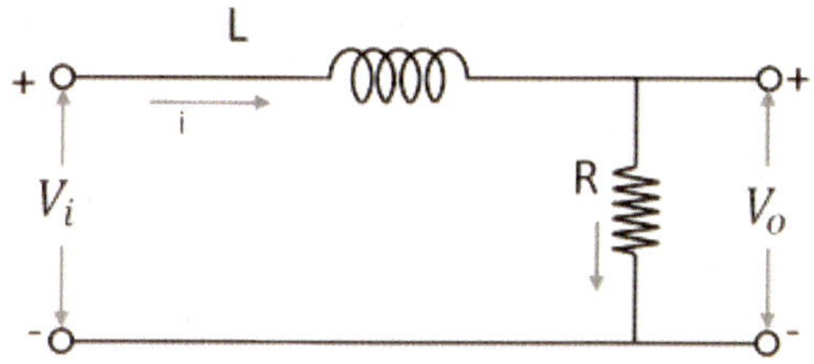

Low Pass RL Circuit

The capacitor filter or RC filter and the inductor filter or RL filter both act as low pass filters:

- The RC filter – As the capacitor is placed in shunt, the AC it allows is grounded. This by passes all the high frequency components while allows DC at the output.
- The RL filter – As the inductor is placed in series, the DC is allowed to the output. The inductor blocks AC which is not allowed at the output.

The symbol for a low pass filter (LPF) is as given below:

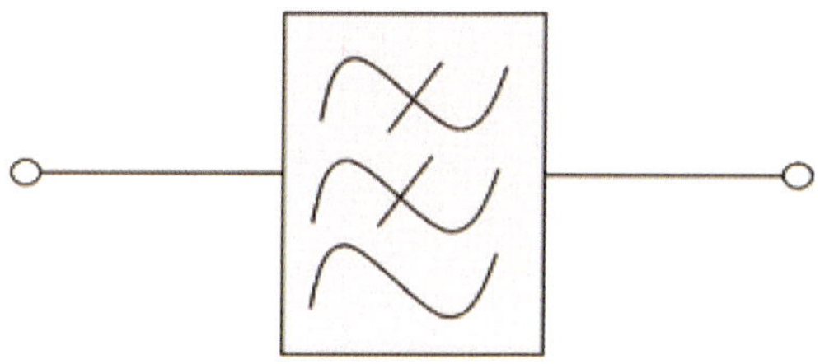
Symbol for LPF

Frequency Response

The frequency response of a practical filter is as shown here under and the frequency response of an ideal LPF when the practical considerations of electronic components are not considered will be as follows:

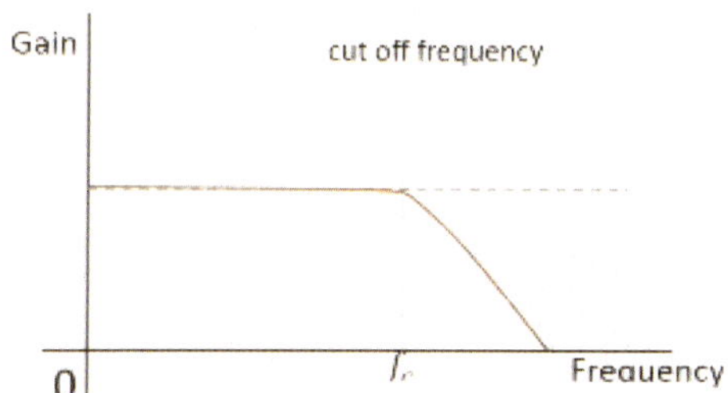

Frequency response for LPF

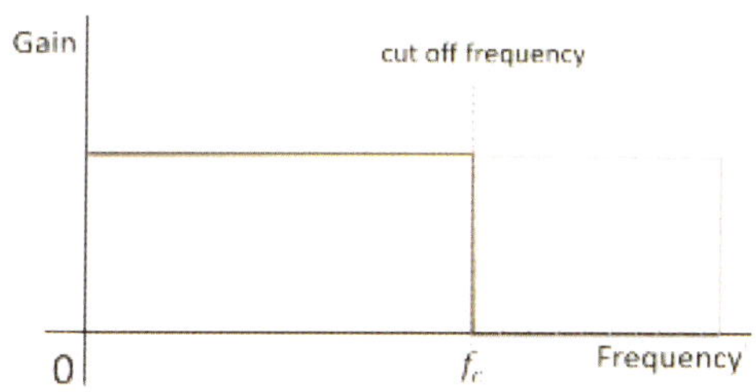

Ideal Frequency response for LPF

The cut off frequency for any filter is the critical frequency fcfc for which the filter is intended to attenuate (cut) the signal. An ideal filter has a perfect cut-off whereas a practical one has few limitations.

The RLC Filter

After knowing about the RC and RL filters, one may have an idea that it would be good to add these two circuits in order to have a better response. The following figure shows how the RLC circuit looks like.

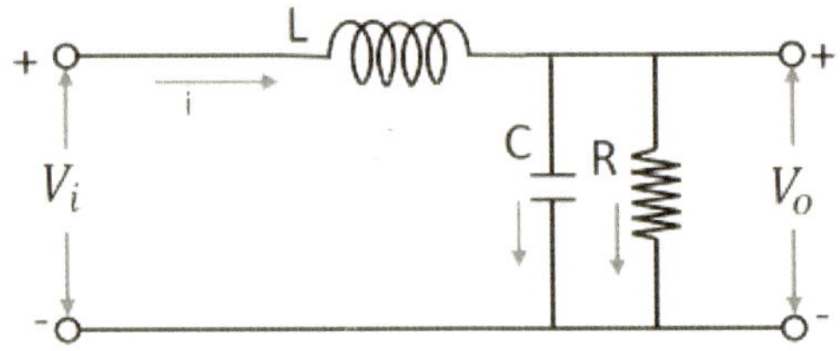

Low Pass RLC Circuit

The signal at the input goes through the inductor which blocks AC and allows DC. Now, that output is again passed through the capacitor in shunt, which grounds the remaining AC component if any, present in the signal, allowing DC at the output. Thus we have a pure DC at the output. This is a better low pass circuit than both of them.

High Pass Filter

A Filter circuit which allows a set of frequencies that are above a specified value can be termed as a High pass filter. This filter passes the higher frequencies. The circuit diagram of a high pass filter using RC and RL are as shown below:

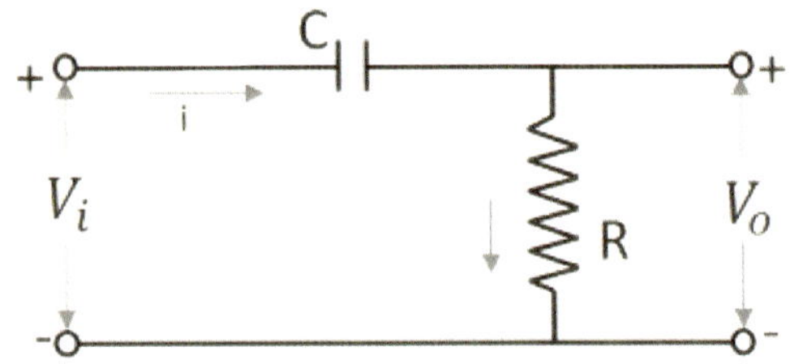

High Pass RC Circuit

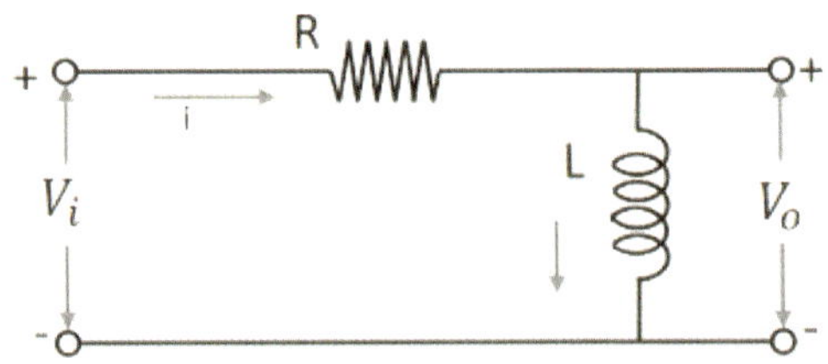

High Pass RL Circuit

The capacitor filter or RC filter and the inductor filter or RL filter both act as high pass filters.

The RC Filter

As the capacitor is placed in series, it blocks the DC components and allows the AC components to the output. Hence the high frequency components appear at the output across the resistor.

The RL Filter

As the inductor is placed in shunt, the DC is allowed to be grounded. The remaining AC component, appears at the output. The symbol for a high pass filter (HPF) is as given below:

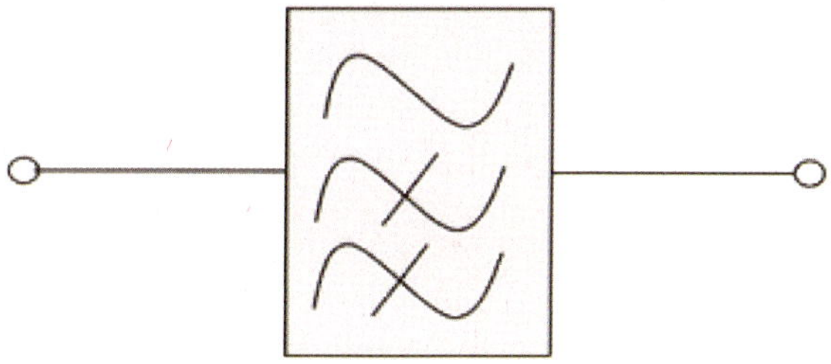

Symbol for HPF

Frequency Response

The frequency response of a practical filter is as shown here under and the frequency response of an ideal HPF when the practical considerations of electronic components are not considered will be as follows:

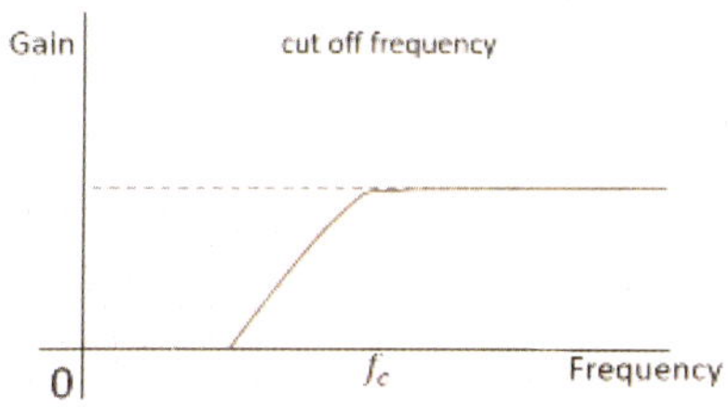

Frequency response for HPF

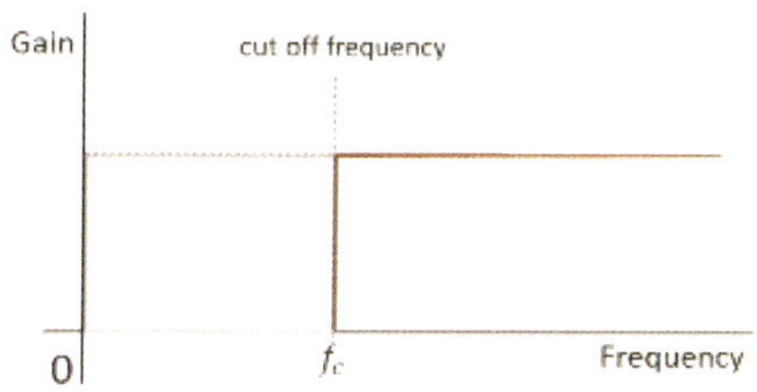

Ideal Frequency response for HPF

The cut-off frequency for any filter is the critical frequency fcfc for which the filter is intended to attenuate (cut) the signal. An ideal filter has a perfect cut-off whereas a practical one has few limitations.

The RLC Filter

After knowing about the RC and RL filters, one may have an idea that it would be good to add these two circuits in order to have a better response. The following figure shows how the RLC circuit looks like.

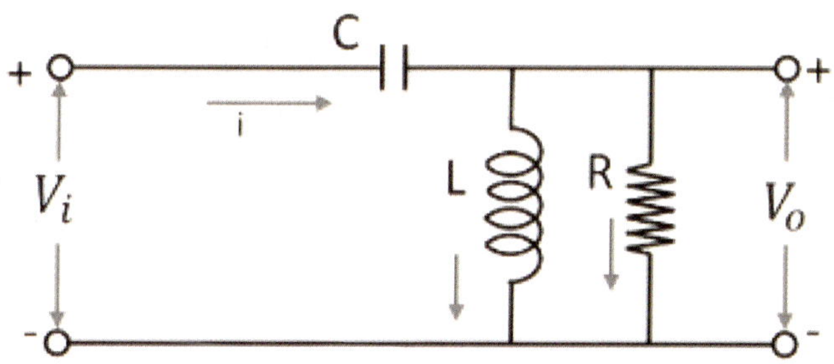

High Pass RLC Circuit

The signal at the input goes through the capacitor which blocks DC and allows AC. Now, that output is again passed through the inductor in shunt, which grounds the remaining DC component if any, present in the signal, allowing AC at the output. Thus we have a pure AC at the output. This is a better high pass circuit than both of them.

Band Pass Filter

A Filter circuit which allows a set of frequencies that are between two specified values can be termed as a Band pass filter. This filter passes a band of frequencies.

As we need to eliminate few of the low and high frequencies, to select a set of specified frequencies, we need to cascade a HPF and a LPF to get a BPF. This can be understood easily even by observing the frequency response curves.

The circuit diagram of a band pass filter is as shown below:

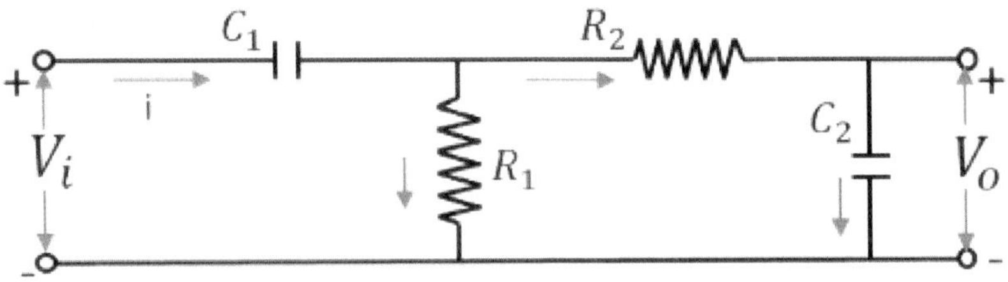

Circuit diagram for Band Pass Filter

The above circuit can also be constructed using RL circuits or RLC circuits. The above one is a RC circuit chosen for simple understanding.

The symbol for a band pass filter (BPF) is as given below.

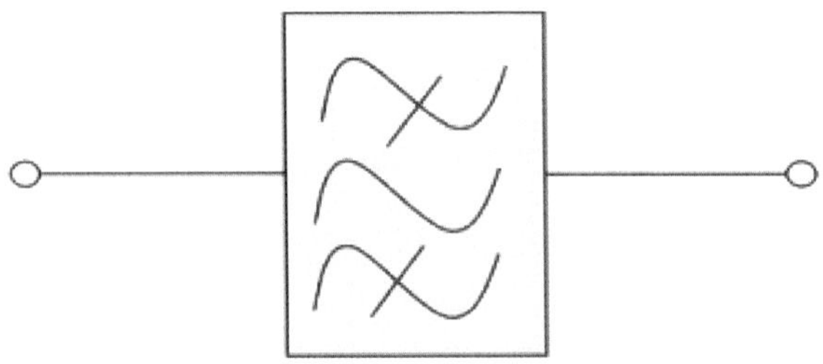

Symbol for BPF

Frequency Response

The frequency response of a practical filter is as shown here under and the frequency response of an ideal BPF when the practical considerations of electronic components are not considered will be as follows. The cut-off frequency for any filter is the critical frequency *fc* for which the filter is intended to attenuate (cut) the signal. An ideal filter has a perfect cut-off whereas a practical one has few limitations.

Band Stop Filter

A Filter circuit which blocks or attenuates a set of frequencies that are between two specified values can be termed as a Band Stop filter. This filter rejects a band of frequencies and hence can also be called as Band Reject Filter.

As we need to eliminate few of the low and high frequencies, to select a set of specified frequencies, we need to cascade a LPF and a HPF to get a BSF. This can be understood easily even by observing the frequency response curves.

The circuit diagram of a band stop filter is as shown below:

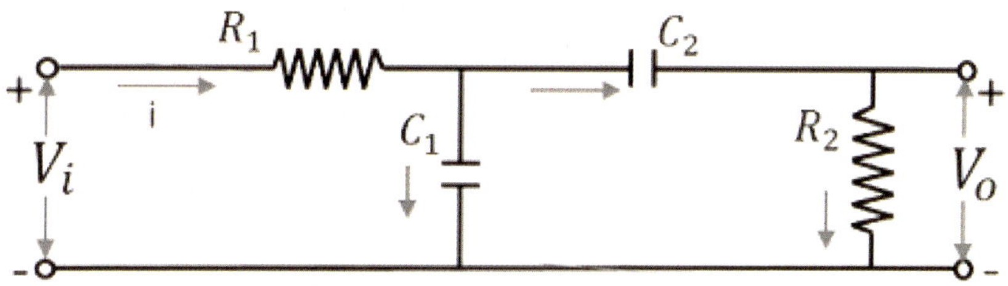

Circuit diagram for Band Stop Filter

The above circuit can also be constructed using RL circuits or RLC circuits. The above one is a RC circuit chosen for simple understanding.

The symbol for a band stop filter (BSF) is as given below:

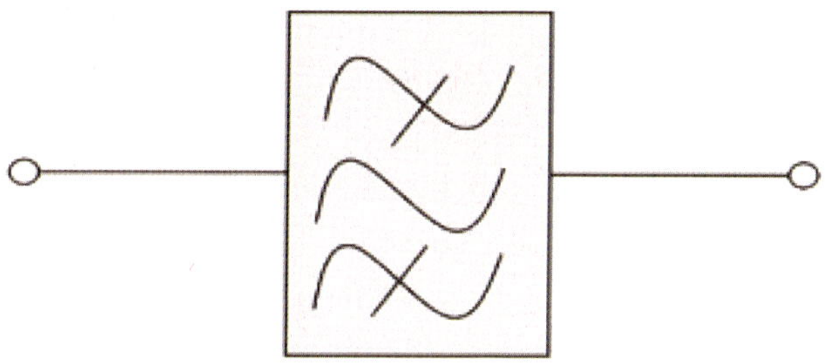

Symbol for BPF

Frequency Response

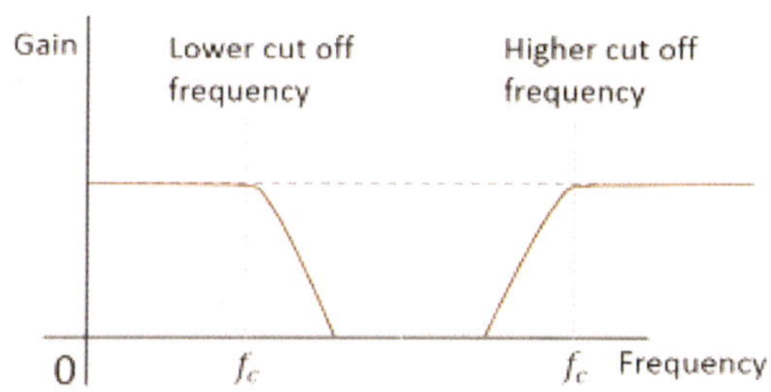

Frequency response for BSF

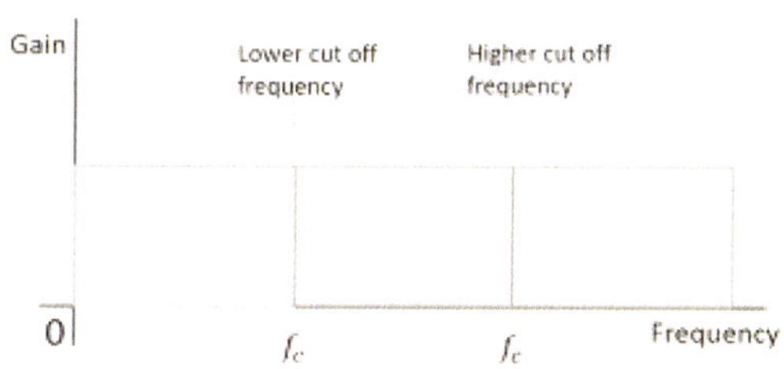

Ideal Frequency response for BSF

The frequency response of a practical filter is as shown here under and the frequency response of an ideal BSF when the practical considerations of electronic components are not considered will be as follows. The cut-off frequency for any filter is the critical frequency *fc* for which the filter is intended to attenuate (cut) the signal. An ideal filter has a perfect cut-off whereas a practical one has few limitations.

Low-pass and high-pass filter circuits are used as special circuits in many applications. Low-pass filter (LPF) can work as an Integrator, whereas the high-pass filter (HPF) can work as a Differentiator. These two mathematical functions are possible only with these circuits which reduce the efforts of an electronics engineer in many applications.

Low Pass Filter as Integrator

At low frequencies, the capacitive reactance tends to become infinite and at high frequencies the reactance becomes zero. Hence at low frequencies, the LPF has finite output and at high frequencies the output is nil, which is same for an integrator circuit. Hence low pass filter can be said to be worked as an integrator.

For the LPF to behave as an integrator,

$$\tau \gg T$$

Where, τ=RC the time constant of the circuit

Then the voltage variation in C is very small.

$$i = \frac{V_0}{R}$$

$$V_i \cong iR$$

$$Since \frac{1}{C}\int i\,dt + iR$$

$$i = \frac{V_i}{R}$$

$$Since, V_0 = \frac{1}{C}\int idt = \frac{1}{RC}\int V_i dt = \frac{1}{\tau}\int V_i d$$

$$Output \propto \int input$$

Hence a LPF with large time constant produces an output that is proportional to the integral of an input.

Frequency Response

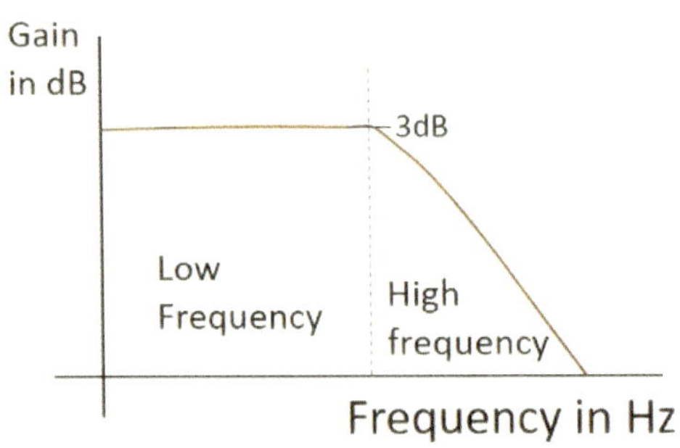

The Frequency response of a practical low pass filter, when it works as an Integrator is as shown in figure.

Output Waveform

If the integrator circuit is given a sinewave input, the output will be a cosine wave. If the input is a square wave, the output wave form changes its shape and appears as in the figure below:

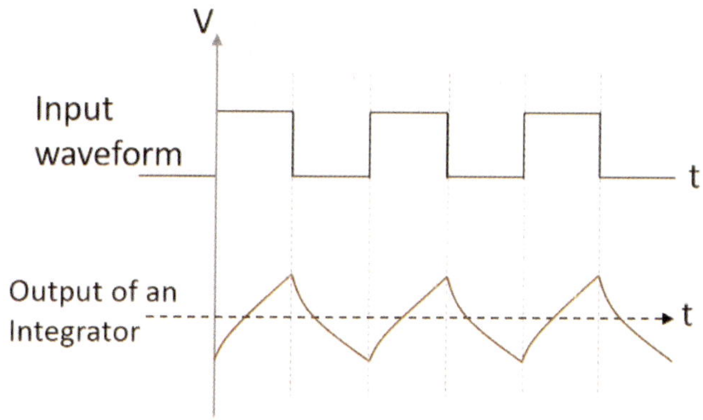

High Pass Filter as Differentiator

At low frequencies, the output of a differentiator is zero whereas at high frequencies, its output is of some finite value. This is same as for a differentiator. Hence the high pass filter is said to be behaved as a differentiator.

If time constant of the RC HPF is very much smaller than time period of the input signal, then circuit behaves as a differentiator. Then the voltage drop across R is very small when compared to the drop across C.

$$V_i = \frac{1}{C}\int i\,dt + iR$$

But $iR = V_o$ is small,

$$\text{since, } V_i = \frac{1}{C}\int i\,dt$$

$$i = \frac{V_0}{R}$$

$$\text{Since, } V_i = \frac{1}{\tau}\int V_0\,dt$$

Differentiating on both sides,

$$\frac{dVi}{dt} = \frac{V_0}{\tau}$$

$$V_0 = \frac{dVi}{dt}$$

$$\text{Since, } V_0 \ \alpha \ \frac{dVi}{dt}$$

The output is proportional to the differential of the input signal.

Frequency Response

The frequency response of a practical high pass filter, when it works as a Differentiator is as shown below:

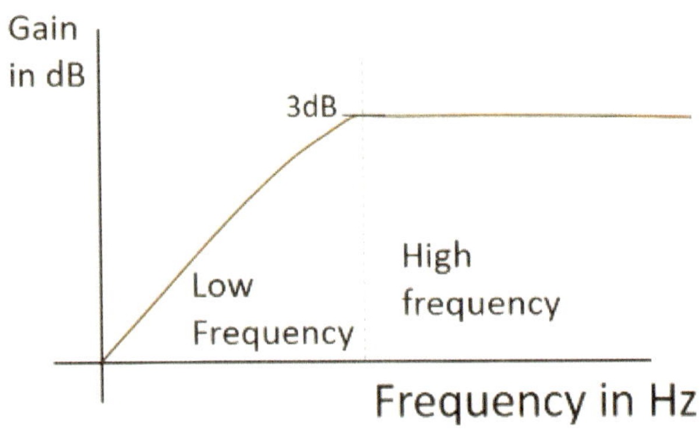

Output Wave Form

If the differentiator circuit is given a sinewave input, the output will be a cosine wave. If the input is a square wave, the output wave form changes its shape and appears as in the figure below:

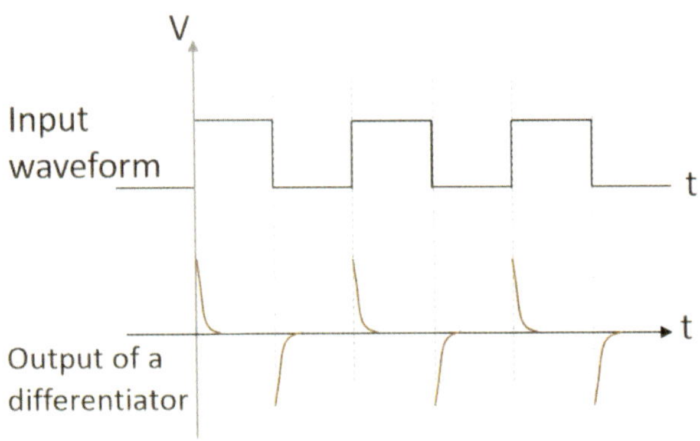

These two circuits are mostly used in various electronic applications. A differentiator circuit produces a constant output voltage when the input applied tends to change steadily. An integrator circuit produces a steadily changing output voltage when the input voltage applied is constant.

Along with resistors, the non-linear elements like diodes are used in nonlinear wave shaping circuits to get required altered outputs. Either the shape of the wave is attenuated or the dc level of the wave is altered in the Non-linear wave shaping.

The process of producing non-sinusoidal output wave forms from sinusoidal input, using non-linear elements is called as nonlinear wave shaping.

Diode Clipping Circuits

The Diode Clipper, also known as a *Diode Limiter*, is a wave shaping circuit that takes an input waveform and clips or cuts off its top half, bottom half or both halves together.

This clipping of the input signal produces an output waveform that resembles a flattened version of the input. For example, the half-wave rectifier is a clipper circuit, since all voltages below zero are eliminated.

But Diode Clipping Circuits can be used a variety of applications to modify an input waveform using signal and Schottky diodes or to provide over-voltage protection using zener diodes to ensure that the output voltage never exceeds a certain level protecting the circuit from high voltage spikes. Then diode clipping circuits can be used in voltage limiting applications.

Types of Clippers

The clipper circuits are generally categorized into three types: series clippers, shunt clippers and dual (combination) clippers. In series clippers, the diode is connected in series with the output load resistance. In shunt clippers, the diode is connected in parallel with the output load resistance.

The series clippers are again classified into four types: series positive clipper, series positive clipper with bias, series negative clipper and series negative clipper with bias. The shunt (parallel) clippers are again classified into four types: shunt positive clipper, shunt positive clipper with bias, shunt negative clipper, and shunt negative clipper with bias.

The various types of clippers are as follows:

- Series positive clipper;
- Series positive clipper with bias;
- Series negative clipper;
- Series negative clipper with bias;
- Shunt positive clipper;
- Shunt positive clipper with bias;
- Shunt negative clipper;
- Shunt negative clipper with bias;
- Dual (combination) clipper;
- Series positive clipper.

Series Positive Clipper

In series positive clipper, the positive half cycles of the input AC signal is removed.

If the diode is arranged in such a way that the arrowhead of the diode points towards the input and the diode is in series with the output load resistance, then the clipper is said to be a series positive clipper.

In the circuit diagram, the diode D is connected in series with the output load resistance RL and the arrowhead of the diode is pointing towards the input. So the circuit is said to be a series positive clipper.

The vertical line in the diode symbol represents the cathode (n-side) and the opposite end represents the anode (p-side).

During positive half cycle:

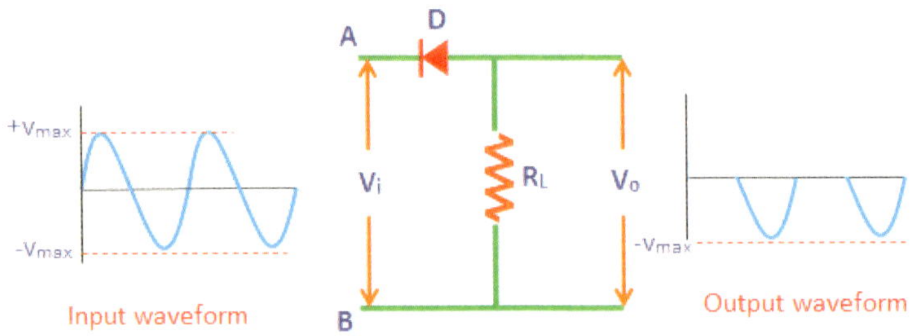

Series positive clipper

During the positive half cycle, terminal A is positive and terminal B is negative. That means the positive terminal A is connected to n-side and the negative terminal B is connected to p-side of the diode. As we already know that if the positive terminal is connected to n-side and the negative terminal is connected to p-side then the diode is said to be reverse biased. Therefore, the diode D is reverse biased during the positive half cycle.

During reverse biased condition, no current flows through the diode. So the positive half cycle is blocked or removed at the output.

During Negative Half Cycle

During the negative half cycle, terminal A is negative and terminal B is positive. That means the negative terminal A is connected to n-side and the positive terminal B is connected to p-side of the diode. As we already know that if the negative terminal is connected to n-side and the positive terminal is connected to p-side then the diode is said to be forward biased. Therefore, the diode D is forward biased during the negative half cycle.

During forward biased condition, electric current flows through the diode. So the negative half cycle is allowed at the output.

Thus, a series of positive half cycles are completely removed at the output.

We know that a clipper either clips a portion of half cycle or clips a complete half cycle. In this case, complete half cycles are removed.

Thus, a series positive clipper removes the series of positive half cycles.

Series Positive Clipper with Bias

Sometimes it is desired to remove a small portion of positive or negative half cycles. In such cases, the biased clippers are used.

The construction of the series positive clipper with bias is almost similar to the series positive clipper. The only difference is an extra element called battery is used in series positive clipper with bias.

- Series positive clipper with positive bias
  - During positive half cycle

During the positive half cycle, terminal A is positive and terminal B is negative. That means the positive terminal is connected to n-side and the negative terminal is connected to p-side. As we already know that if the positive terminal is connected to n-side and the negative terminal is connected to p-side then the diode is said to be reverse biased. Therefore, the diode is reverse biased by the input supply voltage V_i.

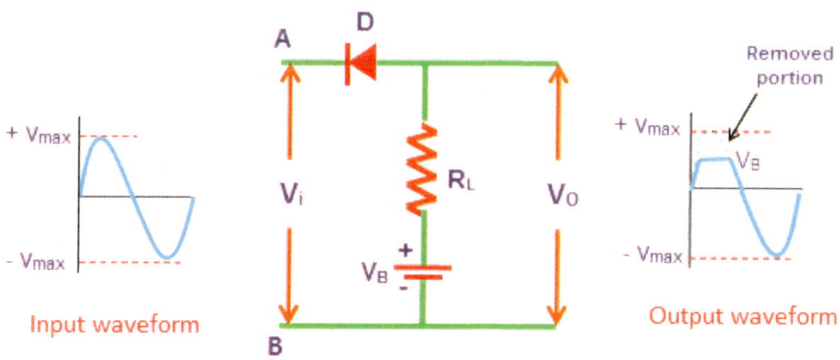

Series positive clipper with positive bias

However, we are supplying the voltage from another source called battery. As shown in the figure, the positive terminal of the battery is connected to p-side and the negative terminal of the battery is connected to n-side of the diode. Therefore, the diode is forward biased by the battery voltage V_B.

That means the diode is reverse biased by the input supply voltage (V_i) and forward biased by the battery voltage (V_B).

Initially, the input supply voltage V_i is less than the battery voltage V_B ($V_i < V_B$). So the battery voltage dominates the input supply voltage. Hence, the diode is forward biased by the battery voltage and allows electric current through it. As a result, the signal appears at the output.

When the input supply voltage V_i becomes greater than the battery voltage V_B, the diode D is reverse biased. So no current flows through the diode. As a result, input signal does not appear at the output.

Thus, the clipping (removal of a signal) takes place during the positive half cycle only when the input supply voltage becomes greater than the battery voltage.

  - During negative half cycle

During the negative half cycle, terminal A is negative and terminal B is positive. That means the diode D is forward biased due to the input supply voltage. Furthermore, the battery is also connected in such a way that the positive terminal is connected to p-side and the negative terminal is connected to n-side. So the diode is forward biased by both battery voltage VB and input supply voltage Vi.

That means, during the negative half cycle, it doesn't matter whether the input supply voltage is greater or less than the battery voltage, the diode always remains forward biased. So the complete negative half cycle appears at the output.

Thus, the series positive clipper with positive bias removes a small portion of positive half cycles.

- Series positive clipper with negative bias
  - During positive half cycle

During the positive half cycle, the diode D is reverse biased by both input supply voltage Vi and battery voltage VB. So no signal appears at the output during the positive half cycle. Therefore, the complete positive half cycle is removed.

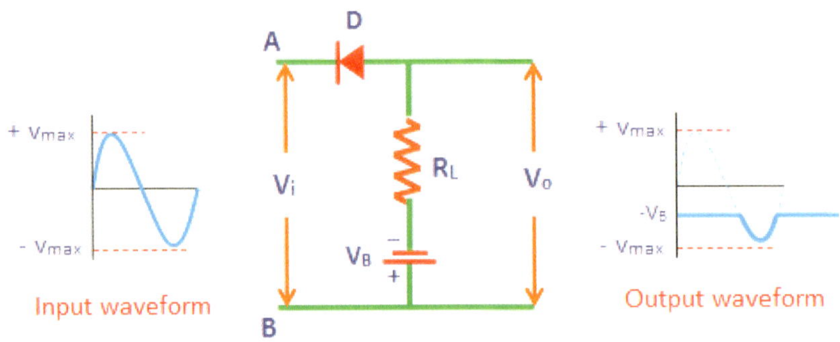

Series positive clipper with negative bias

- During negative half cycle

During the negative half cycle, the diode is forward biased by the input supply voltage Vi and reverse biased by the battery voltage VB. However, initially, the battery voltage VB dominates the input supply voltage Vi. So the diode remains to be reverse biased until the Vi becomes greater than VB. When the input supply voltage Vi becomes greater than the battery voltage VB, the diode is forward biased by the input supply voltage Vi. So the signal appears at the output.

Series Negative Clipper

In series negative clipper, the negative half cycles of the input AC signal is removed at the output. The circuit construction of the series negative clipper is shown in the figure.

If the diode is arranged in such a way that the arrowhead of the diode points towards the output and the diode is in series with the output load resistance, then the clipper is said to be a series negative clipper. In simple words, in a series negative clipper, the diode is connected in a direction opposite to that of the series positive clipper.

The vertical line in the diode symbol represents the cathode (n-side) and the opposite end represents the anode (p-side).

- During positive half cycle:

During the positive half cycle, terminal A is positive and terminal B is negative. That means the positive terminal A is connected to p-side and the negative terminal B is connected to n-side of the diode. As we already know that if the positive terminal is connected to p-side and the negative terminal is connected to n-side then the diode is said to be forward biased. Therefore, the diode D is forward biased during the positive half cycle.

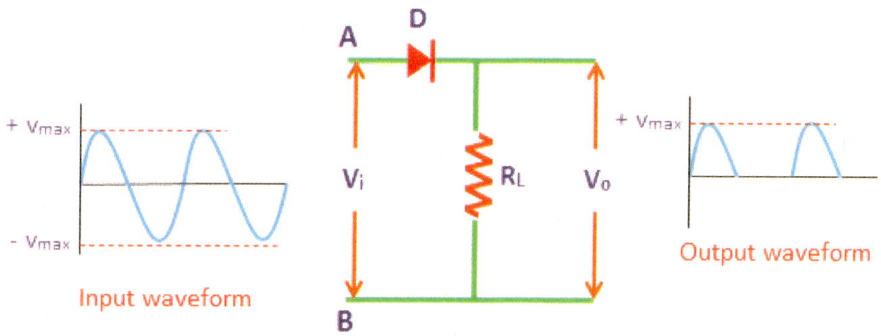

Series negative clipper

During forward biased condition, electric current flows through the diode. So the positive half cycle is allowed at the output. Therefore, a series of positive half cycles appears at the output.

- During negative half cycle:

During the negative half cycle, the terminal A is negative and the terminal B is positive. That means the negative terminal A is connected to p-side and the positive terminal B is connected to n-side of the diode. As we already know that if the negative terminal is connected to p-side and the positive terminal is connected to n-side then the diode is said to be reverse biased. Therefore, the diode D is reverse biased during the negative half cycle.

During reverse biased condition, no current flows through the diode. So the negative half cycle is completely blocked or removed at the output. In other words, a series of negative half cycles are removed at the output.

Thus, the series negative clipper removes the series of negative half cycles.

Series Negative Clipper with Bias

Sometimes it is desired to remove a small portion of positive or negative half cycles of the input AC signal. In such cases, the biased clippers are used.

The construction of the series negative clipper with bias is almost similar to the series negative clipper. The only difference is an extra element called battery is used in series negative clipper with bias.

- Series negative clipper with positive bias
    - During positive half cycle

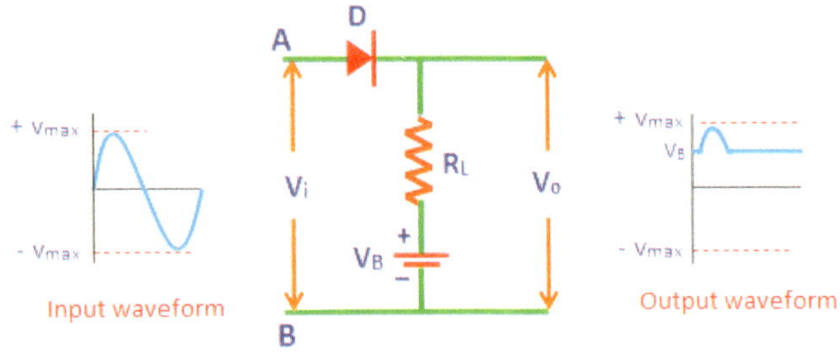

Series negative clipper with positive bias

During the positive half cycle, terminal A is positive and terminal B is negative. That means the positive terminal A is connected to p-side and the negative terminal B is connected to n-side. As we already know that if the positive terminal is connected to p-side and the negative terminal is connected to n-side then the diode is said to be forward biased. However, we are also supplying the voltage from another source called battery. As shown in the figure, the positive terminal of the battery is connected to n-side and the negative terminal of the battery is connected to p-side of the diode.

That means the diode is forward biased by input supply voltage Vi and reverse biased by battery voltage VB. Initially, the battery voltage is greater than the input supply voltage. Hence, the diode is reverse biased and does not allow electric current. Therefore, no signal appears at the output.

When the input supply voltage Vi becomes greater than the battery voltage VB, the diode is forward biased and allows electric current. As a result, the signal appears at the output.

- During negative half cycle

During the negative half cycle, the diode is reverse biased by both input supply voltage Vi and battery voltage VB. So it doesn't matter whether the input supply voltage is greater or less than the battery voltage VB, the diode always remains reverse biased. Therefore, during the negative half cycle, no signal appears at the output.

- Series negative clipper with negative bias
  - During positive half cycle

During the positive half cycle, the diode D is forward biased by both input supply voltage Vi and the battery voltage VB. So it doesn't matter whether the input supply voltage is greater or less than battery voltage VB, the diode always remains forward biased. Therefore, during the positive half cycle, the signal appears at the output.

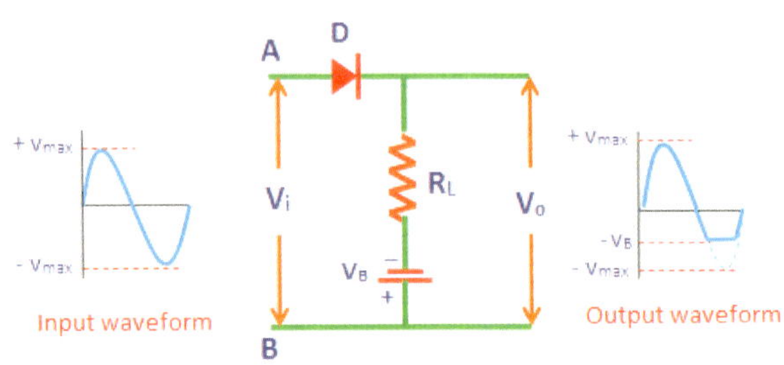

Series negative clipper with negative bias

- During negative half cycle:

During the negative half cycle, the diode D is reverse biased by the input supply voltage Vi and forward biased by the battery voltage VB. Initially, the input supply voltage Vi is less than the battery voltage VB. So the diode is forward biased by the battery voltage VB. As a result, the signal appears at the output.

When the input supply voltage Vi becomes greater than the battery voltage VB, the diode will become reverse biased. As a result, no signal appears at the output.

Shunt Positive Clipper

In shunt clipper, the diode is connected in parallel with the output load resistance. The operating principles of the shunt clipper are nearly opposite to the series clipper.

The series clipper passes the input signal to the output load when the diode is forward biased and blocks the input signal when the diode is reverse biased.

The shunt clipper on the other hand passes the input signal to the output load when the diode is reverse biased and blocks the input signal when the diode is forward biased.

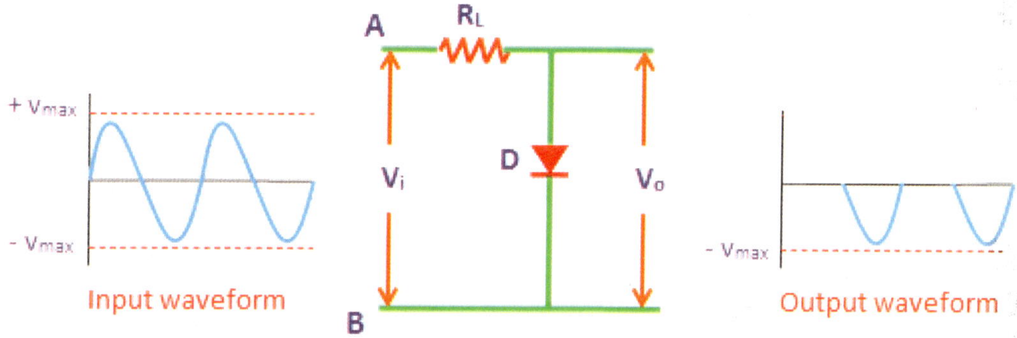

Shunt positive clipper

In shunt positive clipper, during the positive half cycle the diode is forward biased and hence no output is generated. On the other hand, during the negative half cycle the diode is reverse biased and hence the entire negative half cycle appears at the output.

Shunt Positive Clipper with Bias

- Shunt positive clipper with positive bias:

During the positive half cycle, the diode is forward biased by the input supply voltage Vi and reverse biased by the battery voltage VB. However, initially, the input supply voltage Vi is less than the battery voltage VB. Hence, the battery voltage VB makes the diode to be reverse biased. Therefore, the signal appears at the output. However, when the input supply voltage Vi becomes greater than the battery voltage VB, the diode D is forward biased by the input supply voltage Vi. As a result, no signal appears at the output.

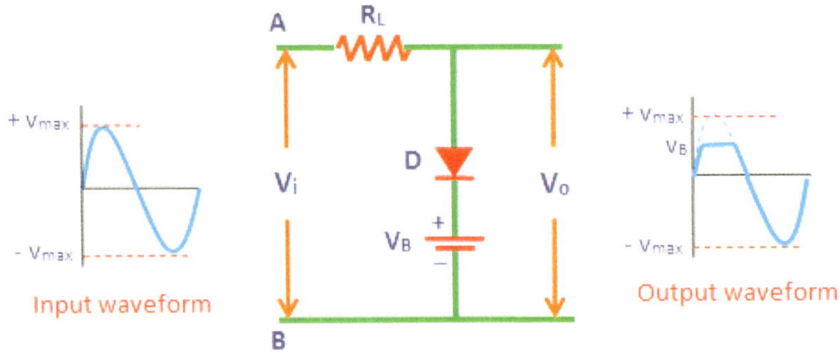

Shunt positive clipper with positive bias

During the negative half cycle, the diode is reverse biased by both input supply voltage and battery voltage. So it doesn't matter whether the input supply voltage is greater or lesser than the battery voltage, the diode always remains reverse biased. As a result, a complete negative half cycle appears at the output.

- Shunt positive clipper with negative bias:

During the positive half cycle, the diode is forward biased by both input supply voltage Vi and battery voltage VB. Therefore, no signal appears at the output during the positive half cycle.

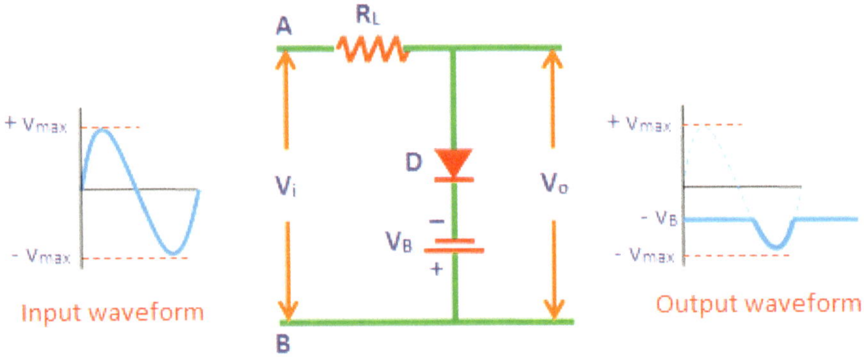

Shunt positive clipper with negative bias

During the negative half cycle, the diode is reverse biased by the input supply voltage and forward biased by the battery voltage. However, initially, the input supply voltage Vi is less than the battery voltage VB. So the battery voltage makes the diode to be forward biased. As a result, no signal appears at the output. However, when the input supply voltage Vi becomes greater than the battery voltage VB, the diode is reverse biased by the input supply voltage Vi. As a result, the signal appears at the output.

Shunt Negative Clipper

In shunt negative clipper, during the positive half cycle the diode is reverse biased and hence the entire positive half cycle appears at the output. On the other hand, during the negative half cycle the diode is forward biased and hence no output signal is generated.

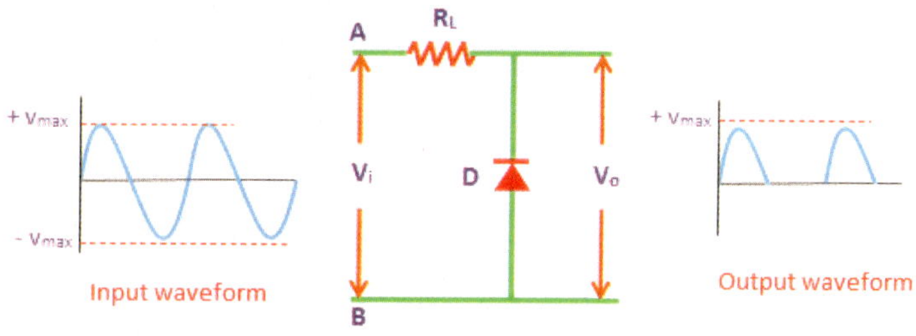

Shunt negative clipper

Shunt Negative Clipper with Bias

- Shunt negative clipper with positive bias

During the positive half cycle, the diode is reverse biased by the input supply voltage Vi and forward biased by the battery voltage VB. However, initially, the input supply voltage is less than the battery voltage. So the diode is forward biased by the battery voltage. As a result, no signal appears at the output. However, when the input supply voltage becomes greater than the battery voltage then the diode is reverse biased by the input supply voltage. As a result, the signal appears at the output.

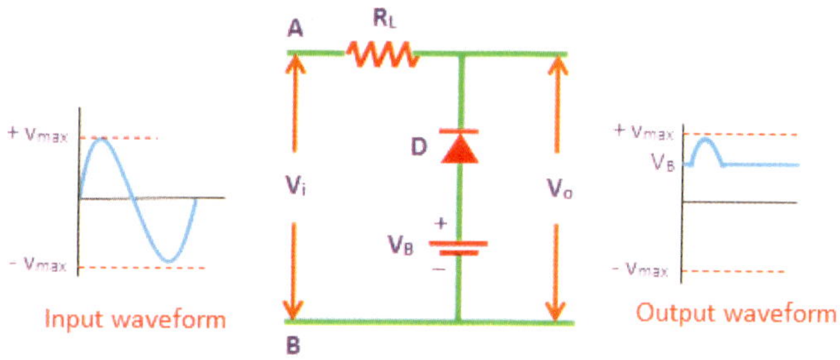

Shunt negative clipper with positive bias

During the negative half cycle, the diode is forward biased by both input supply voltage Vi and battery voltage VB. So the complete negative half cycle is removed at the output.

- Shunt negative clipper with negative bias

During the positive half cycle, the diode is reverse biased by both input supply voltage Vi and battery voltage VB. As a result, the complete positive half cycle appears at the output.

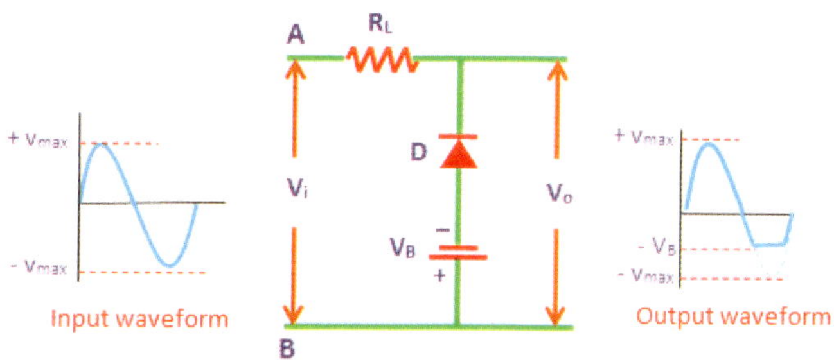

Shunt negative clipper with negative bias

During the negative half cycle, the diode is forward biased by the input supply voltage Vi and reverse biased by the battery voltage VB. However, initially, the input supply voltage is less than the battery voltage. So the diode is reverse biased by the battery voltage. As a result, the signal appears at the output. However, when the input supply voltage becomes greater than the battery voltage, the diode is forward biased by the input supply voltage. As a result, the signal does not appear at the output.

Dual (combination) Clipper

Sometimes it is desired to remove a small portion of both positive and negative half cycles. In such cases, the dual clippers are used.

The dual clippers are made by combining the biased shunt positive clipper and biased shunt negative clipper.

Let us consider a dual clipper circuit in which a sinusoidal ac voltage is applied to the input terminals of the circuit.

- During positive half cycle

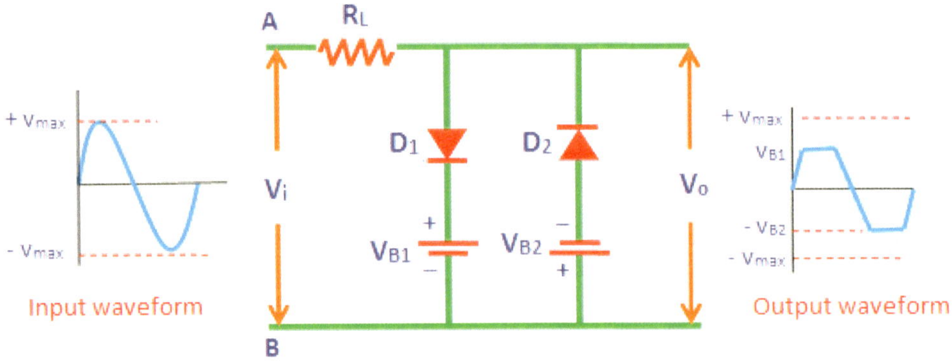

Dual (Combination) clipper

During the positive half cycle, the diode D1 is forward biased by the input supply voltage Vi and reverse biased by the battery voltage VB1. On the other hand, the diode D2 is reverse biased by both input supply voltage Vi and battery voltage VB2.

Initially, the input supply voltage is less than the battery voltage. So the diode D1 is reverse biased by the battery voltage VB1. Similarly, the diode D2 is reverse biased by the battery voltage VB2. As a result, the signal appears at the output. However, when the input supply voltage Vi becomes greater than the battery voltage VB1, the diode D1 is forward biased by the input supply voltage. As a result, no signal appears at the output.

- During negative half cycle

During the negative half cycle, the diode D1 is reverse biased by both input supply voltage Vi and battery voltage VB1. On the other hand, the diode D2 is forward biased by the input supply voltage Vi and reverse biased by the battery voltage VB2.

Initially, the battery voltage is greater than the input supply voltage. Therefore, the diode D1 and diode D2 are reverse biased by the battery voltage. As a result, the signal appears at the output.

When the input supply voltage becomes greater than the battery voltage VB2, the diode D2 is forward biased. As a result, no signal appears at the output.

Drawbacks of Series and Shunt Diode Clippers

- In series clippers, when the diode is in 'OFF' position, there will be no transmission of the input signal to output. But in the case of high-frequency signals transmission occurs through diode capacitance which is undesirable. This is the drawback of using the diode as a series element in such clippers.
- In shunt clippers, when the diode is in the 'off condition, transmission of input signal should take place to output. But in the case of high-frequency input signals, diode capacitance affects the circuit operation adversely and the signal gets attenuated (that is, it passes through diode capacitance to ground).

Applications of Clippers

- Clippers are commonly used in power supplies.
- Used in TV transmitters and Receivers.
- They are employed for different wave generation such as square, rectangular, or trapezoidal waves.
- Series clippers are used as noise limiters in FM transmitters.

Transistor

Transistor is a semiconductor device that can both conduct and insulate. A transistor can act as a switch and an amplifier. It converts audio waves into electronic waves and resistor, controlling electronic current. Transistors have very long life, smaller in size, can operate on lower voltage supplies for greater safety and required no filament current. The first transistor was fabricated with germanium. A transistor performs the same function as a vacuum tube triode, but using semiconductor junctions instead of heated electrodes in a vacuum chamber. It is the fundamental building block of modern electronic devices and found everywhere in modern electronic systems.

A transistor is a three terminal device. Namely:

- Base: This is responsible for activating the transistor.
- Collector: This is the positive lead.
- Emitter: This is the negative lead.

The basic idea behind a transistor is that it lets you control the flow of current through one channel by varying the intensity of a much smaller current that's flowing through a second channel.

Types of Transistors

There are two types of transistors in present; they are bipolar junction transistor (BJT), field effect transistors (FET). A small current is flowing between the base and the emitter; base terminal can control a larger current flow between the collector and the emitter terminals. For a field-effect transistor, it also has the three terminals, they are gate, source, and drain, and a voltage at the gate can control a current between source and drain. The simple diagrams of BJT and FET are shown in figure below:

Bipolar Junction Transistor(BJT)

Field Effect Transistors(FET)

As you can see, transistors come in a variety of different sizes and shapes. One thing all of these transistors have in common is that they each have three leads.

Symbols, Pins, and Construction

Transistors are fundamentally three-terminal devices. On a bi-polar junction transistor (BJT), those pins are labeled collector (C), base (B), and emitter (E). The circuit symbols for both the NPN and PNP BJT are below:

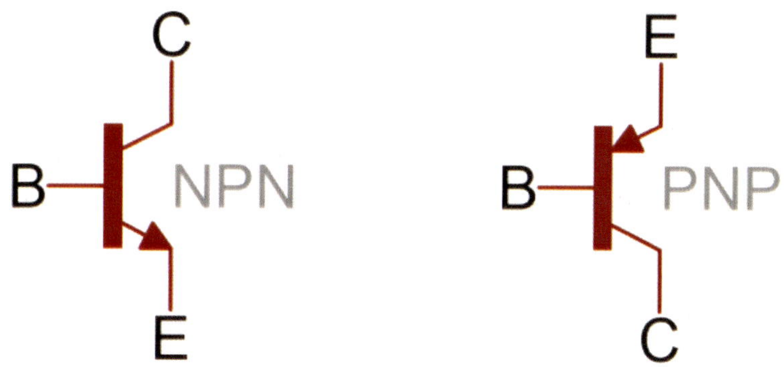

The only difference between an NPN and PNP is the direction of the arrow on the emitter. The arrow on an NPN points out, and on the PNP it points in.

Transistor Construction

Transistors rely on semiconductors to work their magic. A semiconductor is a material that's not quite a pure conductor (like copper wire) but also not an insulator (like air). The conductivity of a semiconductor – how easily it allows electrons to flow – depends on variables like temperature or the presence of more or less electrons.

A Transistor as Two Diodes

Transistors are kind of like an extension of another semiconductor component: diodes. In a way transistors are just two diodes with their cathodes (or anodes) tied together:

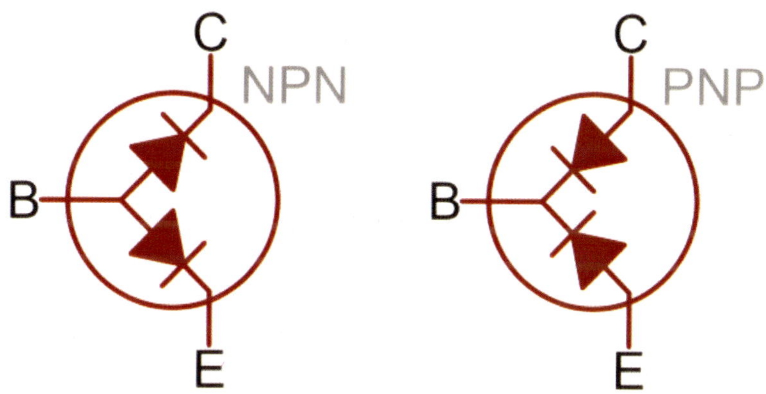

The diode connecting base to emitter is the important one here; it matches the direction of the arrow on the schematic symbol, and shows you which way current is intended to flow through the transistor.

The diode representation is a good place to start, but it's far from accurate. Don't base your understanding of a transistor's operation on that model (and definitely don't try to replicate it on a breadboard, it won't work).

Transistor Structure and Operation

Transistors are built by stacking three different layers of semiconductor material together. Some of those layers have extra electrons added to them (a process called "doping"), and others have electrons removed (doped with "holes" – the absence of electrons). A semiconductor material with *extra* electrons is called an n-type (*n* for negative because electrons have a negative charge) and a material with electrons removed is called a p-type (for positive). Transistors are created by either stacking an *n* on top of a *p* on top of an *n*, or *p* over *n* over *p*.

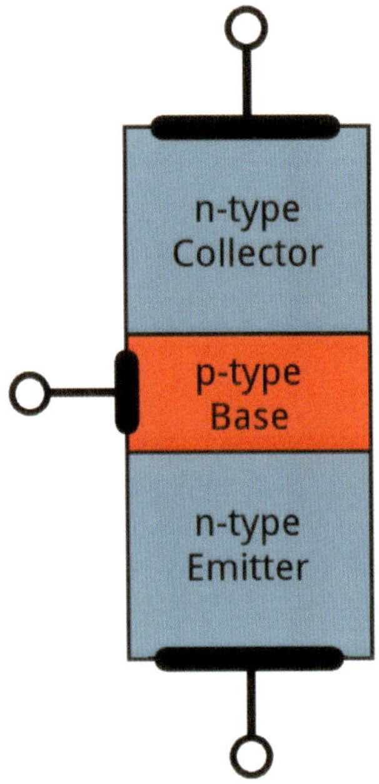

Simplified diagram of the structure of an NPN. Notice the origin of any acronyms

With some hand waving, we can say electrons can easily flow from n regions to p regions, as long as they have a little force (voltage) to push them. But flowing from a *p* region to an *n* region is really hard (requires a *lot* of voltage). But the special thing about a transistor – the part that makes our two-diode model obsolete – is the fact that electrons *can*easily flow from the p-type base to the n-type collector as long as the base-emitter junction is forward biased(meaning the base is at a higher voltage than the emitter).

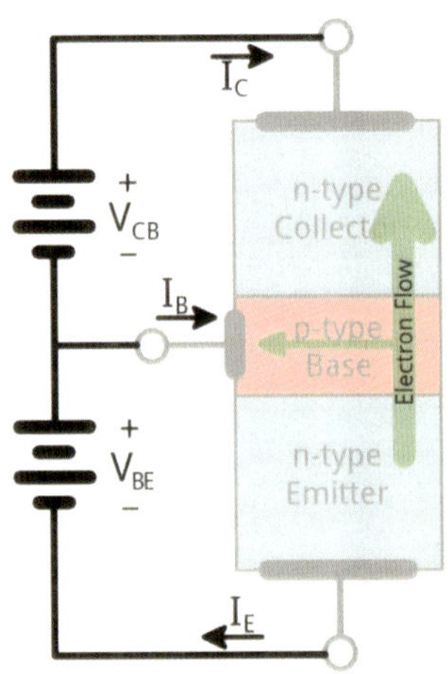

The NPN transistor is designed to pass electrons from the emitter to the collector (so conventional current flows from collector to emitter). The emitter "emits" electrons into the base, which controls the number of electrons the emitter emits. Most of the electrons emitted are "collected" by the collector, which sends them along to the next part of the circuit.

A PNP works in a same but opposite fashion. The base still controls current flow, but that current flows in the opposite direction – from emitter to collector. Instead of electrons, the emitter emits "holes" (a conceptual absence of electrons) which are collected by the collector.

The transistor is kind of like an electron valve. The base pin is like a handle you might adjust to allow more or less electrons to flow from emitter to collector.

Modes of Biasing

Following are the different modes of transistor base biasing:

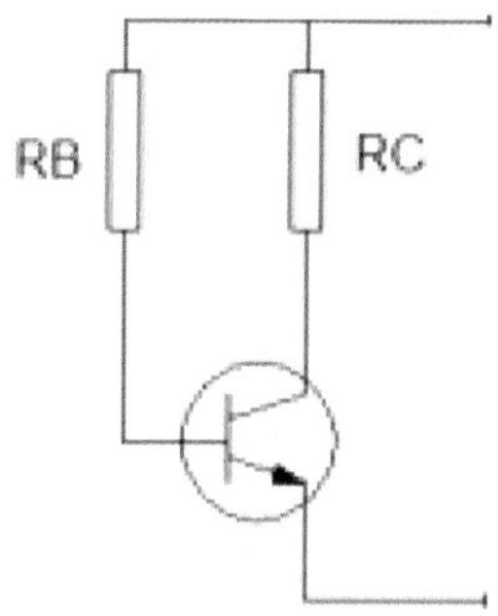

Current biasing

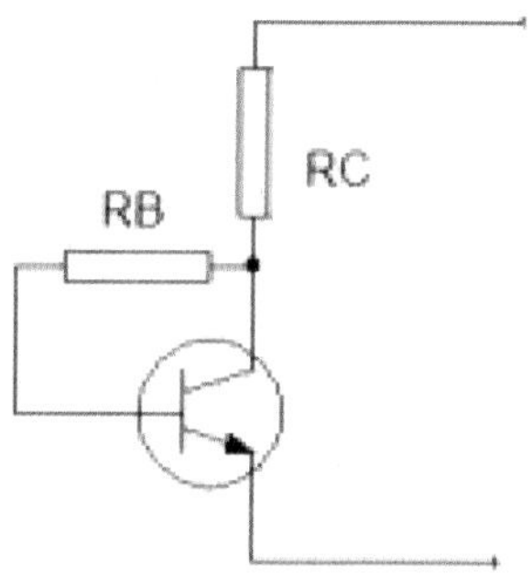

Feedback biasing

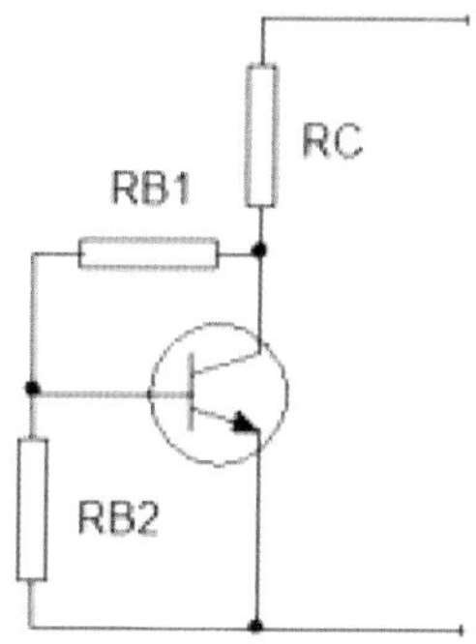

Double feedback biasing

Current Biasing

As shown in the figure above, two resistors RC and RB are used to set the base bias. These resistors establish the initial operating region of the transistor with a fixed current bias.

The transistor forward biases with a positive base bias voltage through RB. The forward base-Emitter voltage drop is 0.7 volts. Therefore the current through RB is $I_B = (V_{cc} - V_{BE}) / I_B$.

Feedback Biasing

Figure above shows the transistor biasing by the use of a feedback resistor. The base bias is obtained from the collector voltage. The collector feedback ensures that the transistor is always biased in the active region. When the collector current increases, the voltage at the collector drops. This reduces the base drive which in turn reduces the collector current. This feedback configuration is ideal for transistor amplifier designs.

Double Feedback Biasing

Figure above shows how the biasing is achieved using double feedback resistors. By using two resistors RB1 and RB2 increases the stability with respect to the variations in Beta by increasing the current flow through the base bias resistors. In this configuration, the current in RB1 is equal to 10 % of the collector current.

Voltage Dividing Biasing

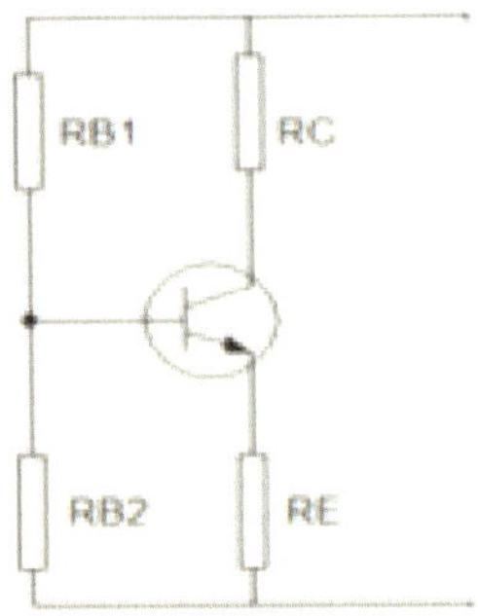

Voltage divider biasing

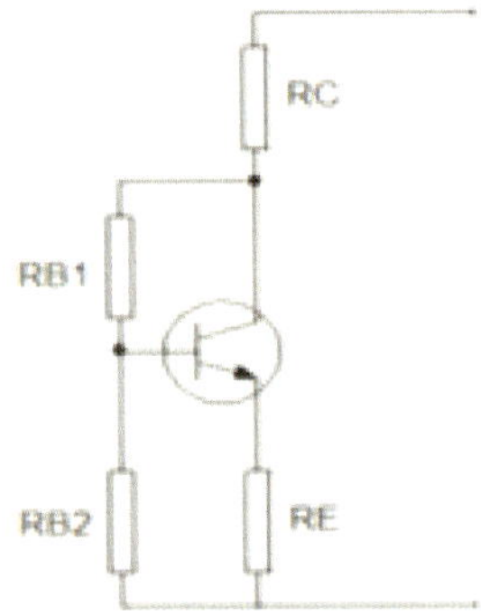

Double base biasing

Figure above shows the Voltage divider biasing in which two resistors RB1 and RB2 are connected to the base of the transistor forming a voltage divider network. The transistor gets biases by the voltage drop across RB2. This kind of biasing configuration is used widely in amplifier circuits.

Double Base Biasing

Figure above shows a double feedback for stabilization. It uses both Emitter and Collector base feedback to improve the stabilization through controlling the collector current. Resistor values should be selected so as to set the voltage drop across the Emitter resistor 10% of the supply voltage and the current through RB1, 10% of the collector current.

Advantages of Transistor

- Smaller mechanical sensitivity.
- Lower cost and smaller in size, especially in small-signal circuits.
- Low operating voltages for greater safety, lower costs and tighter clearances.
- Extremely long life.
- No power consumption by a cathode heater.
- Fast switching.

Bipolar Junction Transistor

BJT was invented in 1948 by William Shockley, Brattain, and John Bardeen which has remolded not only the world of electronics but also in our day to day life. The Bipolar junction transistors uses both charge carriers that are electron and holes. Indifference the unipolar transistors such as field effect transistors use only one kind of charge carriers. For the operation purpose BJT uses two semiconductor type n-type and p- type between two junctions. The main basic function of a BJT is to amplify current it will allow BJTs are used as amplifiers or switches to produce wide applicability in electronic equipment include mobile phones, industrial control, television, and radio transmitters. There are two different types of BJTs are available, they are NPN, and PNP.

The Bipolar junction transistor is a solid state device and in the BJTs the current flow in two terminals, they are emitter and collector and the amount of current controlled by the third terminal i.e. base terminal. It is different to the other type of transistor i.e. Field effect transistor which is the output current is controlled by the input voltage. The basic symbol of the BJTs n-type and p-type is shown below:

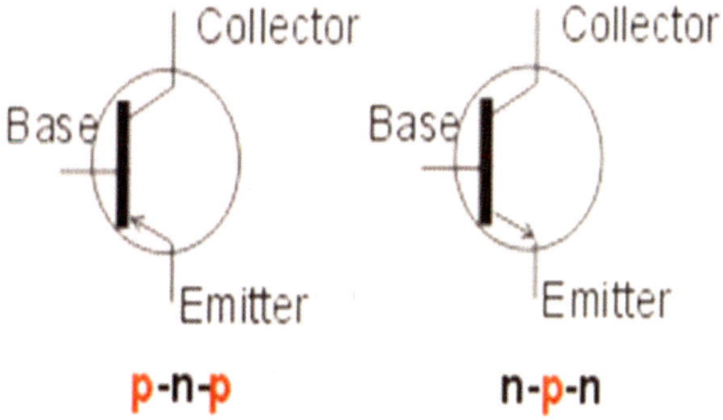

Bipolar Junction Transistors

Types

As we have seen a semiconductor offer less resistance to flow current in one direction and high resistance is another direction and we can call transistor as the device mode of the semiconductor. The bipolar junction transistors consist of two types of transistors. Which, given us,

- Point contact
- Junction transistor

By comparing two transistors the junction transistors are used more than point type transistors. Further the junction transistors are classified into two types which are given below. There are three electrodes for each junction transistor they are emitter, collector, and base,

- PNP junction transistors
- NPN junction transistors.

PNP Junction Transistor

In PNP transistor, the emitter is more positive with base and also with respect to the collector. The PNP transistor is a three terminal device which is made from the semiconductor material. The three terminals are collector, base, and the emitter and the transistor are used for switching and amplifying applications. The operation of PNP transistor is shown below.

Generally the collector terminal is connected to the positive terminal and the emitter to a negative supply with resistor either the emitter or collector circuit. To the base terminal the voltage is applied and it operates transistor as an ON/OFF state. The transistor is in OFF state when the base voltage is same as the emitter voltage. The transistor mode is in ON state when the base voltage decreases with respect to the emitter. By using this property the transistor can act on both applications like switch and amplifier. The basic diagram of PNP transistor is shown below.

NPN Junction Transistor

The NPN transistor is exactly opposite to the PNP transistor. The NPN transistor contains three terminals which are same as the PNP transistor which are emitter, collector, and base.

Generally the positive supply is given to the collector terminal and the negative supply to the emitter terminal with a resistor either the emitter or collector or emitter circuit. To the base terminal the voltage is applied and it operated as an ONN/OFF state of a transistor. The transistor is in OFF state when the base voltage is same as the emitter. If the base voltage is increased with respect to the emitter then the transistor mode is in ON state. By using this condition the transistor can act like both applications which are amplifier and switch. The basic symbol and the NPN configuration diagram as shown below:

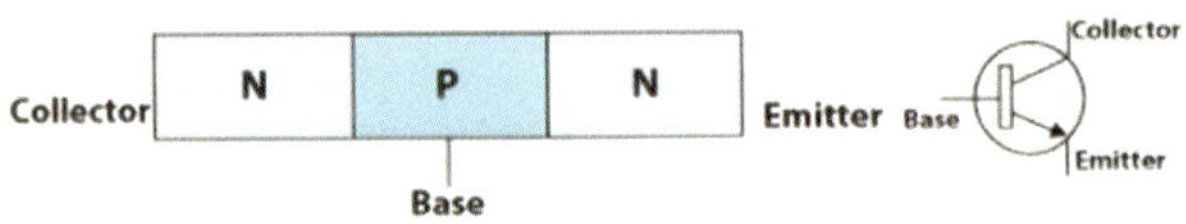

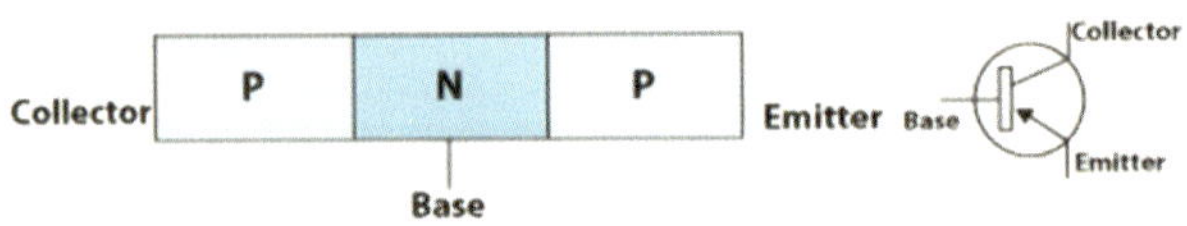

PNP & NPN Junction Transistor

Hetero Bipolar Junction

The Hetero bipolar junction transistor is also a type is the bipolar junction transistor. It uses different semiconductor materials to the emitter and base region and produces hetero junction. The HBT can handle the singles of very high frequencies of several hundred GHz generally it is used in ultrafast circuits and mostly used in radio frequency. Its applications are used in cellular phones, and RF power amplifiers.

Working Principle of BJT

The BE junction is a forward bias and the CB is a reverse bias junction. The width of the depletion region of the CB junction is higher than the BE junction. The forward bias at the BE junction decreases the barrier potential and produces electrons to flow from the emitter to the base and the base is a thin and lightly doped it has very few holes and less amount of electrons from the emitter about 2% it recombine in the base region with holes and from the base terminal it will flow out. This initiates the base current flow due to combination of electrons and holes. The left over large number of electrons will pass the reverse bias collector junction to initiate the collector current. By using KCL we can observe the mathematical equation,

$$I_E = I_B + I_C$$

The base current is very less as compared to emitter and collector current,

$I_E \sim I_C$

Here the operation of PNP transistor is same as the NPN transistor the only difference is only holes instead of electrons. The below diagram shows the PNP transistor of the active mode region:

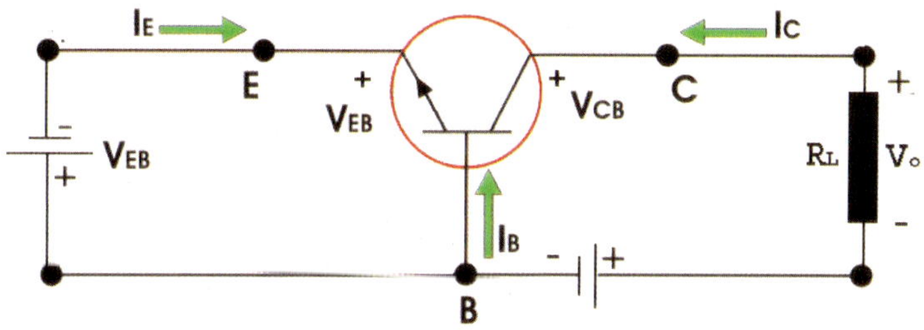

Working Principle of BJT

Bipolar Junction Transistors Characteristics

The three parts of a BJT are collector, emitter and base. Before knowing about the bipolar junction transistor characteristics, we have to know about the modes of operation for this type of transistors. The modes are :

1. Common Base (CB) mode
2. Common Emitter (CE) mode
3. Common Collector (CC) mode.

All three types of modes are shown below.

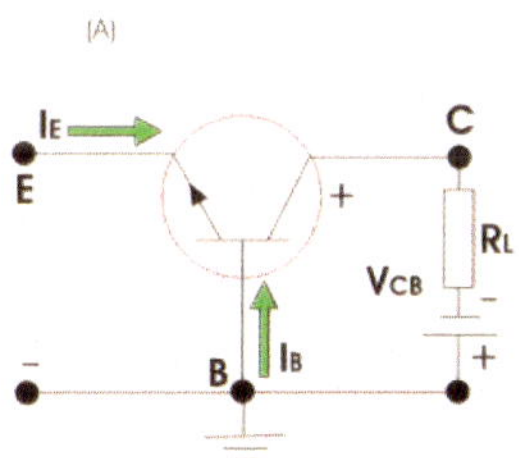

Common Base (CB) mode

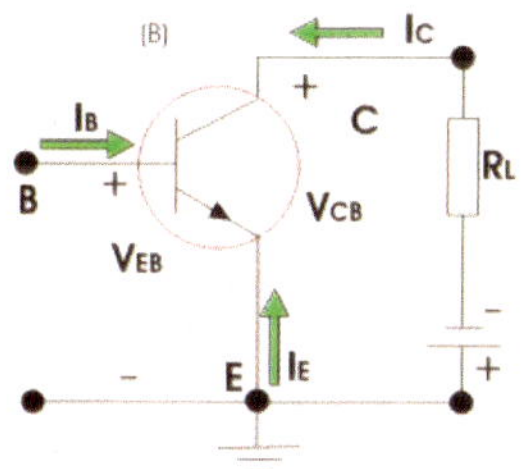

Common Emitter (CE) mode

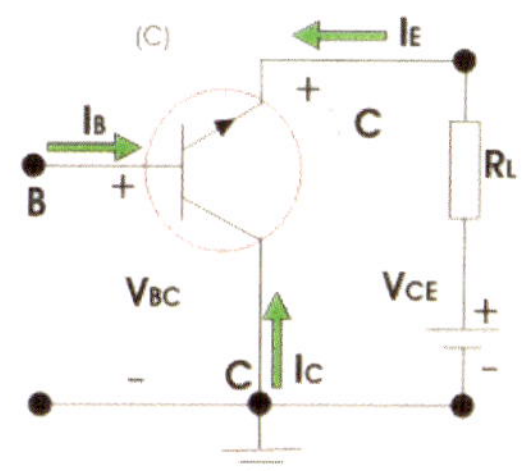

Common Collector (CC) mode

Now coming to the characteristics of BJT there are different characteristics for different modes of operation. Characteristics is nothing but the graphical forms of relationships among different current and voltage variables of the transistor. The characteristics for p-n-p transistors are given for different modes and different parameters.

Common Base Characteristics

Input Characteristics

For p-n-p transistor, the input current is the emitter current (I_E) and the input voltage is the collector base voltage (V_{CB}).

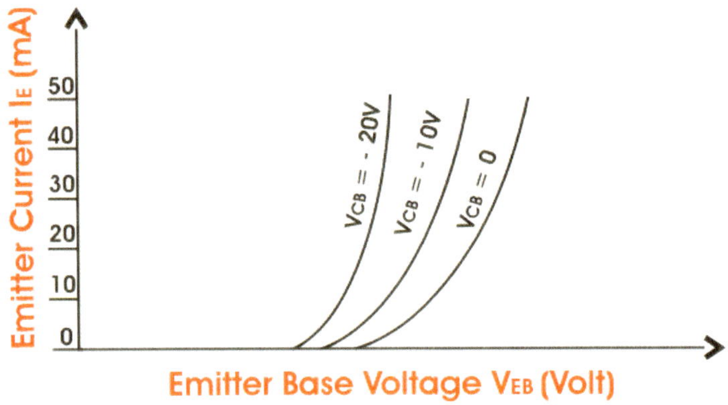

As the emitter - base junction is forward biased, therefore the graph of I_E Vs V_{EB} is similar to the forward characteristics of a p-n diode. I_E increases for fixed V_{EB} when V_{CB} increases.

Output Characteristics

The output characteristics shows the relation between output voltage and output current I_C is the output current and collector-base voltage and the emitter current I_E is the input current and works as the parameters. The figure below shows the output characteristics for a p-n-p transistor in CB mode.

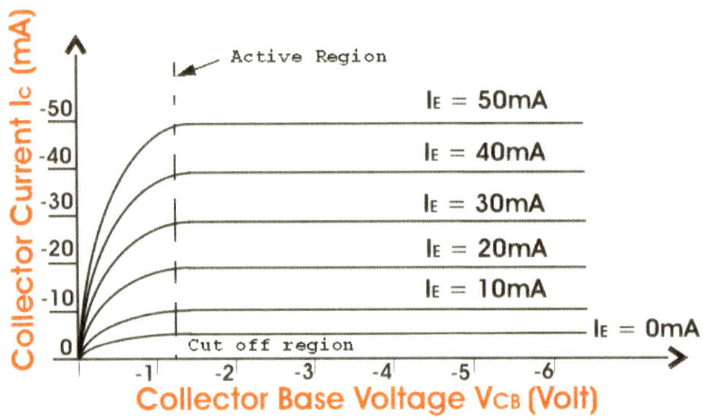

As we know for p-n-p transistors I_E and V_{EB} are positive and I_C, I_B, V_{CB} are negative. These are three regions in the curve, active region saturation region and the cut off region. The active region is the region where the transistor operates normally. Here the emitter junction is reverse biased. Now the saturation region is the region where both the emitter collector junctions are forward biased. And finally the cut off region is the region where both emitter and the collector junctions are reverse biased.

Common Emitter Characteristics

Input Characteristics

I_B (Base Current) is the input current, V_{BE} (Base - Emitter Voltage) is the input voltage for CE (Common Emitter) mode. So, the input characteristics for CE mode will be the relation between I_B and V_{BE} with V_{CE} as parameter. The characteristics are shown below,

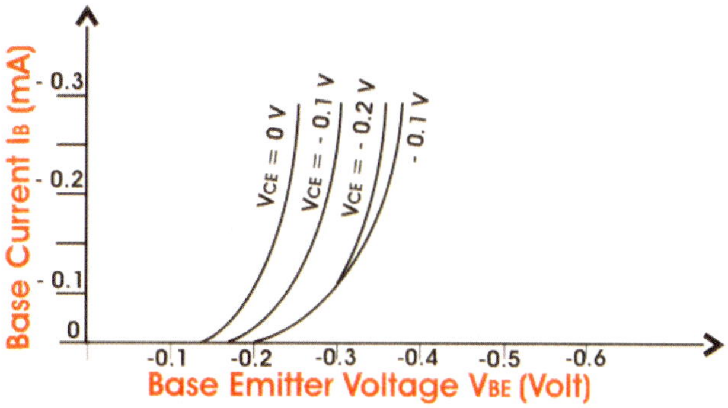

The typical CE input characteristics are similar to that of a forward biased of p-n diode. But as V_{CB} increases the base width decreases.

Output Characteristics

Output characteristics for CE mode is the curve or graph between collector current (I_C) and collector - emitter voltage (V_{CE}) when the base current I_B is the parameter. The characteristics is shown below in the figure.

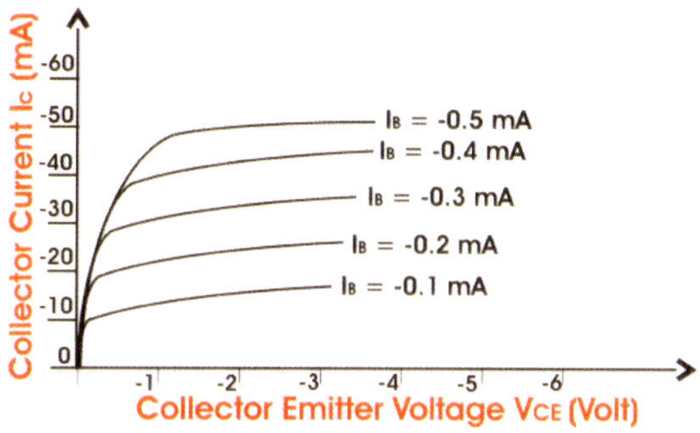

Like the output characteristics of common - base transistor CE mode has also three regions named:

1. Active region,
2. Cut-off regions,
3. Saturation region.

The active region has collector region reverse biased and the emitter junction forward biased. For cut-off region the emitter junction is slightly reverse biased and the collector current is not totally cut-off. And finally for saturation region both the collector and the emitter junction are forward biased.

Application of BJT

BJT's are used in discrete circuit designed due to availability of many types, and obviously because of its high transconductane and output resistance which is better than MOSFET. BJT's are suitable for high frequency application also. That's why they are used in radio frequency for wireless systems. Another application of BJT can be stated as small signal amplifier, metal proximity photocell, etc.

Bipolar Junction Transistor Amplifier

To understand the concept of Bipolar Junction Transistor Amplifier, we should look through the diagram of a p-n-p transistor first.

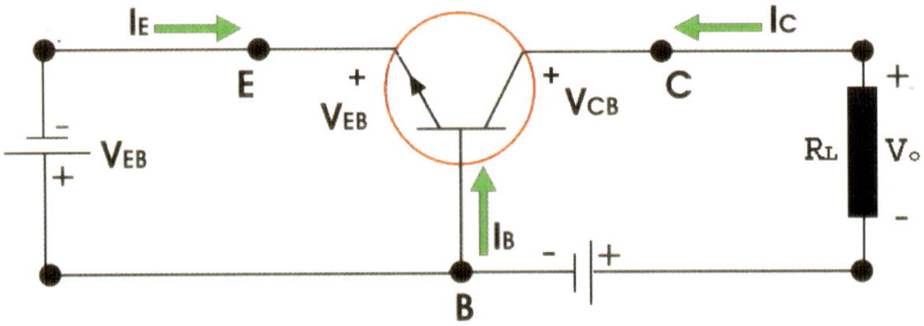

Now as the input voltage is changed a little, say ΔV_i of the emitter-base voltage changes the barrier height and the emitter current by ΔI_E. This change in emitter current develops a voltage drop ΔV_O across the load resistance R_L, where,

$$\Delta V_o = -R_L \Delta I_c$$

ΔV_O gives the output voltage of the amplifier. There is a negative sign because of the collector current gives a voltage drop across R_L with polarity opposite to the reference polarity. The voltage gain A_V for the amplifier is given the ratio between the output voltages ΔV_O to the input voltage ΔV_i, so, is called the current gain ratio of the transistor.

$$A_v = \frac{\Delta V_o}{\Delta V_i} = \frac{R_L \Delta I_c}{r_c \Delta I_E}$$

$$\frac{I_c}{I_E} = A_I$$

From the figure diagram shown above we can see that an increase in the emitter voltage reduces the forward bias at the emitter junction thus decreases the collector current. It indicates that the output voltage and the input voltage are in phase. Now, finally the power gain Ap of the transistor is the ratio between the output power and the input power,

$$A_p = \frac{(\Delta I_c)^2 R_L}{(\Delta I_E)^2 r_c} = A_I . A_V$$

Field-effect Transistor

Field effect transistor or FET is a transistor, where output current is controlled by electric field. FET sometimes, is called uni-polar transistor as it involves single carrier type operation. The basic types of FET transistor is completely different from BJT transistor basics. FET is three terminal semiconductor devices, with source, drain and gate terminals.

The charge carries are electrons or holes, which flow from the source to drain though an active channel. This flow of electrons from source to drain is controlled.

Types of FET

FETs are of two types- JFETs or MOSFETs.

Junction FET

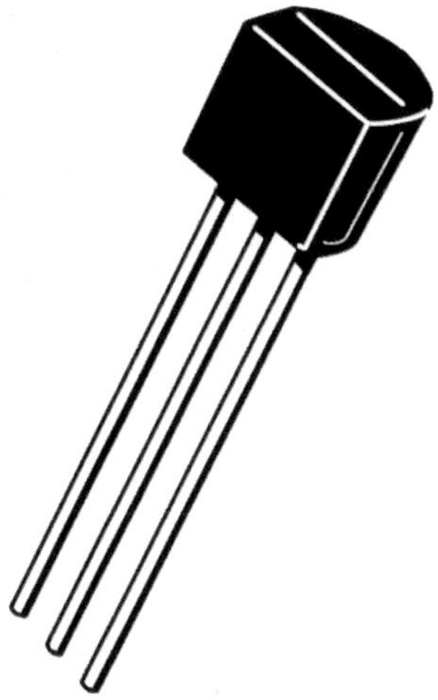

A Junction FET

The Junction FET transistor is a type of field effect transistor that can be used as an electrically controlled switch. The electric energy flows through an active channel between sources to drain terminals. By applying a reverse bias voltage to gate terminal, the channel is strained so the electric current is switched off completely.

The junction FET transistor is available in two polarities which are mentioned below.

N- Channel JFET

N channel JFET

N channel JFET consists of an n type bar at the sides of which two p type layers are doped. The channel of electrons constitutes the N channel for the device. Two ohmic contacts are made at both ends of the N-channel device, which are connected together to form the gate terminal.

The source and drain terminals are taken from the other two sides of the bar. The potential difference between source and drain terminals is termed as Vdd and potential difference between source and gate terminal is termed as Vgs. The charge flow is due to flow of electrons from source to drain.

Whenever a positive voltage is applied across drain and source terminals, electrons flows from the source 'S' to drain 'D' terminal , where as conventional drain current Id flows through the drain to source. As current flows through the device, it is in on state.

When a negative polarity voltage is applied to the gate terminal, a depletion region is created in the channel. The channel width is reduced, hence increasing the channel resistance between the source and drain. Since the gate source junction is reverse biased and no current flows in the device, it is in off condition.

So basically if voltage applied at the gate terminal is increased, less amount of current will flow from the source to drain.

The N channel JFET has greater conductivity than the P channel JFET. So the N channel JFET is more efficient conductor compared to P channel JFET.

P-Channel JFET

P channel JFET consists of a P type bar, at two sides of which n type layers are doped. The gate terminal is formed by joining the ohmic contacts at both the sides. Like in an N channel JFET, the source and drain terminals are taken from the other two sides of the bar. A P type channel, consisting of holes as charge carriers, is formed between source and drain terminal.

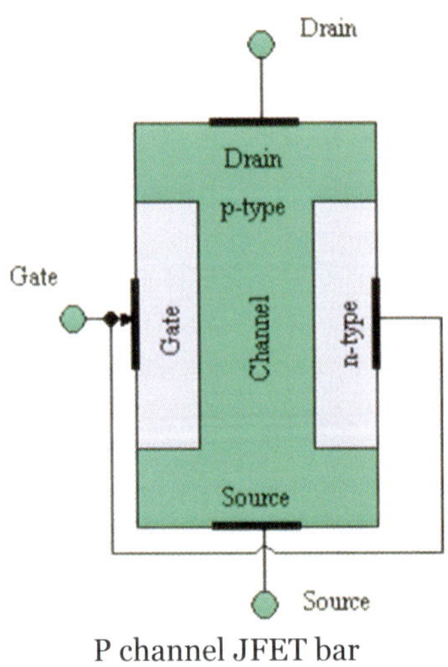

P channel JFET bar

A negative voltage applied to the drain and source terminals ensure the flow of current from source to drain terminal and the device operates in ohmic region. A positive voltage applied to the gate terminal ensures the reduction of channel width, thus increasing the channel resistance. More positive is the gate voltage; less is the current flowing through the device.

Characteristics of p Channel Junction FET Transistor

Given below is characteristic curve of p channel Junction Field Effect transistor and different modes of operation of the transistor:

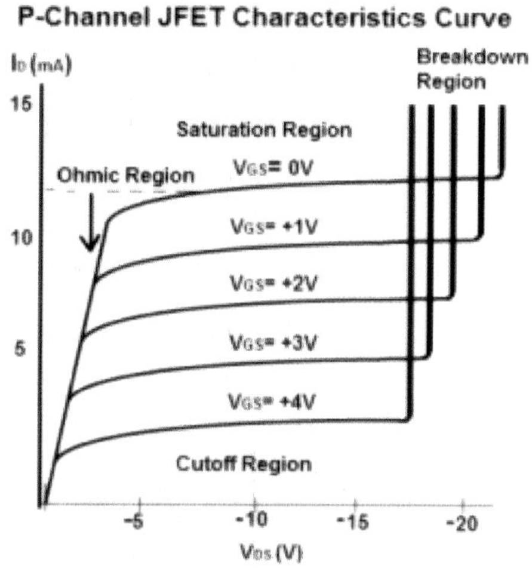

Characteristics of p channel junction FET transistor

- Cutoff region: When the voltage applied to the gate terminal is enough positive for the channel width to be minimum, no current flows. This causes the device to be in cut off region.
- Ohmic region: The current flowing through the device is linearly proportional to the applied voltage until a breakdown voltage is reached. In this region, the transistor shows some resistance to the flow of current.
- Saturation region: When the drain source voltage reaches a value such that the current flowing through the device is constant with the drain source voltage and varies only with the gate source voltage, the device is said to be in saturation region.
- Break down region: When the drain source voltage reaches a value which causes the depletion region to break down, causing an abrupt increase in the drain current, the device is said to be in breakdown region. This breakdown region is reached earlier for lower value of drain source voltage when gate source voltage is more positive.

MOSFET Transistor

MOSFET transistor as its name suggests is a p type (n type) semiconductor bar (with two heavily doped n type regions diffused into it) with a metal oxide layer deposited on its surface and holes taken out of the layer to form source and drain terminals. A metal layer is deposited on the oxide layer to form the gate terminal. One of the basic applications of field effect transistor is using a MOSFET as a switch.

MOSFET transistor

This type of FET transistor has three terminals, which are source, drain, and gate. The voltage applied to the gate terminal controls the flow of current from source to drain. The presence of an insulating layer of metal oxide results in the device having high input impedance.

Types of MOSFET Transistor based on Operation Modes

A MOSFET transistor is the most commonly used type of field effect transistor. MOSFET operation is achieved in two modes, based upon which MOSFET transistors are classified. MOSFET operation in enhancement mode consists of a gradual formation of a channel whereas in depletion mode MOSFET, it consists of an already diffused channel. An advanced application of MOSFET is CMOS.

Enhancement MOSFET Transistor

When a negative voltage is applied to the gate terminal of MOSFET, the positive charge carrying carriers or holes get accumulated more near the oxide layer. A channel is formed from the source to drain terminal.

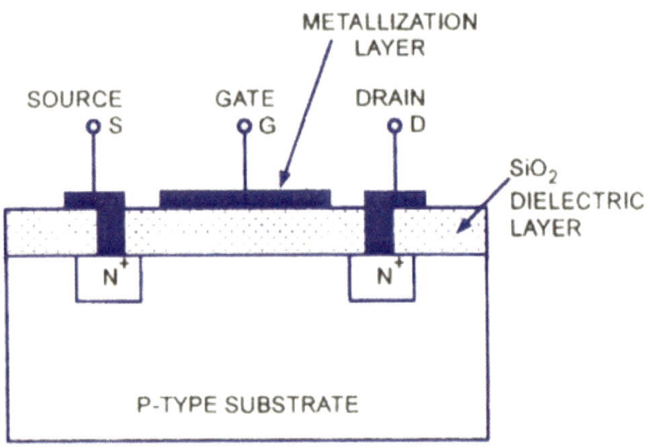

Enhancement MOSFET Transistor

As the voltage is made more negative, the channel width increases and current flows from source to drain terminal. Thus as the flow of current 'enhances' with applied gate voltage, this device is called Enhancement type MOSFET.

Depletion Mode MOSFET Transistor

A depletion mode MOSFET consists of a channel diffused between the drain to source terminal. In absence of any gate voltage, current flows from source to drain because of the channel.

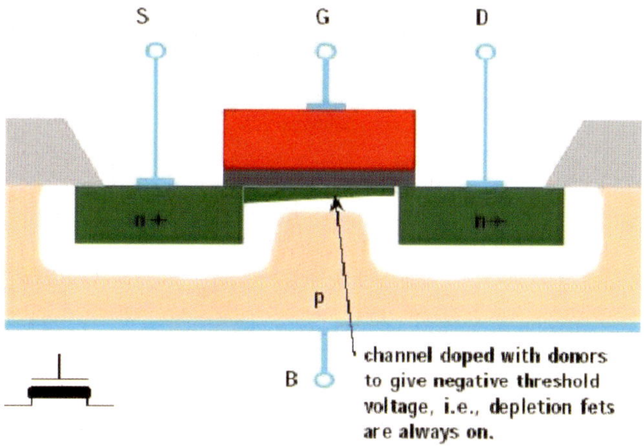

Depletion mode MOSFET transistor

When this gate voltage is made negative, positive charges get accumulated in the channel. This causes a depletion region or region of immobile charges in the channel and hinders the flow of current. Thus as the flow of current is affected by the formation of depletion region, this device is called depletion mode MOSFET.

Applications involving MOSFET as a Switch

Controlling Speed of BLDC Motor

MOSFET can be used as a switch to operate a DC motor. Here a transistor is used to trigger the MOSFET. PWM signals from a microcontroller are used to switch on or off the transistor.

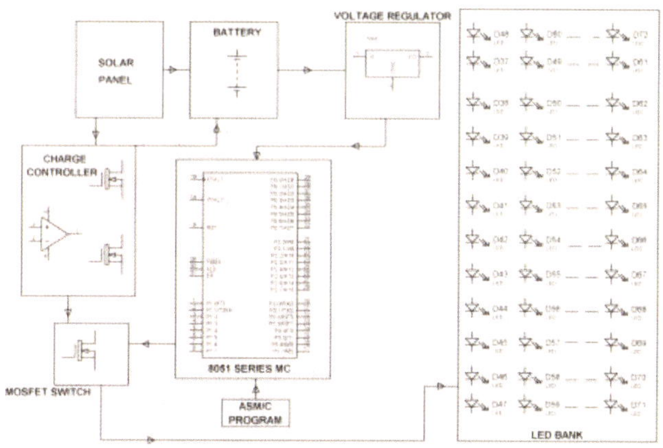

Controlling speed of BLDC motor

A logic low signal from the microcontroller pin results in the OPTO Coupler to operate, generating a high logic signal at its output. The PNP transistor is cut off and accordingly the MOSFET gets triggered and is switched ON. The drain and source terminals are shorted and the current flows to the motor windings such that it starts rotating. PWM signals ensure speed control of the motor.

Driving an Array of LEDs

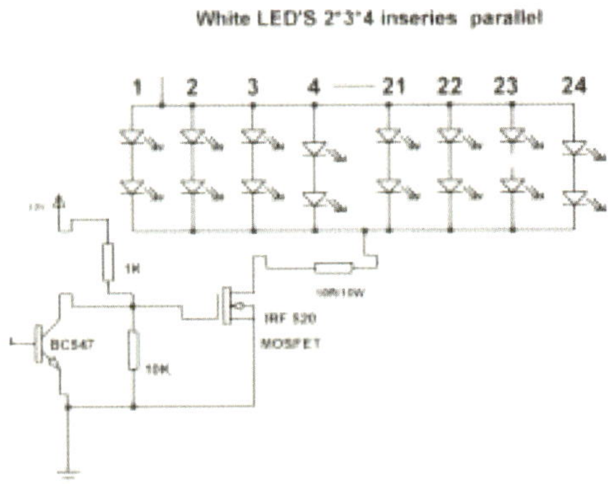

Driving an array of LEDs

MOSFET operation as a switch involves application of controlling the intensity of an array of LEDs. Here a transistor, driven by signals from external source like microcontroller, is used to drive the MOSFET. When the transistor is switched off, the MOSFET gets the supply and is switched ON, thus providing proper biasing to the LED array.

Switching Lamp using MOSFET

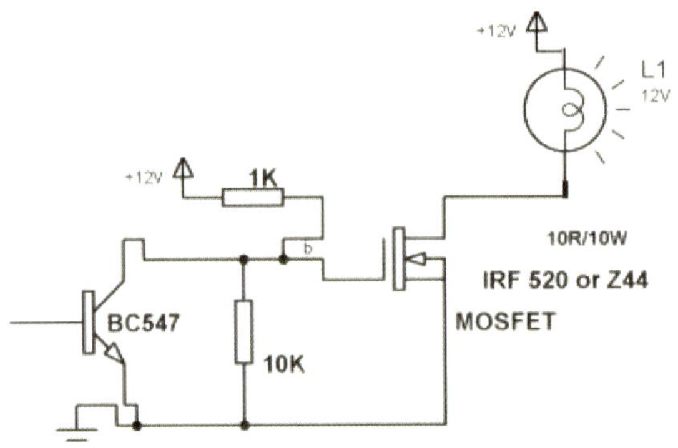

MOSFET can be used as a switch to control the switching of lamps. Here also, the MOSFET is triggered using a transistor switch. PWM signals from an external source like a microcontroller is used to control the conduction of transistor and accordingly the MOSFET switches on or off, thus control the switching of the lamp.

Transistor Biasing

Transistor Biasing is the process of setting a transistors DC operating voltage or current conditions to the correct level so that any AC input signal can be amplified correctly by the transistor.

A transistors steady state of operation depends a great deal on its base current, collector voltage, and collector current and therefore, if a transistor is to operate as a linear amplifier, it must be properly biased to have a suitable operating point.

Establishing the correct operating point requires the proper selection of bias resistors and load resistors to provide the appropriate input current and collector voltage conditions. The correct biasing point for a bipolar transistor, either NPN or PNP, generally lies somewhere between the two extremes of operation with respect to it being either "fully-ON" or "fully-OFF" along its load line. This central operating point is called the "Quiescent Operating Point", or Q-point for short.

When a bipolar transistor is biased so that the Q-point is near the middle of its operating range, that is approximately halfway between cut-off and saturation, it is said to be operating as a Class-A amplifier. This mode of operation allows the output current to increase and decrease around the amplifiers Q-point without distortion as the input signal swings through a complete cycle. In other words, the output current flows for the full 360° of the input cycle.

So how do we set this Q-point biasing of a transistor? – The correct biasing of the transistor is achieved using a process know commonly as Base Bias.

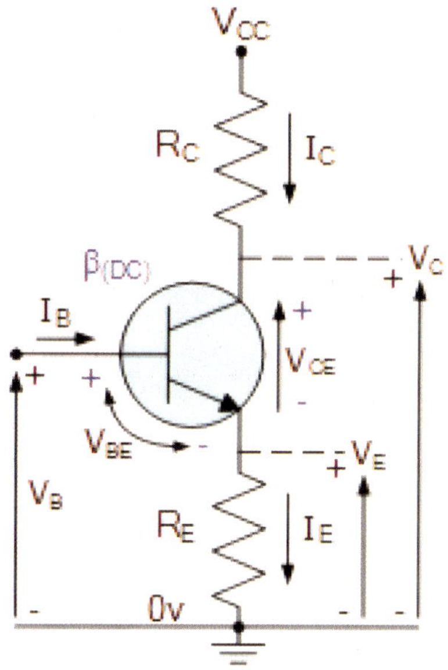

But before we start looking at the possible different *transistor biasing* arrangements, lets first reminder ourselves of the basic transistor circuit and associated voltages and currents as shown on the left.

The function of the "DC Bias level" or "no input signal level" is to correctly set the transistors Q-point by setting its Collector current (I_C) to a constant and steady state value without an input signal applied to the transistors Base.

This steady-state or DC operating point is set by the values of the circuits DC supply voltage (Vcc) and the value of the biasing resistors connected the transistors Base terminal.

Since the transistors Base bias currents are steady-state DC currents, the appropriate use of coupling and bypass capacitors will help block bias current setup for one transistor stage affecting the bias conditions of the next. Base bias networks can be used for Common-base (CB), common-collector (CC) or common-emitter (CE) transistor configurations.

Base Biasing a Common Emitter Amplifier

One of the most frequently used biasing circuits for a transistor circuit is with the self-bias of the emitter-bias circuit where one or more biasing resistors are used to set up the initial DC values of transistor currents, (I_B), (I_C) and (I_E).

The two most common forms of transistor biasing are: *Beta Dependent* and *Beta Independent*. Transistor bias voltages are largely dependent on transistor beta, (β) so the biasing set up for one transistor may not necessarily be the same for another transistor. Transistor biasing can be achieved either by using a single feed back resistor or by using a simple voltage divider network to provide the required biasing voltage.

The following are five examples of transistor Base bias configurations from a single supply (V_{cc}).

Fixed Base Biasing a Transistor

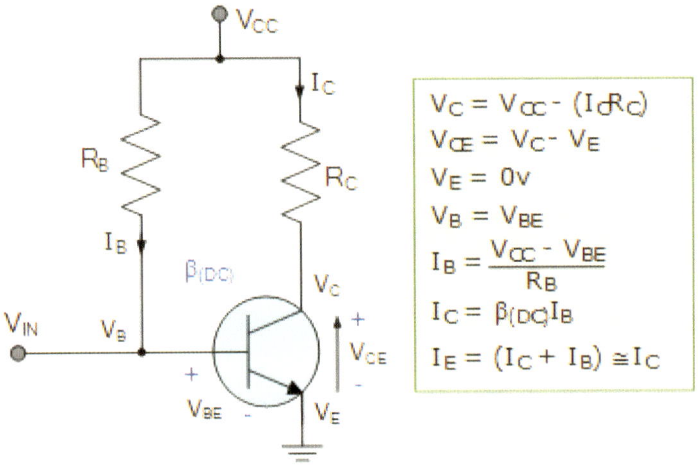

The circuit shown is called as a "fixed base bias circuit", because the transistors base current, I_B remains constant for given values of V_{cc}, and therefore the transistors operating point must also remain fixed. This two resistor biasing network is used to establish the initial operating region of the transistor using a fixed current bias.

This type of transistor biasing arrangement is also beta dependent biasing as the steady-state condition of operation is a function of the transistors beta β value, so the biasing point will vary over a wide range for transistors of the same type as the characteristics of the transistors will not be exactly the same.

The emitter diode of the transistor is forward biased by applying the required positive base bias voltage via the current limiting resistor R_B. Assuming a standard bipolar transistor, the forward base-emitter voltage drop will be 0.7V. Then the value of R_B is simply: $(V_{CC} - V_{BE})/I_B$ where, I_B is defined as I_C/β.

With this single resistor type of biasing method the biasing voltages and currents do not remain stable during transistor operation and can vary enormously. Also the temperature of the transistor can adversely effect the operating point.

Collector Feedback Biasing a Transistor

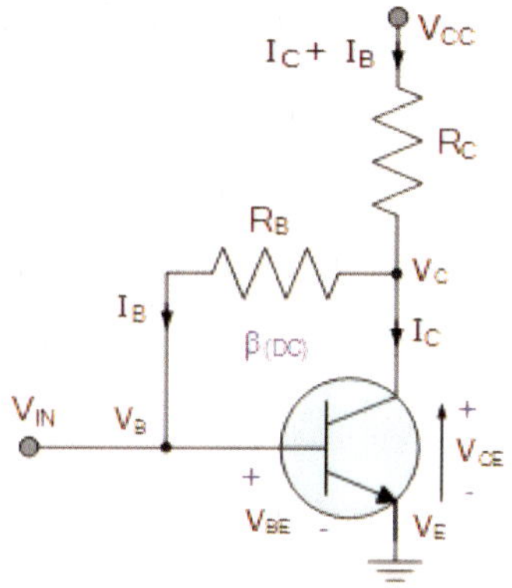

$$V_C = V_{CC} - R_C(I_C + I_B)$$
$$V_E = 0v$$
$$V_B = V_{BE}$$
$$I_B = \frac{V_C - V_B}{R_B}$$
$$I_C = \beta_{(DC)} I_B$$
$$I_E = (I_C + I_B) \cong I_C$$

This self biasing collector feedback configuration is another beta dependent biasing method that requires only two resistors to provide the necessary DC bias for the transistor. The collector to base feedback configuration ensures that the transistor is always biased in the active region regardless of the value of Beta (β) as the DC base bias voltage is derived from the collector voltage, V_C providing good stability.

In this circuit, the base bias resistor, R_B is connected to the transistors collector C, instead of to the supply voltage rail, V_{cc}. Now if the collector current increases, the collector voltage drops, reducing the base drive and thereby automatically reducing the collector current to keep the transistors Q-point fixed. Then this method of collector feedback biasing produces negative feedback as there is feedback from the output to the input through resistor, R_B.

The biasing voltage is derived from the voltage drop across the load resistor, R_L. So if the load current increases there will be a larger voltage drop across R_L, and a corresponding reduced collector voltage, V_C which will cause a corresponding drop in the base current, I_B which in turn, brings I_C back to normal.

The opposite reaction will also occur when transistors collector current becomes less. Then this method of biasing is called self-biasing with the transistors stability using this type of feedback bias network being generally good for most amplifier designs.

Dual Feedback Transistor Biasing

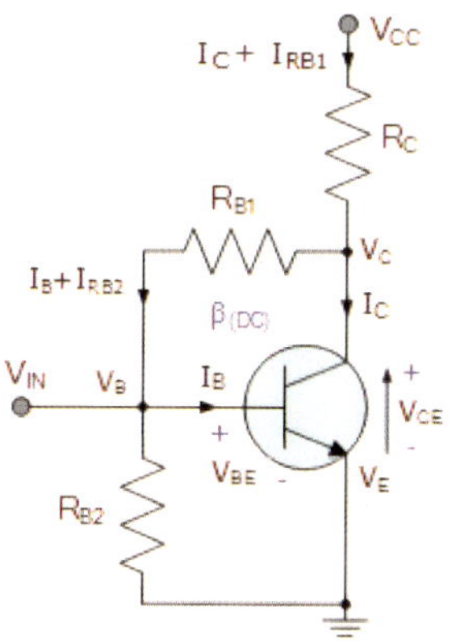

$$V_C = V_{CC} - R_C(I_C + I_{B1})$$
$$V_E = 0v$$
$$V_B = V_{BE}$$
$$I_{RB2} = \frac{V_B}{R_{B2}}$$
$$I_{RB1} = I_B + I_{RB2} = \frac{V_C - V_B}{R_{B1}}$$
$$I_C = \beta_{(DC)} I_B$$
$$I_E = (I_C + I_B) \cong I_C$$

Adding an additional resistor to the base bias network of the previous configuration improves stability even more with respect to variations in Beta, (β) by increasing the current flowing through the base biasing resistors.

The current flowing through R_{B1} is generally set at a value equal to about 10% of collector current, I_C. Obviously it must also be greater than the base current required for the minimum value of Beta, β.

One of the advantages of this type of self biasing configuration is that the resistors provide both automatic biasing and R*f* feedback at the same time.

Transistor Biasing with Emitter Feedback

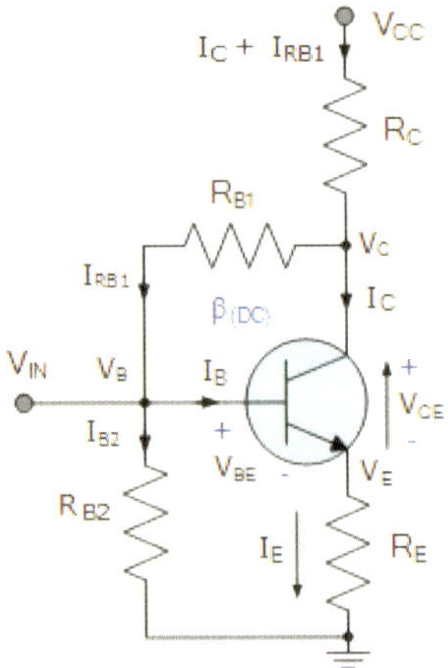

$$V_C = V_{CC} - R_C(I_C + I_{B1})$$
$$V_E = I_E R_E = V_B - V_{BE}$$
$$V_{CE} = V_C - V_E$$
$$V_B = V_{BE} + V_E$$
$$I_{RB2} = \frac{V_B}{R_{B2}}$$
$$I_{RB1} = I_B + I_{RB2} = \frac{V_C - V_B}{R_{B1}}$$
$$I_C = \beta_{(DC)} I_B$$
$$I_E = (I_C + I_B) \cong I_C$$

This type of transistor biasing configuration, often called self-emitter biasing, uses both emitter and base-collector feedback to stabilize the collector current even more as resistors R_{B1} and R_E as well as the base-emitter junction of the transistor are all effectively connected in series with the supply voltage, V_{CC}.

The downside of this emitter feedback configuration is that the output has reduced gain because of the base resistor connection as the collector voltage determines the current flowing through the feedback resistor, R_{B1} producing what is called "degenerative feedback".

The current flowing from the emitter, I_E (which is a combination of $I_C + I_B$) causes a voltage drop to appear across R_E in such a direction, that it reverse biases the base-emitter junction.

So if the emitter current increases, voltage drop $I*R_E$ also increases. Since the polarity of this voltage reverse biases the base-emitter junction, I_B automatically decrease. Therefore the emitter current increase less than it would have done had there been no self biasing resistor.

Resistor values are generally set so that the voltage drop across emitter resistor R_E is approximately 10% of V_{CC} and the current flowing through resistor R_{B1} is 10% of the collector current I_C.

This type of transistor biasing configuration works best at relatively low power supply voltages.

Voltage Divider Transistor Biasing

The common emitter transistor is biased using a voltage divider network to increase stability. The name of this biasing configuration comes from the fact that the two resistors R_{B1} and R_{B2} form a voltage or potential divider network across the supply with their center point junction connected the transistors base terminal as shown.

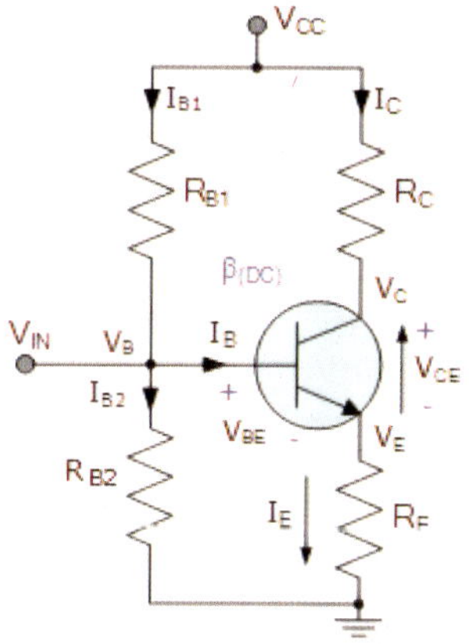

$$V_C = V_{CC} - R_C I_C = (V_E + V_{CE})$$
$$V_E = I_E R_E = V_B - V_{BE}$$
$$V_{CE} = V_C - V_E = V_{CC} - (I_C R_C + I_E R_E)$$
$$V_B = V_{BE} + V_E = V_{RB2} = \left(\frac{R_{B2}}{R_{B1}+R_{B2}}\right) V_{CC}$$
$$I_{B2} = \frac{V_B}{R_{B2}}$$
$$I_{B1} = I_B + I_{B2} = \frac{V_{CC} - V_B}{R_{B1}}$$
$$R_B = \frac{R_{B1} \times R_{B2}}{R_{B1}+R_{B2}} \qquad I_B = \frac{V_B - V_{BE}}{R_B + (1+\beta)R_E}$$
$$I_C = \beta_{(DC)} I_B$$
$$I_E = I_C + I_B = \frac{V_E}{R_E}$$

This voltage divider biasing configuration is the most widely used transistor biasing method, as the emitter diode of the transistor is forward biased by the voltage dropped across resistor R_{B2}. Also, voltage divider network biasing makes the transistor circuit independent of changes in beta as the voltages at the transistors base, emitter, and collector are dependant on external circuit values.

To calculate the voltage developed across resistor R_{B2} and therefore the voltage applied to the base terminal we simply use the voltage divider formula for resistors in series.

Generally the voltage drop across resistor R_{B2} is much less than for resistor R_{B1}. Then clearly the transistors base voltage V_B with respect to ground, will be equal to the voltage across R_{B2}.

The current flowing through resistor R_{B2} is generally set at 10 times the value of the required base current I_B so that it has no effect on the voltage divider current or changes in Beta.

The goal of Transistor Biasing is to establish a known Q-point in order for the transistor to work efficiently and produce an undistorted output signal. Correct biasing of the transistor also establishes its initial AC operating region with practical biasing circuits using either a two or four-resistor bias network.

In bipolar transistor circuits, the Q-point is represented by (V_{CE}, I_C) for the NPN transistors or (V_{EC}, I_C) for PNP transistors. The stability of the base bias network and therefore the Q-point is generally assessed by considering the collector current as a function of both Beta (β) and temperature.

Transistor Amplifiers

Amplifier is a circuit that is used for amplifying a signal. The input signal to an amplifier will be a current or voltage and the output will be an amplified version of the input signal. An amplifier circuit which is purely based on a transistor or transistors is called a transistor amplifier. Transistors amplifiers are commonly used in applications like RF (radio frequency), audio, OFC (optic fibre communication) etc. Anyway the most common application we see in our day to day life is the usage of transistor as an audio amplifier. As you know there are three transistor configurations that are used commonly i.e. common base (CB), common collector (CC) and common emitter (CE). In common base configuration has a gain less than unity and common collector configuration (emitter follower) has a gain almost equal to unity). Common emitter follower has a gain that is positive and greater than unity. So, common emitter configuration is most commonly used in audio amplifier applications.

A good transistor amplifier must have the following parameters; high input impedance, high band width, high gain, high slew rate, high linearity, high efficiency, high stability etc. Input impedance: Input impedance is the impedance seen by the input voltage source when it is connected to the input of the transistor amplifier. In order to prevent the transistor amplifier circuit from loading the input voltage source, the transistor amplifier circuit must have high input impedance.

Bandwidth

The range of frequency that an amplifier can amplify properly is called the bandwidth of that particular amplifier. Usually the bandwidth is measured based on the half power points i.e. the points where the output power becomes half the peak output power in the frequency Vs output graph. In simple words, bandwidth is the difference between the lower and upper half power points. The band width of a good audio amplifier must be from 20 Hz to 20 KHz because that is the frequency range that is audible to the human ear. The frequency response of a single stage RC coupled transistor is shown in the figure below, points tagged P1 and P2 are the lower and upper half power points respectively.

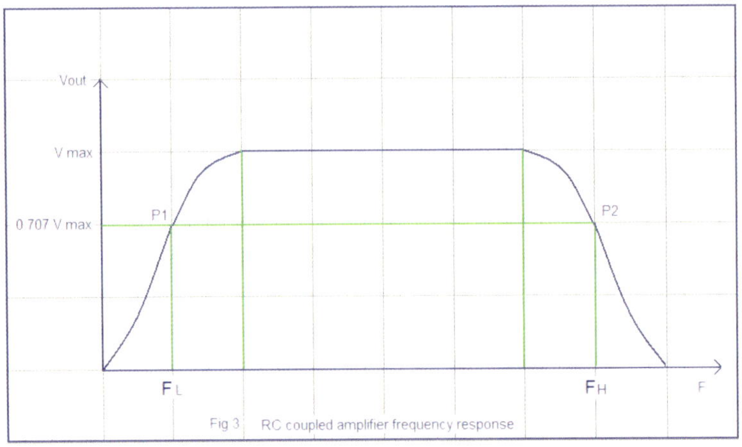

RC coupled amplifier frequency response

Gain

Gain of an amplifier is the ratio of output power to the input power. It represents how much an amplifier can amplify a given signal. Gain can be simply expressed in numbers or in decibel (dB). Gain in number is expressed by the equation $G = P_{out} / P_{in}$. In decibel the gain is expressed by the equation Gain in dB = $10 \log (P_{out} / P_{in})$. Here Pout is the power output and Pin is the power input. Gain can be also expressed in terms of output voltage / input voltage or output current / input current. Voltage gain in decibel can be expressed using the equation, Av in dB = $20 \log (V_{out} / Vin)$ and current gain in dB can be expressed using the equation $Ai = 20 \log (I_{out} / I_{in})$.

Derivation of Gain

$$G = 10 \log (P_{out} / P_{in})$$

Let $P_{out} = V_{out} / R_{out}$ and $P_{in} = V_{in} / R_{in}$. Where V*out* is the output voltage V_{in} is the input voltage, P_{out} is the output power, P_{in} is the input power, R_{in} is the input voltage and R_{out} is the output resistance. Substituting this in the above equation we have,

$$G = 10\log (V_{out^2}/R_{out}) / (V_{in^2}/R_{in})$$

Let $R_{out} = R_{in}$, then the equation 2 becomes,

$$G = 10\log (V_{out^2} / V_{in^2})$$

i.e.

$$G = 20 \log (V_{out} / V_{in}).$$

Efficiency

Efficiency of an amplifier represents how efficiently the amplifier utilizes the power supply. In simple words it is a measure of how much power from the power supply is usefully converted to the output. Efficiency is usually expressed in percentage and the equation is $\zeta = (P_{out}/ P_s) \times 100$. Where, ζ is the efficiency, P*out* is the power output and P*s* is the power drawn from the power supply.

Class A transistor amplifiers have up to 25% efficiency, Class AB has up to 55% can class C has up to 90% efficiency. Class A provides excellent signal reproduction but the efficiency is very low while Class C has high efficiency but the signal reproduction is bad. Class AB stands in between them and so it is used commonly in audio amplifier applications.

Stability

Stability is the capacity of an amplifier to resist oscillations. These oscillations may be high amplitude ones masking the useful signal or very low amplitude, high frequency oscillations in the spectrum. Usually stability problems occur during high frequency operations, close to 20KHz in case of audio amplifiers. Adding a Zobel network at the output, providing negative feedback etc improves the stability.

Slew Rate

Slew rate of an amplifier is the maximum rate of change of output per unit time. It represents how quickly the output of an amplifier can change in response to the input. In simple words, it represents the speed of an amplifier. Slew rate is usually represented in V/μS and the equation is $SR = dVo/dt$.

Linearity

An amplifier is said to be linear if there is a linear relationship between the input power and the output power. It represents the flatness of the gain. 100% linearity is not possible practically as the amplifiers using active devices like BJTs , JFETs or MOSFETs tend to lose gain at high frequencies due to internal parasitic capacitance. In addition to this the input DC decoupling capacitors sets a lower cutoff frequency.

Noise

Noise refers to unwanted and random disturbances in a signal. In simple words, it can be said to be unwanted fluctuation or frequencies present in a signal. It may be due to design flaws, component failures, external interference, due to the interaction of two or more signals present in a system, or by virtue of certain components used in the circuit.

Output Voltage Swing

Output voltage swing is the maximum range up to which the output of an amplifier could swing. It is measured between the positive peak and negative peak and in single supply amplifiers it is measured from positive peak to the ground. It usually depends on the factors like supply voltage, biasing, and component rating.

Common Emitter RC Coupled Amplifier

The common emitter RC coupled amplifier is one of the simplest and elementary transistor amplifier that can be made. Don't expect much boom from this little circuit, the main purpose of this circuit is pre-amplification i.e to make weak signals strong enough for further processing or amplification. If designed properly, this amplifier can provide excellent signal characteristics. The circuit diagram of a single stage common emitter RC coupled amplifier using transistor is shown in figure below.

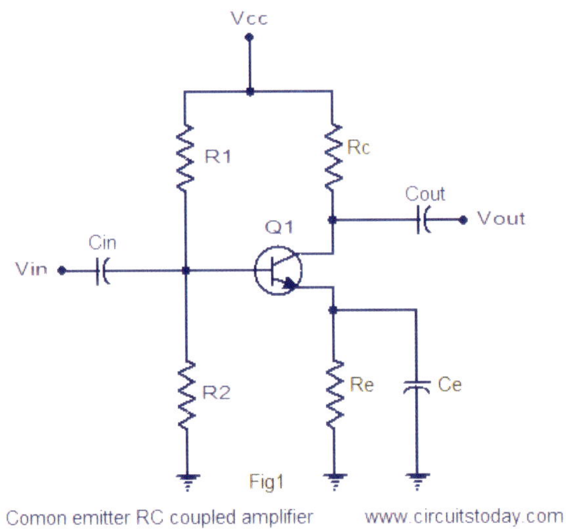

RC coupled amplifier

Capacitor Cin is the input DC decoupling capacitor which blocks any DC component if present in the input signal from reaching the Q1 base. If any external DC voltage reaches the base of Q1, it will alter the biasing conditions and affects the performance of the amplifier.

R1 and R2 are the biasing resistors. This network provides the transistor Q1's base with the necessary bias voltage to drive it into the active region. The region of operation where the transistor is completely switched of is called cut-off region and the region of operation where the transistor is completely switched ON (like a closed switch) is called saturation region. The region in between cut-off and saturation is called active region. Refer figure below for better understanding. For a transistor amplifier to function properly, it should operate in the active region. Let us consider this simple situation where there is no biasing for the transistor. As we all know, a silicon transistor requires 0.7 volts for switch ON and surely this 0.7 V will be taken from the input audio signal by the transistor. So all parts of there input wave form with amplitude ≤ 0.7V will be absent in the output waveform. In the other hand if the transistor is given with a heavy bias at the base ,it will enter into saturation (fully ON) and behaves like a closed switch so that any further change in the base current due to the input audio signal will not cause any change in the output. The voltage across collector and emitter will be 0.2V at this condition (Vce sat = 0.2V). That is why proper biasing is required for the proper operation of a transistor amplifier.

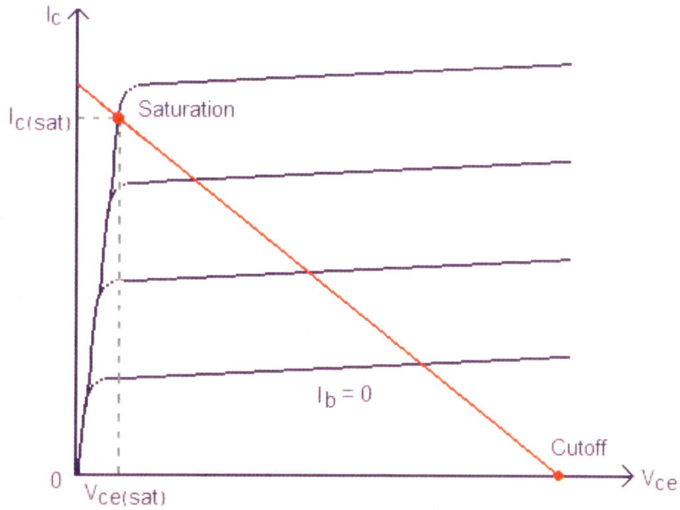

BJT output c haracteristics

C_{out} is the output DC decoupling capacitor. It prevents any DC voltage from entering into the succeeding stage from the present stage. If this capacitor is not used the output of the amplifier (V_{out}) will be clamped by the DC level present at the transistors collector.

R_c is the collector resistor and R_e is the emitter resistor. Values of R_c and R_e are so selected that 50% of V_{cc} gets dropped across the collector & emitter of the transistor. This is done to ensure that the operating point is positioned at the center of the load line. 40% of V_{cc} is dropped across R_c and 10% of V_{cc} is dropped across R_e. A higher voltage drop across R_e will reduce the output voltage swing and so it is a common practice to keep the voltage drop across R_e = 10% V_{cc} . C_e is the emitter by-pass capacitor. At zero signal condition (i.e, no input) only the quiescent current (set by the biasing resistors R_1 and R_2 flows through the R_e). This current is a direct current of magnitude few milli amperes and C_e does nothing. When input signal is applied, the transistor amplifies it and as a result a corresponding alternating current flows through the R_e. The job of C_e is to bypass this alternating component of the emitter current. If C_e is not there , the entire emitter current will flow through R_e and that causes a large voltage drop across it. This voltage drop gets added to the V_{be} of the transistor and the bias settings will be altered. It reality, it is just like giving a heavy negative feedback and so it drastically reduces the gain.

Design of RC Coupled Amplifier

The design of a single stage R_C coupled amplifier is shown below.

The nominal vale of collector current I_c and hfe can be obtained from the datasheet of the transistor.

Design of R_e and C_e

Let voltage across R_e; $VR_e = 10\%Vcc$

Voltage across R_c; $VR_c = 40\%\ V_{cc}$.

The remaining 50% will drop across the collector-emitter .

From above 2 equations, $R_c = 0.4\ (V_{cc}/I_c)$ and $R_e = 01(V_{cc}/I_c)$.

Design of R_1 and R_2

Base current $I_b = I_c/h_{fe}$.

Let $I_c \approx I_e$.

Let current through R_1; $IR_1 = 10Ib$.

Also voltage across R_2 ; VR_2 must be equal to $Vb_e + VR_e$. From this VR_2 can be found.

There fore $VR_1 = V_{cc} - VR_2$. Since VR_1 ,VR_2 and IR_1 are found we can find R_1 and R_2 using the following equations.

$R_1 = VR_1/IR_1$ and $R_2 = VR_2/IR_1$.

Finding C_e

Impedance of emitter by-pass capacitor should be one by tenth of R_e.

i.e, $XC_e = 1/10\ (R_e)$.

Also $XC_e = 1/2\prod FC_e$.

F can be selected to be 100Hz.

From this Ce can be found.

Finding C_{in}

Impedance of the input capacitor(C_{in}) should be one by tenth of the transistors input impedance (R_{in}).

i.e, $XC_{in} = 1/10\ (R_{in})$

$R_{in} = R_1$ parallel R_2 parallel $(1 + (h_{fe}\ r_e))$

$r_e = 25mV/I_e$.

$Xc_{in} = 1/2\prod FC_{in}$.

From this C_{in} can be found.

Finding Cout

Impedance of the output capacitor (C_{out}) must be one by tenth of the circuit's output resistance (R_{out}).

i.e, $XC_{out} = 1/10\ (R_{out})$.

$R_{out} = R_c$.

$XC_{out} = 1/\ 2\prod FC_{out}$.

From this C_{out} can be found.

Setting the Gain

Introducing a suitable load resistor R_L across the transistor's collector and ground will set the gain.

Expression for the voltage gain (A_v) of a common emitter transistor amplifier is as follows.

$A_v = -(r_c/r_e)$

$r_e = 25mV/I_e$

and $r_c = R_c$ parallel R_L.

References

- Zenerdiode-definition-vicharacteristics-breakdowns, semiconductor-diodes, electronic-devices-and-circuits: physics-and-radio-electronics.com, Retrieved 16 May 2018
- Zener-diode-circuit-working-applications: elprocus.com, Retrieved 26 July 2018
- Electronic-circuits-nonlinear-wave-shapping, electronic-circuits: tutorialspoint.com, Retrieved 26 April 2018
- Transistors-basics-types-baising-modes: elprocus.com, Retrieved 16 April 2018
- Bipolar-junction-transistors-working-principle-and-applications: elprocus.com, Retrieved 15 March 2018
- Transistor-amplifier: circuitstoday.com, Retrieved 28 June 2018

4

Capacitance and Capacitors

Any change in an electric charge relative to a change in the electric potential is expressed in terms of a ratio called the capacitance. A passive electrical component, which stores electrical potential energy is known as a capacitor. The topics elaborated in this chapter on the fundamental concepts of capacitors and capacitance include mutual and self capacitance, parasitic capacitance and quantum capacitance, which will help in developing a comprehensive understanding of this physical quantity and its measurement.

Capacitance

Capacitance is the ability to store electric charge. In its simplest form a capacitor consists of two parallel plates or electrodes that are separated from each other by an insulating dielectric. It is found that when a battery or any other voltage source is connected to the two plates as shown a current flows for a short time as it charges up. It is found that one plate of the capacitor receives an excess of electrons, while the other has too few. In this way the capacitor plate or electrode with the excess of electrons becomes negatively charged, while the capacitor electrode becomes positively charged.

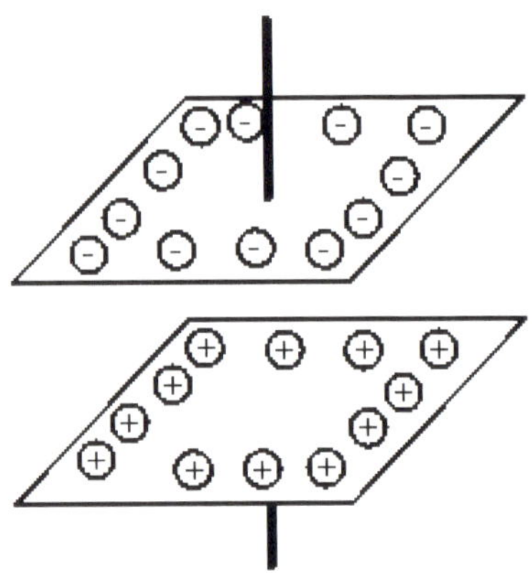

Charge stored between two plates of a capacitor

If the battery is removed the capacitor will retain its charge. However if a resistor is placed across the plates, a current will flow in the resistor until the capacitor becomes discharged.

Units of Capacitance: Farad

It is necessary to quantify a capacitor in terms of its ability to store charge. The basic unit of capacitance is the Farad, named after Michael Faraday.

The definition of a Farad is: A capacitor has a capacitance of one Farad when a potential difference of one volt will charge it with one coulomb of electricity (i.e. one Amp for one second).

In view of the fact that a capacitor with a capacitance of one Farad is too large for most electronics applications, components with much smaller values of capacitance are normally used. Three prefixes (multipliers) are used, μ (micro), n (nano) and p (pico):

PREFIX	MULTIPLIER	
μ	10^{-6} (millionth)	1000000μF = 1F
n	10^{-9} (thousand-millionth)	1000nF = 1μF
p	10^{-12} (million-millionth)	1000pF = 1nF

Capacitance depends on four things:

1. The area of the plates
2. The distance between the plates
3. The type of dielectric material
4. Temperature.

Of these four, temperature has the least effect in most capacitors.The value of most capacitors is fairly stable over a "normal" range of temperatures.

Capacitor values may be fixed or variable. Most variable capacitors have a very small value a few tens or hundreds of pF) The value is varied by either:

- Changing the area of the plates.
- Changing the thickness of the dielectric.

Capacitance (C) is directly proportional to the area of the two plates that directly overlap, the greater the overlapping area, the greater the capacitance.

Capacitance is inversely proportional to the distance between the plates. i.e. if the plates move apart, the capacitance reduces.

The Dielectric

The electrons on one plate of the capacitor affect the electrons on the other plate by causing the orbits of the electrons within the dielectric material (the insulating layer between the plates) to distort. The amount of distortion depends on the nature of the dielectric material and this is measured by the permittivity of the material.

Permittivity

Permittivity is quoted for any particular material as relative permittivity, which is a measure of how efficient a dielectric material is. It is a number without units which indicates how much greater the permittivity of the material is than the permittivity of air (or a vacuum), which is given a permittivity of 1 (one). For example, if a dielectric material such as mica has a relative permittivity of 6, this means the capacitor will have a permittivity, and so a capacitance, six times that of one whose dimensions are the same, but whose dielectric is air.

Dielectric Strength

Another important aspect of the dielectric is the dielectric strength. This indicates the ability of the dielectric to withstand the voltage placed across it when the capacitor is charged. Ideally the dielectric must by as thin as possible, so giving the maximum capacitance for a given size of component. However, the thinner the dielectric layer, the more easily its insulating properties will break down. The dielectric strength therefore governs the maximum working voltage of a capacitor.

Maximum Working Voltage

It is very important when using capacitors that the maximum working voltage indicated by the manufacturer is not exceeded. Otherwise there will be a great danger of a sudden insulation breakdown within the capacitor. As it is likely that a maximum voltage existed across the capacitor at this time (hence the breakdown) large currents will flow with a real risk of fire or explosion in some circuits.

Charge on a Capacitor

The charge (Q) on a capacitor depends on a combination of the above factors, which can be given together as the Capacitance (C) and the voltage applied (V). For a component of a given capacitance, the relationship between voltage and charge is constant. Increasing the applied voltage results in a proportionally increased charge. This relationship can be expressed in the formula;

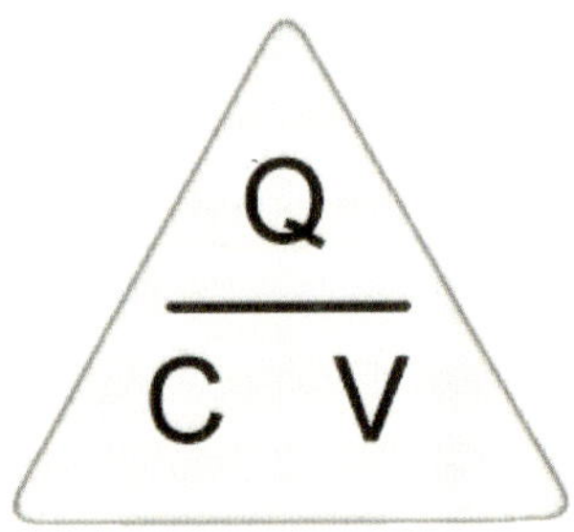

$Q = CV$

or

$C = Q/V$

or

$V = Q/C$

Where, V is the voltage applied, in Volts.

C is the capacitance in Farads.

Q is the quantity of charge in Coulombs.

So any of these quantities can be found provided the other two are known.

This equation can then be developed to calculate the work required for charging a capacitor, and hence the energy stored in it.

$$W = \int_0^Q V\, dQ = \int_0^Q \frac{q}{c} dQ = \frac{1}{2} CV^2$$

Electrical Elastance

Electrical Elastance is the inverse of capacitance. The SI unit is the reciprocal farad. Although the term daraf (farad spelled backwards) is sometimes used, this is not approved by SI. Electrical inductor-capacitor-resistor circuits follow differential equations that can be interpreted as representing mass-spring-damper systems. If voltage is taken to be force and current velocity, elastance corresponds to the elastic constant of a spring.

Mutual and Self Capacitance

Self Capacitance sensing only requires one pin of the MCU. The Capacitive Touch Sensor Unit (CTSU) measures the capacitance between the pin and ground. Self capacitance sensing is best used for single touch sensors such as buttons, sliders and wheels.

In the case of self-capacitance, each capacitive sensor is a conductive pad laid on a PCB, surrounded by a ground pattern.

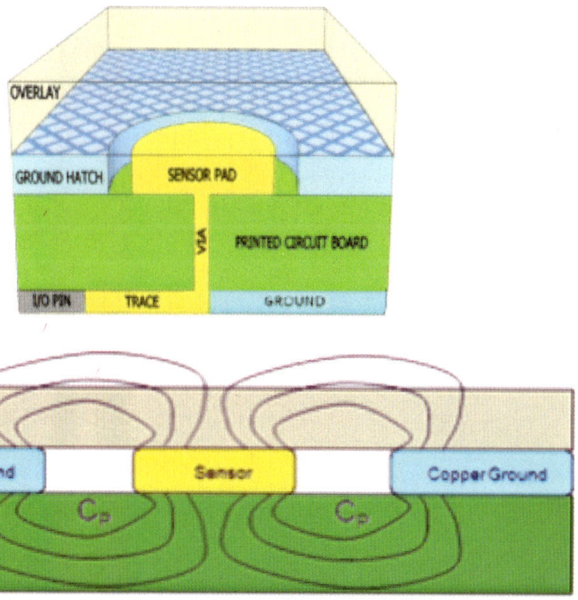

Sensor pad on a PCB

This sensor forms a parasitic capacitance Cp with the surrounding ground pattern, and the electric field lines can be seen in the area above the sensor. When a conductor like a finger enters the area above this sensor, it alters the electric field lines and effectively adds a finger capacitance Cf to the sensor.

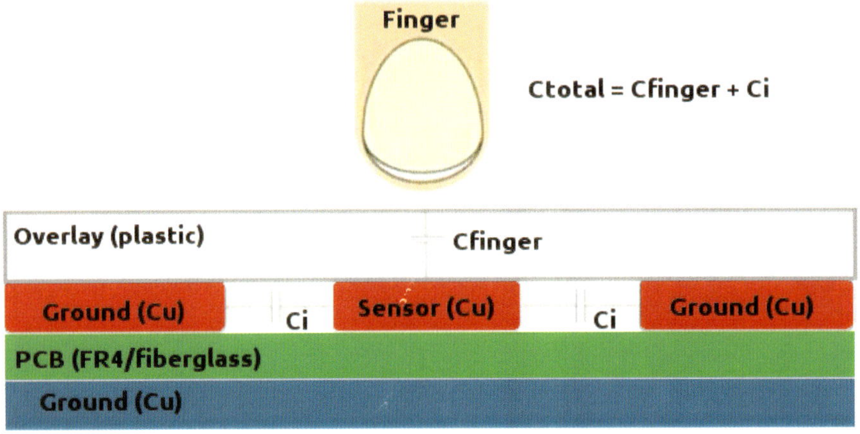

Finger capacitance added to the sensor

This results in an increase in capacitance of the sensor from C_p to $C_p + C_f$.

By continuously measuring the capacitance of the sensor(s) in the system and looking for a sudden change in capacitance, a microcontroller can determine when the finger was placed on the sensor. Here, the absolute value, or the parasitic capacitance of the sensor does not matter. The microcontroller just looks for a sudden change in capacitance and if this change is above a particular threshold, a finger presence is reported.

Mutual Capacitance sensing requires two pins of the MCU. The Capacitive Touch Sensor Unit (CTSU) measures the capacitance between the transmitter (Tx) and receiver (Rx). This method reduces the influence of parasitic capacitance which is problematic in Self capacitance measurements. Mutual capacitance sensing is best used for key pads matrices, touch screens and proximity sensing. It also has the unique property of not being influenced by the presence of water. It is often used when water rejection in an application is required, for example medical, home appliance etc.

This is the latest addition to the catalogue of capacitive sensing techniques. Such capacitive panels have two conductive layers stacked together with a very thin separation.

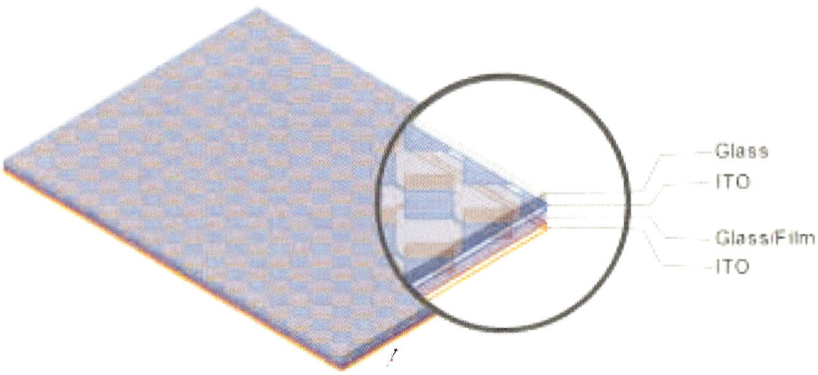

Dual layer ITO - Mutul cap

In this technology, when a finger touches the panel, the mutual capacitance between the row and column is reduced. This reduction in capacitance is used to identify the presence of a finger.

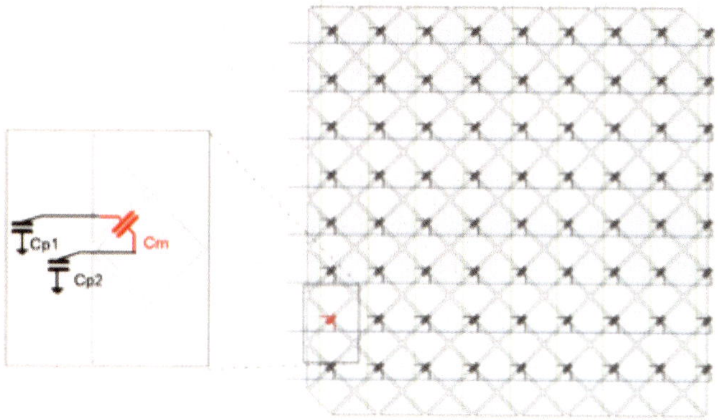

Mutual capacitance between Rx-Tx

As every intersection has its own mutual capacitance and can be independently tracked, this method provides a distinct advantage for detecting multiple fingers.

Which method is right for your application?

Table: Sensing technologies compartion

Method	Linearity	Accuracy	Size Scalability	Optical Clarity	Damage Resistant	Multitouch
Infrared	★★★★★	★★★	★★★★★	★★★★★	★★★	Yes (expensive)
Surface Acoustic Wave (SAW)	★★★★	★★★★	★★	★★★	★★★★★	No
Surface Capacitance	★★	★★	★★	★★★★★	★★★★★	No
Resistive	★★★★	★★★★	★★★★	★★	★	Yes (expensive)
Projected Capacitance	★★★★★	★★★★	★★★	★★★★★	★★★★★	Yes

All the technologies mentioned in table have their merits and demerits for touchscreen-based applications. However, when it comes to implementing buttons or slider for a front panel where two buttons are separated by a significant distance, then the only economical solution is projected capacitance. This is because it works with PCB traces; all other technologies require that a special panel has to be designed.

Mutual Capacitance vs Self-Capacitance Touchscreen Technology

Capacitive touchscreens are often preferred over their resistive counterpart because of their superior image quality. As you may already know, capacitive devices identify touch by measuring changes in capacitance. Because the human body is an excellent conductor of electricity, touching the surface of a capacitive device causes a disturbance in its electrical field (capacitance), which the device uses to determine when and where the touch occurred. There are several different types of capacitive touchscreen devices, however, two of which being mutual capacitance and self-capacitance.

Mutual capacitance devices are designed with a grid-like array of traces. A typical devices using this technology, for instance, may feature a 16 by 14 array of traces; thus, creating 224 capacitors. The device works by applying a uniform electrical field over these capacitors. And like other capacitive devices, touching the surface disturbs the electrical field, at which point the device can identify the location of the user's touch.

Capacitance changes in mutual capacitance devices can be measured at every point of the grid. Using the example previously mentioned, there would be 224 possible points of contact on a device that uses 224 capacitors. One of the key benefits of using mutual capacitance is its support for multi-touch function. With multi-touch function, the device can identify when two or more points of simultaneous contact are made. A common example of multi-touch is the pinch-to-zoom feature found on many smartphones and tablet computers. However, multi-touch can be used for a wide variety of other applications.

On the other hand, a self-capacitance touchscreen device features capacitors that act independently of one another. Like mutual capacitance, it also features a grid-like array of traces. However, the rows and columns function independently. In other words, changes in capacitance is measured on each column or row, typically by a current meter.

Self-capacitance touchscreen devices offer a higher signal strength than their mutual capacitance counterpart, resulting in a higher level of sensitivity. The downside, however, is that they are only able to identify a single point of contact at any time. This means you can not use multi-touch features with a self-capacitance device.

Of course, there are other types of capacitance technology, such as surface capacitance. With this configuration, one side of the insulator layer has a conductive material applied to it. The device then emits a voltage across this layer to create an electrical field, which it uses to identify when and where touch occurs.

Parasitic Capacitance

Parasitic capacitance is also know as stray capacitance.

Stray capacitance is unintended and unwanted capacitance in a circuit. Capacitance doesn't exist only within capacitors. In fact, any two surfaces at different electric potential, and that are close enough together to generate an electric field have capacitance, and thus act like a capacitor. Such effects are often present within circuits (for example between conductive runs or component leads), even though they are not intended. This unintended capaictance is referred to as stray capacitance, and it can result in a disruption of normal current flow within a circuit.

Designers of circuits try to minimize stray capacitance as much as possible. They do this by keeping the leads of electronic components very short and grouping components in such a way to eliminate capacitive coupling.

A good example of this is an inductor. When buying an inductor fresh from a manufacturer, the leads of the inductor are usually quite long.

The leads can extend for several inches on inductors, as can be seen from the illustration above.

However, plugging an inductor with the leads this long into a circuit can create some issues in the circuit.

When you have inductors that have very long leads close to each other, these leads are, essentially, wires. When you have wires close to each other in circuits, they can produce a capacitive effect. Even a small amount of wire can have considerable capacitance. When you have this, the capacitance can serve to impede or block low-frequency signals. This is because capacitance or a device that acts as an capacitor has high impedance to low-frequency signals. Thus, it is difficult for low-frequency signals to pass through a circuit that exhibits capacitive qualities. When you add unwanted capacitance into a circuit, the circuit can now act (again unwanted) to block low-frequency signals. Thus, an inductor, which is a device that has low reactance, or impedance, at low frequencies, may now impede low-frequency signals from passing through. If this is radio circuit or an audio circuit, entire ranges of frequencies may be blocked, which of course can cause poor transmission of numerous frequencies. This is why it is desired that no stray capacitance exist in a circuit.

Therefore, this is the reason inductor leads must be kept short, less than 1.5mm in length ideally, to effectively stop capacitive effects, which can limit an inductor's ability to pass low-frequency signals.

Surface mount inductors are even better to use in circuits because their lead-less terminals being placed directly on the power plane of a circuit stops almost all capacitance. This is because surface mount components have no leads, but just terminals. This will limit the inductance that inductors can emit and will allow them to pass low frequency signals better in a circuit.

Types of Parasitic Capacitances

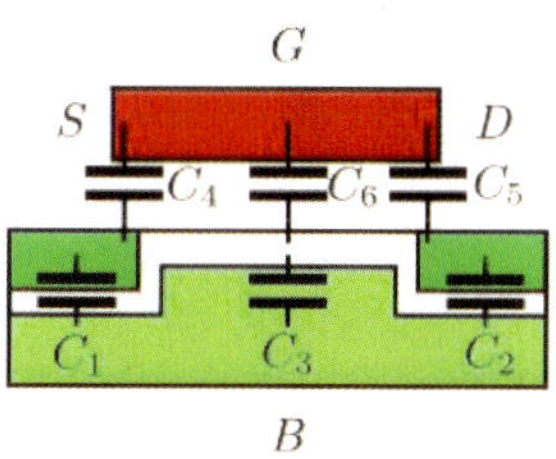

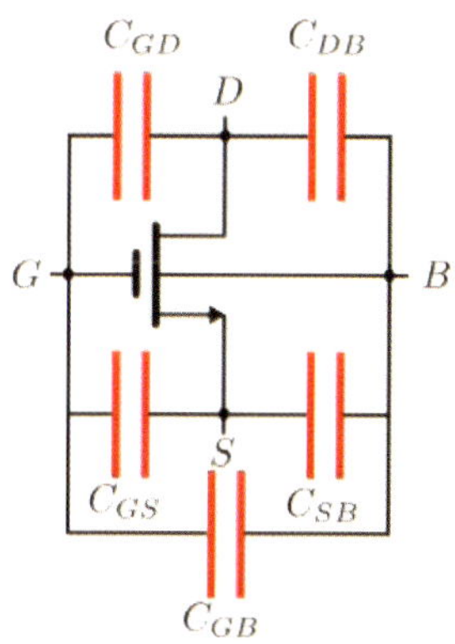

C_1 and C_2 are capacitances created by the depletion regions between source/drain and bulk. C_3 is the depletion capacitance between the channel and bulk. C_4 and C_5 are capacitances caused by the overlap between the gate and the source/drain diffusions. Finally, C_6 is the oxide capacitance between gate and the channel and is split between drain and source depending on the region of operation of the transistor. We want to map this physical capacitances to the terminals of the transistor, so we can use them in our model.

Junctions Depletion Capacitances

C_1 and C_2 are capacitances created by the depletion regions of the junctions between source/drain and bulk. These junctions are three-dimensional: 3 sides that are away from the channel, one side that is in front of the channel, and the bottom side (the upper side has isolator above it, so it has no depletion).

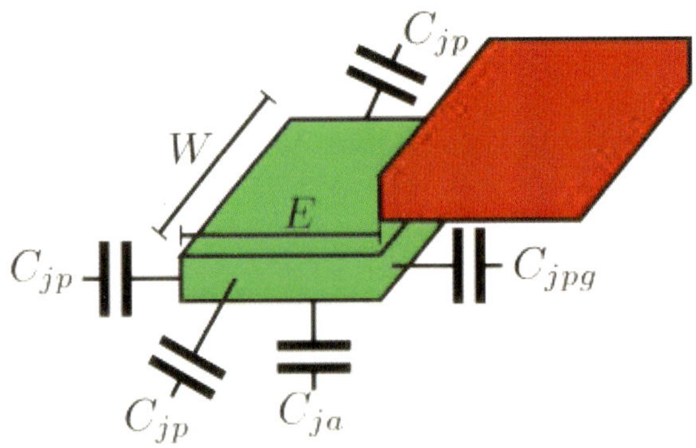

We define 3 normalized capacitances:

- C_{ja} - Bottom plate capacitance per unit area ($fF/\mu m^2$).
- C_{jp} - Sidewalls away from channel capacitance per unit length ($fF/\mu m$). The height is a process parameter and is included in the capacitance.
- C_{jpg} - Sidewall in front of the channel capacitance per unit length ($fF/\mu m$). The height is a process parameter and is included in the capacitance.

In order to calculate C_{SB} or C_{DB}, we need to sum all these capacitances.

$$C_{SB}=CDB=WEC_{ja}+(W+2E)C_{jp}+WC_{jpg}$$

Overlap and Gate-channel Capacitances

C_4 and C_5 are overlap capacitances and are only proportional to the width of the transistor. C_6 is the gate-channel capacitance and its total value is split between drain and source in a way that depends on the region of operation of the transistor.

First, we define the normalized capacitances:

- Cov - Overlap capacitance per unit length ($fF/\mu m$).
- Cox - Gate to channel capacitance per unit area ($fF/\mu m^2$).

C_{GS} and C_{GD} have a base value of the overlap capacitance WC_{ov}. To that we add the gate to channel capacitance WLC_{ox} according to the region of operation:

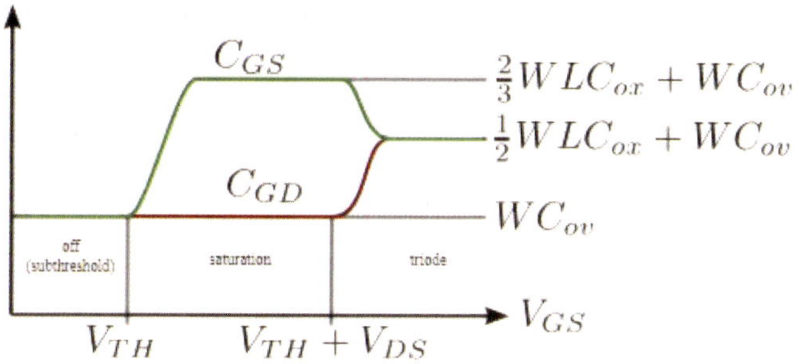

In subthreshold region, there is no gate-channel capacitance because there is no channel. In saturation, the channel is pinched-off and there is no gate-channel capacitance at the drain and only two-thirds go to the source. In triode, the channel is not pinched-off and the gate-channel capacitance is split equally between drain and source.

Channel-bulk Depletion Capacitance

This capacitance is only relevant in subthreshold regime. In strong inversion, the channel is being driven and shields the transistor from this capacitance. However, in subthreshold there is no channel and this capacitance is in series with the oxide capacitance. Furthermore, C_{GS} and C_{DS} are made of only the overlap capacitance and they are small.

Important Points to Reduce Parasitic Capacitance

1. Use higher metals for the net for which parasitic capacitance is important.

In this case let's take the parasitic capacitance between higher metal and the lower metals into account i.e. parasitic capacitance between signal A and other signals (B, C, D, E, F etc.) This is similar to case 2 of figure below. With using higher metal we are actually increasing the distance between the plates (as in below figure signal A is with M6 and B, C, D, etc. are with M1 and in between M1 and M6 there are M2, M3, M4, M5 metals) which ultimately decreases the parasitic capacitance.

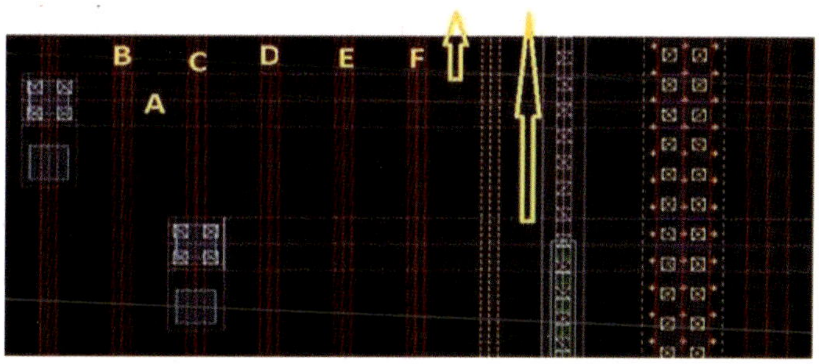

Metal 6 routing over metal 1, so that parasitic capacitance between net A and nets B, C, D, E, F, etc. is less

2. Increase the spacing between the nets in which parasitic capacitance required is less.

In this case most important is the parasitic capacitance between same metals i.e. between M1 and M1 or between M2 and M2 etc. that is case 1 of Figure. So with increasing the space between the nets we are increasing the distance between the plates which ultimately decreases parasitic capacitance. For example in Figure, parasitic capacitance between one bank of signals and another bank of signals is important, so we have placed them at some distance from each other.

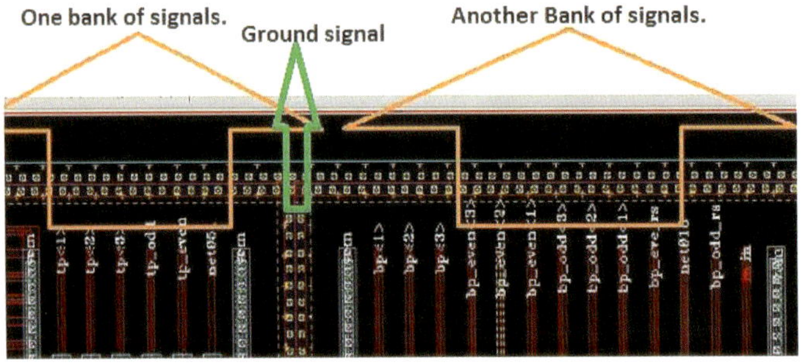

If parasitic capacitance between one bank of signals and another bank of signals is important, we need to place them at some distance from each other

Another way of routing the signals can be as in figure below, i.e. doing routing with alternate metals, which also results in increasing the distance between the nets and ultimately parasitic capacitance. Here we have routed a signal with Metal 3 in between signals A and B which are routed with Metal 1, As a result of this distance between A and B signals have increased which resulted in decreasing parasitic capacitance between A and B.

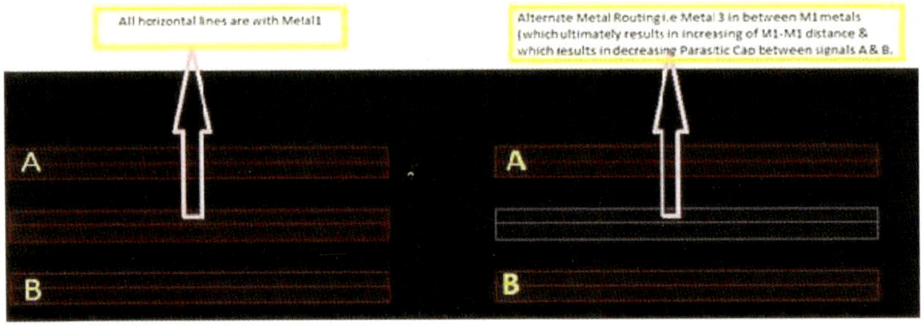

Routing with alternate metals can increase the distance between the nets and ultimately parasitic capacitance

3. Put some other reference signal (with which parasitic capacitance is not so important) in between the nets for which the parasitic capacitance required is less.

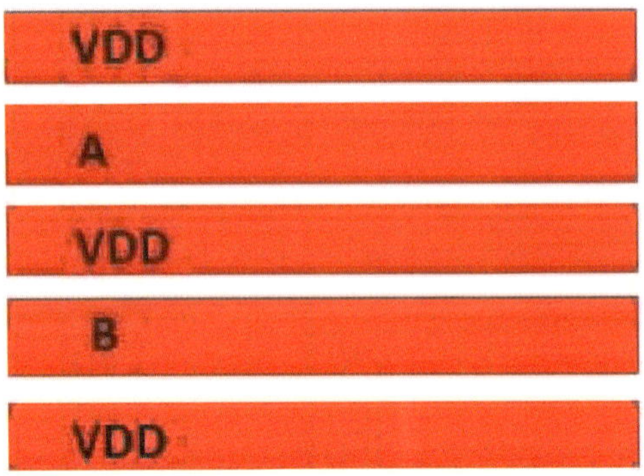

All routes should be with the same Metal and equidistant from each other

As in figure above putting the reference signal in between the nets can also serve our purpose of decreasing parasitic capacitance between the signal nets i.e. between A and B. It not only increases distance between the signal nets but the reference signal also tries to neutralize the parasitic capacitance between the signal nets. This technique is known as shielding from the sides. If your shielding has a low impedance path (e.g. good grounding), the coupled noise flows into this path, and will reduce the impact on vulnerable signals.

4. Avoid too much parallel routing of metals.

Here parallel routing means one metal on another metal completely as shown in figure below.

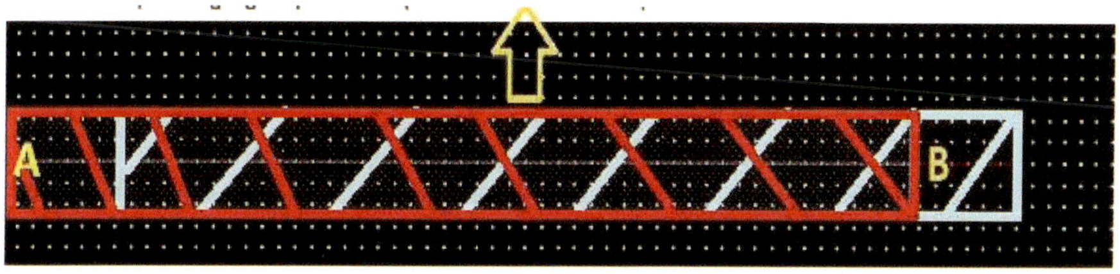

Signal A (Metal 2) is routed over signal B (Metal 1) in parallel, and causes a large area between the plates A and B, which in turn causes higher parasitic capacitance between the plates

With parallel routing there is a maximum area between the two metals so there is maximum capacitance between the two as well. So to avoid that we can do routing as in figure below.

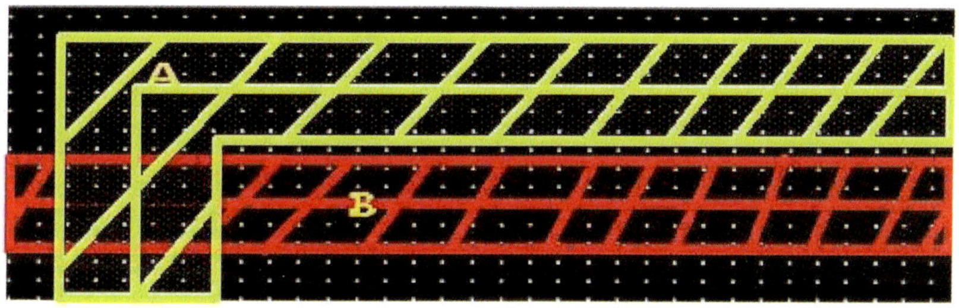

Parasitic capacitance is reduced with this technique

With the effective area overlap between the metal plates decreased, parasitic capacitance has also decreased.

Another technique is by adding some reference signal in between the two signal lines.

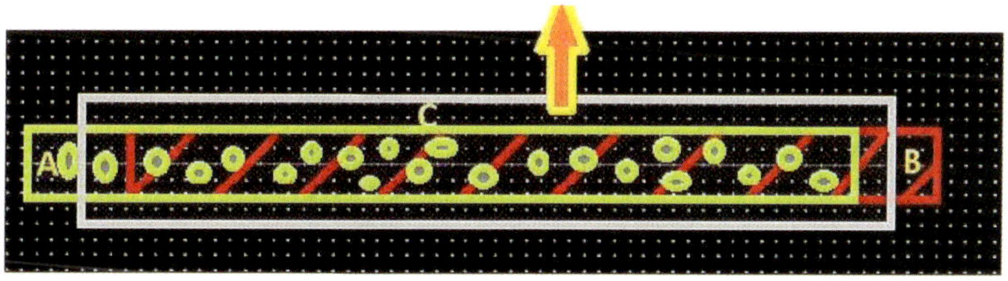

Signal C (Metal 2) is inserted in between Signals A (Metal 3) and B (Metal 1) so as to decrease the parasitic capacitance between A and B

With the figure above method of routing, the effective parasitic capacitance between two signal nets (A and B) decreases. Signal C acts as shielding for signals A and B. If sometimes our signal is too critical that small noise can harm it then we can do coaxial shielding also, which is covering all the four sides of the signal with a reference signal. It can be shown as in figure below that

Metal 2 (shielded signal) is shielded from all the four sides (i.e. top with M3, bottom with M1, sides with M2).

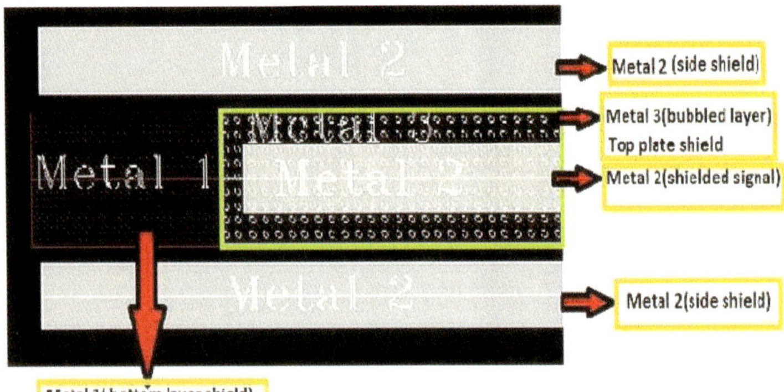

Fig 9

Metal 2 (shielded signal) is shielded from all the four sides (i.e. top with M3, bottom with M1, sides with M2)

Quantum Capacitance

The concept of "quantum capacitance" was used by Luryi in order to develop an equivalent circuit model for devices that incorporate a highly conducting two-dimensional (2D) electron gas. Recently, this term has also been used in the modeling of one-dimensional (1D) systems, such as carbon nanotube (CN) devices.

Equilibrium Quantum Capacitance

In order to derive analytical expressions, it is assumed that our device is in quasiequilibrium, and that the carrier distribution functions are rigidly shifted by the local electrostatic potential. If the density of states (DOS) is symmetric with respect to the Fermi level, E_F, as in graphene, then we can write the charge density, Q, due to electrons and holes in the semiconductor, as

$$Q = q\int_0^\infty g(E)\left[f\left(E+\frac{E_G}{2}+qV_a\right)-f\left(E+\frac{E_G}{2}-qVa\right)\right], dE\ ,$$

where, q is the magnitude of the electronic charge, E is the energy, *g(E)* is the 1D or 2D DOS, *f(E)* is the Fermi-Dirac distribution function, V_a is the local electrostatic potential, E_G is the bandgap, and E_F is taken to be mid-gap when $V_a = 0$. The quantum capacitance, C_Q, is defined as,

$$C_Q = \frac{\partial Q}{\partial V_a}\ ,$$

and has units of F/m^2 and F/m in the 2D and 1D cases, respectively.

Two Dimensions

In the two-dimensional case, if we employ the effective-mass approximation with parabolic bands, the DOS is given by,

$$g(E) = \frac{m}{\pi\hbar^2}\nu(E)$$

where, ν(E) is the number of contributing bands at a given energy, m is the effective mass, and $\hbar$ is Dirac's constant. If we combine this with two Eqs. mentioned before the above one, and exchange the order of differentiation and integration, we get,

$$C_Q = \frac{mq^2}{4\pi\hbar^2 kT}\int_0^\infty \nu(E)\left[\frac{E + \frac{E_G}{2} - qV_a}{2kT}\right] + sech^2\left(\frac{E + \frac{E_G}{2} - qV_a}{2kT}\right) \mathrm{d}E,$$

where k is Boltzmann's constant, and T is temperature. If ν is a constant, we can perform the integration to get,

$$C_Q = \frac{\nu mq^2}{2\pi\hbar^2}\left[2 - \frac{sinh\left(\frac{E_G}{2kT}\right)}{\cosh\left(\frac{\frac{E_G}{2} - qV_a}{2kT}\right) cosh\left(\frac{\frac{E_G}{2} - qV_a}{2kT}\right)}\right],$$

which reduces to,

$$C_Q = \frac{\nu mq^2}{\pi\hbar^2},$$

when $E_G = 0$ in agreement with Ref. 1, where metallic properties were assumed. Note that this function is quantized in the metallic case, but continuous for a semiconductor. For E_G greater than about 15kT, however, the function makes a rapid transition from a small value to that given by Equation when, V_a crosses $E_G/2$, and is thus effectively quantized.

One Dimension

In the one-dimensional, effective-mass case, we have,

$$g(E) = \frac{\nu(E)}{\pi\hbar}\sqrt{\frac{2m}{E}}.$$

The explicit energy dependence of this DOS complicates the evaluation of our integral for C_Q. The approach suggested in Ref. 2, i.e., using the fact that the derivative of *f(E)* is peaked about E_F in order to approximate this integral using a Sommerfeld expansion,5 cannot be done in general, due to the presence of singularities in the 1D DOS.

The capacitance is given by,

$$C_Q = \frac{q^2}{2kTh}\sqrt{\frac{m}{2}}\int_0^{\infty}\frac{\nu(E)}{\sqrt{E}}\left[\operatorname{sech}^2\left(\frac{E+\frac{E_G}{2}-qV_a}{2kT}\right)\right]+\operatorname{sech}^2\left(\frac{E+\frac{E_G}{2}+qV_a}{2kT}\right)dE,$$

where h is Planck's constant. For sufficiently large $|V_a|$, we can completely neglect one of the sech2 (·) terms. As a simple example, if V_a = 0.1 V for a material with $E_G \simeq$ 1 eV, the contribution to the integral from the first term is roughly four orders of magnitude greater than the second. This approximation is equivalent to neglecting hole charge for positive V_a, and electron charge for negative V_a. The solid line in figure shows the equilibrium CQ as a function of V_a for a semiconductor with two valence and conduction bands: at 0.2 and 0.6 eV away from the Fermi level. An effective mass of $0.06m_0$ is assumed, where mo is the free-electron mass. The van Hove singularities, at each band edge, result in corresponding peaks in C_Q.

For a linear energy-wavevector relation, such as that near the Fermi level in graphene or a metallic CN, the DOS is constant. This is the case considered by Burke, 3 and is valid when V_a is such that f(E) is approximately zero before the first van Hove singularity is encountered in the integral. Since the higher energy bands are not relevant to the integration under such a condition, ν is constant, and the DOS is given by,

$$g(E) = \frac{2\nu}{hv_F},$$

where v_F is the Fermi velocity. The result is,

$$C_Q = \frac{2\nu q^2}{hv_F},$$

Note that in Eq. $C_Q = \frac{q^2}{2kTh}\sqrt{\frac{m}{2}}\int_0^{\infty}\frac{\nu(E)}{\sqrt{E}}\left[\operatorname{sech}^2\left(\frac{E+\frac{E_G}{2}-qV_a}{2kT}\right)\right]+\operatorname{sech}^2\left(\frac{E+\frac{E_G}{2}+qV_a}{2kT}\right)dE,$ C_Q does not manifest itself as a multiple of some discrete amount, so "quantum capacitance" is not an appropriate description for a 1D semiconductor, unlike in the metallic 2D and metallic 1D CN cases, where the capacitance is truly quantized.

Capacitor

Capacitors are simple passive device that can store an electrical charge on their plates when connected to a voltage source.

The capacitor is a component which has the ability or "capacity" to store energy in the form of an electrical charge producing a potential difference (Static Voltage) across its plates, much like a small rechargeable battery.

There are many different kinds of capacitors available from very small capacitor beads used in resonance circuits to large power factor correction capacitors, but they all do the same thing, they store charge.

In its basic form, a capacitor consists of two or more parallel conductive (metal) plates which are not connected or touching each other, but are electrically separated either by air or by some form of a good insulating material such as waxed paper, mica, ceramic, plastic or some form of a liquid gel as used in electrolytic capacitors. The insulating layer between a capacitors plates is commonly called the Dielectric.

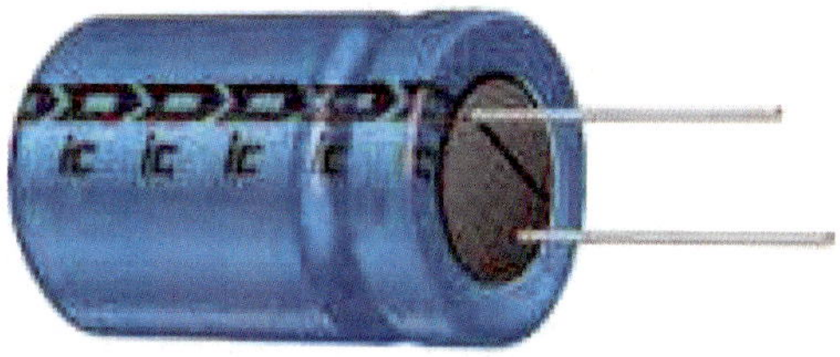

A Typical Capacitor

Due to this insulating layer, DC current can not flow through the capacitor as it blocks it allowing instead a voltage to be present across the plates in the form of an electrical charge.

The conductive metal plates of a capacitor can be either square, circular or rectangular, or they can be of a cylindrical or spherical shape with the general shape, size and construction of a parallel plate capacitor depending on its application and voltage rating.

When used in a direct current or DC circuit, a capacitor charges up to its supply voltage but blocks the flow of current through it because the dielectric of a capacitor is non-conductive and basically an insulator. However, when a capacitor is connected to an alternating current or AC circuit, the flow of the current appears to pass straight through the capacitor with little or no resistance.

There are two types of electrical charge, positive charge in the form of Protons and negative charge in the form of Electrons. When a DC voltage is placed across a capacitor, the positive (+ve) charge quickly accumulates on one plate while a corresponding and opposite negative (-ve) charge accumulates on the other plate. For every particle of +ve charge that arrives at one plate a charge of the same sign will depart from the -ve plate.

Then the plates remain charge neutral and a potential difference due to this charge is established between the two plates. Once the capacitor reaches its steady state condition an electrical current is unable to flow through the capacitor itself and around the circuit due to the insulating properties of the dielectric used to separate the plates.

The flow of electrons onto the plates is known as the capacitors Charging Current which continues to flow until the voltage across both plates (and hence the capacitor) is equal to the applied voltage Vc. At this point the capacitor is said to be "fully charged" with electrons.

The strength or rate of this charging current is at its maximum value when the plates are fully discharged (initial condition) and slowly reduces in value to zero as the plates charge up to a potential difference across the capacitors plates equal to the source voltage.

The amount of potential difference present across the capacitor depends upon how much charge was deposited onto the plates by the work being done by the source voltage and also by how much capacitance the capacitor has and this is illustrated below.

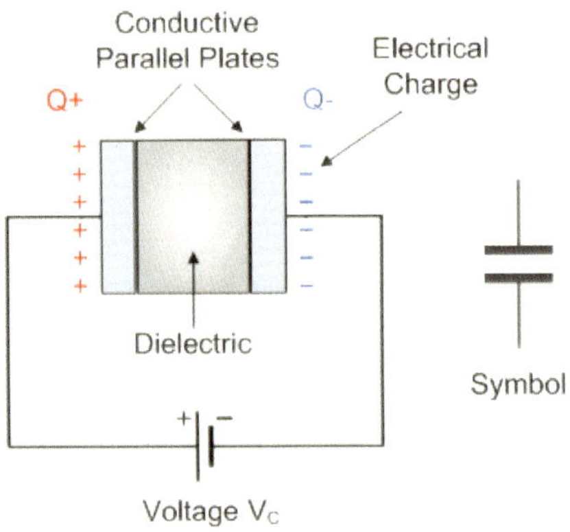

The parallel plate capacitor is the simplest form of capacitor. It can be constructed using two metal or metallised foil plates at a distance parallel to each other, with its capacitance value in Farads, being fixed by the surface area of the conductive plates and the distance of separation between them. Altering any two of these values alters the the value of its capacitance and this forms the basis of operation of the variable capacitors.

Also, because capacitors store the energy of the electrons in the form of an electrical charge on the plates the larger the plates and/or smaller their separation the greater will be the charge that the capacitor holds for any given voltage across its plates. In other words, larger plates, smaller distance, more capacitance.

By applying a voltage to a capacitor and measuring the charge on the plates, the ratio of the charge Q to the voltage V will give the capacitance value of the capacitor and is therefore given as: $C = Q/V$ this equation can also be re-arranged to give the more familiar formula for the quantity of charge on the plates as: $Q = C \times V$

Although we have said that the charge is stored on the plates of a capacitor, it is more correct to say that the energy within the charge is stored in an "electrostatic field" between the two plates. When an electric current flows into the capacitor, charging it up, the electrostatic field becomes more stronger as it stores more energy.

Likewise, as the current flows out of the capacitor, discharging it, the potential difference between the two plates decreases and the electrostatic field decreases as the energy moves out of the plates.

The property of a capacitor to store charge on its plates in the form of an electrostatic field is called the capacitance of the capacitor. Not only that, but capacitance is also the property of a capacitor which resists the change of voltage across it.

The Capacitance of a Capacitor

Capacitance is the electrical property of a capacitor and is the measure of a capacitors ability to store an electrical charge onto its two plates with the unit of capacitance being the Farad (abbreviated to F) named after the British physicist Michael Faraday.

Capacitance is defined as being that a capacitor has the capacitance of One Farad when a charge of One Coulomb is stored on the plates by a voltage of One volt. Note that capacitance, C is always positive in value and has no negative units. However, the Farad is a very large unit of measurement to use on its own so sub-multiples of the Farad are generally used such as micro-farads, nano-farads and pico-farads, for example.

Standard Units of Capacitance

Microfarad (μF) 1μF = 1/1,000,000 = 0.000001 = 10^{-6} F

Nanofarad (nF) 1nF = 1/1,000,000,000 = 0.000000001 = 10^{-9} F

Picofarad (pF) 1pF = 1/1,000,000,000,000 = 0.000000000001 = 10^{-12} F

Then using the information above we can construct a simple table to help us convert between pico-Farad (pF), to nano-Farad (nF), to micro-Farad (μF) and to Farads (F) as shown.

Pico-Farad (pF)	Nano-Farad (nF)	Micro-Farad (μF)	Farads (F)
1,000	1.0	0.001	
10,000	10.0	0.01	
1,000,000	1,000	1.0	
	10,000	10.0	
	100,000	100	
	1,000,000	1,000	0.001
		10,000	0.01
		100,000	0.1
		1,000,000	1.0

Capacitance of a Parallel Plate Capacitor

The capacitance of a parallel plate capacitor is proportional to the area, A in metres2 of the smallest of the two plates and inversely proportional to the distance or separation, d (i.e. the dielectric thickness) given in metres between these two conductive plates.

The generalised equation for the capacitance of a parallel plate capacitor is given as: $C = \varepsilon(A/d)$ where ε represents the absolute permittivity of the dielectric material being used. The permittivity of a vacuum, ε_0 also known as the "permittivity of free space" has the value of the constant 8.84 × 10-12 Farads per metre.

To make the maths a little easier, this dielectric constant of free space, ε_0, which can be written as: $1/(4\pi \times 9{\times}10^9)$, may also have the units of picofarads (pF) per metre as the constant giving: 8.84 for the value of free space. Note though that the resulting capacitance value will be in picofarads and not in farads.

Generally, the conductive plates of a capacitor are separated by some kind of insulating material or gel rather than a perfect vacuum. When calculating the capacitance of a capacitor, we can consider the permittivity of air, and especially of dry air, as being the same value as a vacuum as they are very close.

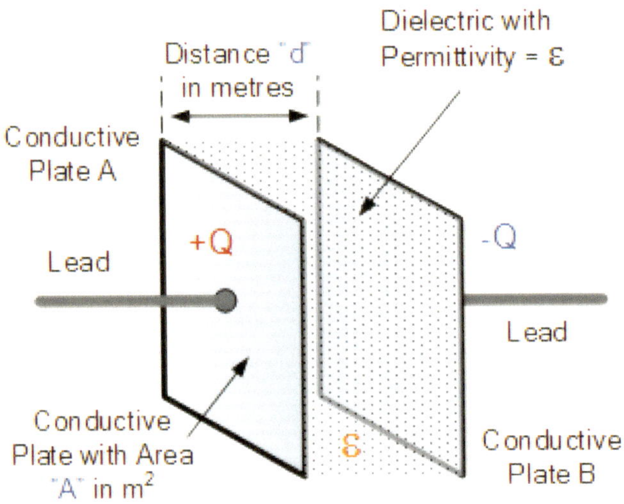

Example

A capacitor is constructed from two conductive metal plates 30cm x 50cm which are spaced 6mm apart from each other, and uses dry air as its only dielectric material. Calculate the capacitance of the capacitor.

Using: $C=\varepsilon_0 \dfrac{A}{d}$

where: $\varepsilon_0 8.84\times10^{-12}$

$A=0.3\times0.5m^2$ and $d=6\times10^{-3}m$

$$C=\frac{8.84\times10^{-12}\times(0.3\times0.5)}{6\times10^{-3}}=0.221nF$$

Then the value of the capacitor consisting of two plates separated by air is calculated as 221pF or 0.221nF.

The Dielectric of a Capacitor

As well as the overall size of the conductive plates and their distance or spacing apart from each other, another factor which affects the overall capacitance of the device is the type of dielectric material being used. In other words the "Permittivity" (ε) of the dielectric.

The conductive plates of a capacitor are generally made of a metal foil or a metal film allowing for the flow of electrons and charge, but the dielectric material used is always an insulator. The various insulating materials used as the dielectric in a capacitor differ in their ability to block or pass an electrical charge.

This dielectric material can be made from a number of insulating materials or combinations of these materials with the most common types used being: air, paper, polyester, polypropylene, Mylar, ceramic, glass, oil, or a variety of other materials.

The factor by which the dielectric material, or insulator, increases the capacitance of the capacitor compared to air is known as the Dielectric Constant, k and a dielectric material with a high dielectric constant is a better insulator than a dielectric material with a lower dielectric constant. Dielectric constant is a dimensionless quantity since it is relative to free space.

The actual permittivity or "complex permittivity" of the dielectric material between the plates is then the product of the permittivity of free space (ε_o) and the relative permittivity (ε_r) of the material being used as the dielectric and is given as:

Complex Permittivity

$$\varepsilon = \varepsilon_0 \times \varepsilon_r$$

In other words, if we take the permittivity of free space, ε_o as our base level and make it equal to one, when the vacuum of free space is replaced by some other type of insulating material, their permittivity of its dielectric is referenced to the base dielectric of free space giving a multiplication factor known as "relative permittivity", ε_r. So the value of the complex permittivity, ε will always be equal to the relative permittivity times one.

Typical units of dielectric permittivity, ε or dielectric constant for common materials are: Pure Vacuum = 1.0000, Air = 1.0006, Paper = 2.5 to 3.5, Glass = 3 to 10, Mica = 5 to 7, Wood = 3 to 8 and Metal Oxide Powders = 6 to 20 etc. This then gives us a final equation for the capacitance of a capacitor as:

$$\text{Capacitance, } C = \frac{\varepsilon_o \varepsilon_r A}{d} \text{ Farads}$$

One method used to increase the overall capacitance of a capacitor while keeping its size small is to "interleave" more plates together within a single capacitor body. Instead of just one set of parallel plates, a capacitor can have many individual plates connected together thereby increasing the surface area, A of the plates.

For a standard parallel plate capacitor as shown above, the capacitor has two plates, labelled A and B. Therefore as the number of capacitor plates is two, we can say that n = 2, where "n" represents the number of plates.

Then our equation above for a single parallel plate capacitor should really be:

$$\text{Capacitance, } C = \frac{\varepsilon_o \varepsilon_r (n-1) A}{d} \text{ Farads}$$

However, the capacitor may have two parallel plates but only one side of each plate is in contact with the dielectric in the middle as the other side of each plate forms the outside of the capacitor. If we take the two halves of the plates and join them together we effectively only have "one" whole plate in contact with the dielectric.

As for a single parallel plate capacitor, n – 1 = 2 – 1 which equals 1 as $C = (\varepsilon_o * \varepsilon_r \times 1 \times A)/d$ is exactly the same as saying: $C = (\varepsilon_o * \varepsilon_r * A)/d$ which is the standard equation above.

Now suppose we have a capacitor made up of 9 interleaved plates, then n = 9 as shown.

Multi-plate Capacitor

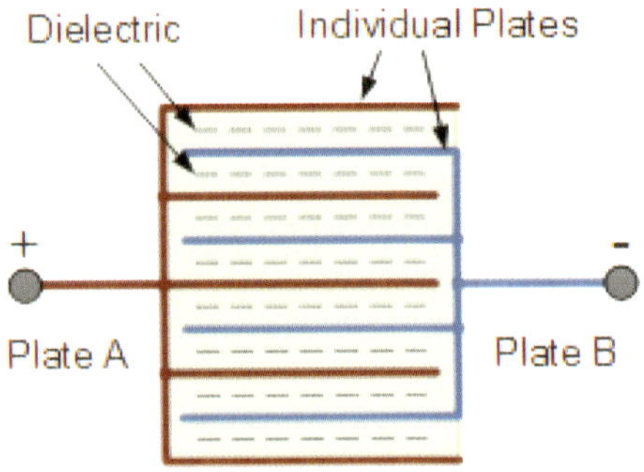

8 mini capacitors in one

Now we have five plates connected to one lead (A) and four plates to the other lead (B). Then BOTH sides of the four plates connected to lead B are in contact with the dielectric, whereas only one side of each of the outer plates connected to A is in contact with the dielectric. Then as above, the useful surface area of each set of plates is only eight and its capacitance is therefore given as:

$$C = \frac{\varepsilon_0 \varepsilon_r (n-1)A}{d} = \frac{\varepsilon_0 \varepsilon_r (9-1)A}{d} = \frac{\varepsilon_0 \varepsilon_r 8A}{d}$$

Modern capacitors can be classified according to the characteristics and properties of their insulating dielectric:

- Low Loss, High Stability such as Mica, Low-K Ceramic, Polystyrene.
- Medium Loss, Medium Stability such as Paper, Plastic Film, High-K Ceramic.
- Polarized Capacitors such as Electrolytic's, Tantalum's.

Voltage Rating of a Capacitor

All capacitors have a maximum voltage rating and when selecting a capacitor consideration must be given to the amount of voltage to be applied across the capacitor. The maximum amount of voltage that can be applied to the capacitor without damage to its dielectric material is generally given in the data sheets as: WV, (working voltage) or as WV DC, (DC working voltage).

If the voltage applied across the capacitor becomes too great, the dielectric will break down (known as electrical breakdown) and arcing will occur between the capacitor plates resulting in a short-circuit. The working voltage of the capacitor depends on the type of dielectric material being used and its thickness.

The DC working voltage of a capacitor is just that, the maximum DC voltage and NOT the maximum AC voltage as a capacitor with a DC voltage rating of 100 volts DC cannot be safely subjected to an alternating voltage of 100 volts. Since an alternating voltage that has an RMS value of 100 volts will have a peak value of over 141 volts! ($\sqrt{2} \times 100$).

Then a capacitor which is required to operate at 100 volts AC should have a working voltage of at least 200 volts. In practice, a capacitor should be selected so that its working voltage either DC or AC should be at least 50 percent greater than the highest effective voltage to be applied to it.

Another factor which affects the operation of a capacitor is Dielectric Leakage. Dielectric leakage occurs in a capacitor as the result of an unwanted leakage current which flows through the dielectric material.

Generally, it is assumed that the resistance of the dielectric is extremely high and a good insulator blocking the flow of DC current through the capacitor (as in a perfect capacitor) from one plate to the other.

However, if the dielectric material becomes damaged due excessive voltage or over temperature, the leakage current through the dielectric will become extremely high resulting in a rapid loss of charge on the plates and an overheating of the capacitor eventually resulting in premature failure of the capacitor. Then never use a capacitor in a circuit with higher voltages than the capacitor is rated for otherwise it may become hot and explode.

Characteristic

The capacitor, as with any other electronic component, comes defined by a series of characteristics. These Capacitor Characteristics can always be found in the data sheets that the capacitor manufacturer provides to us so here are just a few of the more important ones.

- Nominal Capacitance, (C)

The nominal value of the Capacitance, C of a capacitor is the most important of all capacitor characteristics. This value measured in pico-Farads (pF), nano-Farads (nF) or micro-Farads (μF) and is marked onto the body of the capacitor as numbers, letters or coloured bands.

The capacitance of a capacitor can change value with the circuit frequency (Hz) y with the ambient temperature. Smaller ceramic capacitors can have a nominal value as low as one pico-Farad, (1pF) while larger electrolytic's can have a nominal capacitance value of up to one Farad, (1F).

All capacitors have a tolerance rating that can range from -20% to as high as +80% for aluminium electrolytic's affecting its actual or real value. The choice of capacitance is determined by the circuit configuration but the value read on the side of a capacitor may not necessarily be its actual value.

- Working Voltage, (WV)

The Working Voltage is another important capacitor characteristic that defines the maximum continuous voltage either DC or AC that can be applied to the capacitor without failure during its working life. Generally, the working voltage printed onto the side of a capacitors body refers to its DC working voltage, (WVDC).

DC and AC voltage values are usually not the same for a capacitor as the AC voltage value refers to the r.m.s. value and NOT the maximum or peak value which is 1.414 times greater. Also, the specified DC working voltage is valid within a certain temperature range, normally -30°C to +70°C.

Any DC voltage in excess of its working voltage or an excessive AC ripple current may cause failure. It follows therefore, that a capacitor will have a longer working life if operated in a cool environment and within its rated voltage. Common working DC voltages are 10V, 16V, 25V, 35V, 50V, 63V, 100V, 160V, 250V, 400V and 1000V and are printed onto the body of the capacitor.

- Tolerance, (±%)

As with resistors, capacitors also have a Tolerance rating expressed as a plus-or-minus value either in picofarad's (±pF) for low value capacitors generally less than 100pF or as a percentage (±%) for higher value capacitors generally higher than 100pF.

The tolerance value is the extent to which the actual capacitance is allowed to vary from its nominal value and can range anywhere from -20% to +80%. Thus a 100μF capacitor with a ±20% tolerance could legitimately vary from 80μF to 120μF and still remain within tolerance.

Capacitors are rated according to how near to their actual values they are compared to the rated nominal capacitance with coloured bands or letters used to indicated their actual tolerance. The most common tolerance variation for capacitors is 5% or 10% but some plastic capacitors are rated as low as ±1%.

- Leakage Current

The dielectric used inside the capacitor to separate the conductive plates is not a perfect insulator resulting in a very small current flowing or "leaking" through the dielectric due to the influence of the powerful electric fields built up by the charge on the plates when applied to a constant supply voltage.

This small DC current flow in the region of nano-amps (nA) is called the capacitors Leakage Current. Leakage current is a result of electrons physically making their way through the dielectric medium, around its edges or across its leads and which will over time fully discharging the capacitor if the supply voltage is removed.

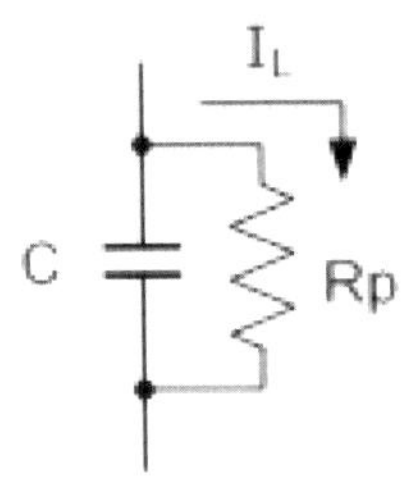

Leakage
Model

When the leakage is very low such as in film or foil type capacitors it is generally referred to as "insulation resistance" (R_p) and can be expressed as a high value resistance in parallel with the capacitor as shown. When the leakage current is high as in electrolytic's it is referred to as a "leakage current" as electrons flow directly through the electrolyte.

Capacitor leakage current is an important parameter in amplifier coupling circuits or in power supply circuits, with the best choices for coupling and/or storage applications being Teflon and the other plastic capacitor types (polypropylene, polystyrene, etc) because the lower the dielectric constant, the higher the insulation resistance.

Electrolytic-type capacitors (tantalum and aluminium) on the other hand may have very high capacitances, but they also have very high leakage currents (typically of the order of about 5-20 μA per μF) due to their poor isolation resistance, and are therefore not suited for storage or coupling applications. Also, the flow of leakage current for aluminium electrolytic's increases with temperature.

- Working Temperature, (T)

Changes in temperature around the capacitor affect the value of the capacitance because of changes in the dielectric properties. If the air or surrounding temperature becomes to hot or to cold the capacitance value of the capacitor may change so much as to affect the correct operation of the circuit. The normal working range for most capacitors is -30° C to +125° C with nominal voltage ratings given for a Working Temperature of no more than +70° C especially for the plastic capacitor types.

Generally for electrolytic capacitors and especially aluminium electrolytic capacitor, at high temperatures (over +85° C the liquids within the electrolyte can be lost to evaporation, and the body of the capacitor (especially the small sizes) may become deformed due to the internal pressure and leak outright. Also, electrolytic capacitors can not be used at low temperatures, below about -10° C, as the electrolyte jelly freezes.

- Temperature Coefficient, (TC)

The Temperature Coefficient of a capacitor is the maximum change in its capacitance over a specified temperature range. The temperature coefficient of a capacitor is generally expressed linearly as parts per million per degree centigrade (PPM/°C), or as a percent change over a particular range of temperatures. Some capacitors are non linear (Class 2 capacitors) and increase their value as the temperature rises giving them a temperature coefficient that is expressed as a positive "P".

Some capacitors decrease their value as the temperature rises giving them a temperature coefficient that is expressed as a negative "N". For example "P100" is +100 ppm/°C or "N200", which is -200 ppm/°C etc. However, some capacitors do not change their value and remain constant over a certain temperature range, such capacitors have a zero temperature coefficient or "NPO". These types of capacitors such as Mica or Polyester are generally referred to as Class 1 capacitors.

Most capacitors, especially electrolytic's lose their capacitance when they get hot but temperature compensating capacitors are available in the range of at least P1000 through to N5000 (+1000 ppm/oC through to -5000 ppm/°C). It is also possible to connect a capacitor with a positive temperature coefficient in series or parallel with a capacitor having a negative temperature coefficient the net result being that the two opposite effects will cancel each other out over a certain range of temperatures. Another useful application of temperature coefficient capacitors is to use them to cancel out the effect of temperature on other components within a circuit, such as inductors or resistors etc.

- Polarization

Capacitor Polarization generally refers to the electrolytic type capacitors but mainly the Aluminium Electrolytic's, with regards to their electrical connection. The majority of electrolytic capacitors are polarized types, that is the voltage connected to the capacitor terminals must have the correct polarity, i.e. positive to positive and negative to negative.

Incorrect polarization can cause the oxide layer inside the capacitor to break down resulting in very large currents flowing through the device resulting in destruction as we have mentioned earlier.

The majority of electrolytic capacitors have their negative, -ve terminal clearly marked with either a black stripe, band, arrows or chevrons down one side of their body as shown, to prevent any incorrect connection to the DC supply.

Some larger electrolytic's have their metal can or body connected to the negative terminal but high voltage types have their metal can insulated with the electrodes being brought out to separate spade or screw terminals for safety.

Also, when using aluminium electrolytic's in power supply smoothing circuits care should be taken to prevent the sum of the peak DC voltage and AC ripple voltage from becoming a "reverse voltage".

- Equivalent Series Resistance, (ESR)

The Equivalent Series Resistance or ESR, of a capacitor is the AC impedance of the capacitor when used at high frequencies and includes the resistance of the dielectric material, the DC resistance of the terminal leads, the DC resistance of the connections to the dielectric and the capacitor plate resistance all measured at a particular frequency and temperature.

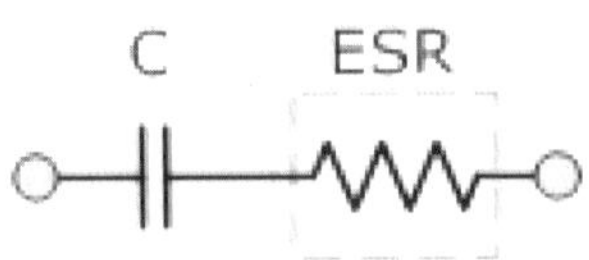

ESR Model

In some ways, ESR is the opposite of the insulation resistance which is presented as a pure resistance (no capacitive or inductive reactance) in parallel with the capacitor. An ideal capacitor would have only capacitance but ESR is presented as a pure resistance (less than 0.1Ω) in series with the capacitor (hence the name Equivalent Series Resistance), and which is frequency dependent making it a "DYNAMIC" quantity.

As ESR defines the energy losses of the "equivalent" series resistance of a capacitor it must therefore determine the capacitor's overall I_2R heating losses especially when used in power and switching circuits.

Capacitors with a relatively high ESR have less ability to pass current to and from its plates to the external circuit because of their longer charging and discharging RC time constant. The ESR of electrolytic capacitors increases over time as their electrolyte dries out. Capacitors with very low ESR ratings are available and are best suited when using the capacitor as a filter.

Capacitors with small capacitance's (less than 0.01μF) generally do not pose much danger to humans. However, when their capacitance's start to exceed 0.1μF, touching the capacitor leads can be a shocking experience.

As a final note, capacitors with small capacitance's (less than 0.01μF) generally do not pose much danger to humans. However, when their capacitance's start to exceed 0.1μF, touching the capacitor leads can be a shocking experience.

Capacitors have the ability to store an electrical charge in the form of a voltage across themselves even when there is no circuit current flowing, giving them a sort of memory with large electrolytic type reservoir capacitors found in television sets, photo flashes and capacitor banks potentially storing a lethal charge.

Capacitor with Air as its Dielectric

$$C=\frac{Q}{V}=\varepsilon\frac{A}{d}$$

Capacitor with a Solid as its Dielectric

$$C=\frac{Q}{V}=\varepsilon_0\varepsilon_r\frac{A}{d}$$

where A is the area of the plates in square metres, m^2 with the larger the area, the more charge the capacitor can store. d is the distance or separation between the two plates. The smaller is this distance, the higher is the ability of the plates to store charge, since the -ve charge on the -Q charged plate has a greater effect on the +Q charged plate, resulting in more electrons being repelled off of the +Q charged plate, and thus increasing the overall charge.

ε_0 (epsilon) is the value of the permittivity for air which is 8.84×10^{-12} F/m, and ε_r is the permittivity of the dielectric medium used between the two plates.

Parallel Plate Capacitor

We have said previously that the capacitance of a parallel plate capacitor is proportional to the surface area A and inversely proportional to the distance, d between the two plates and this is true for dielectric medium of air. However, the capacitance value of a capacitor can be increased by inserting a solid medium in between the conductive plates which has a dielectric constant greater than that of air.

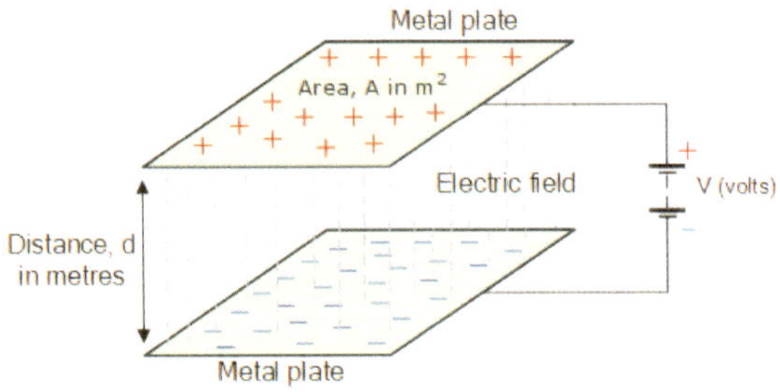

Typical values of epsilon ε for various commonly used dielectric materials are:

Air = 1.0, Paper = 2.5 – 3.5, Glass = 3 – 10, Mica = 5 – 7 etc.

The factor by which the dielectric material, or insulator, increases the capacitance of the capacitor compared to air is known as the Dielectric Constant, (k). "k" is the ratio of the permittivity of the dielectric medium being used to the permittivity of free space otherwise known as a vacuum.

Therefore, all the capacitance values are related to the permittivity of vacuum. A dielectric material with a high dielectric constant is a better insulator than a dielectric material with a lower dielectric constant. Dielectric constant is a dimensionless quantity since it is relative to free space.

Example 1

A parallel plate capacitor consists of two plates with a total surface area of 100 cm². What will be the capacitance in pico-Farads, (pF) of the capacitor if the plate separation is 0.2 cm, and the dielectric medium used is air.

$$C = \varepsilon\frac{A}{d}, \quad \varepsilon=8.85\text{pF/m}$$

$$A=100\text{cm}^2 = 0.01\text{m}^2, d=0.2\text{cm}=0.002\text{m}$$

$$\therefore C=8.85\times10^{-12}\times\frac{0.01\text{m}^2}{0.002\text{m}} = 44\text{pF}$$

then the value of the capacitor is 44pF.

Charging and Discharging of a Capacitor

Consider the following circuit.

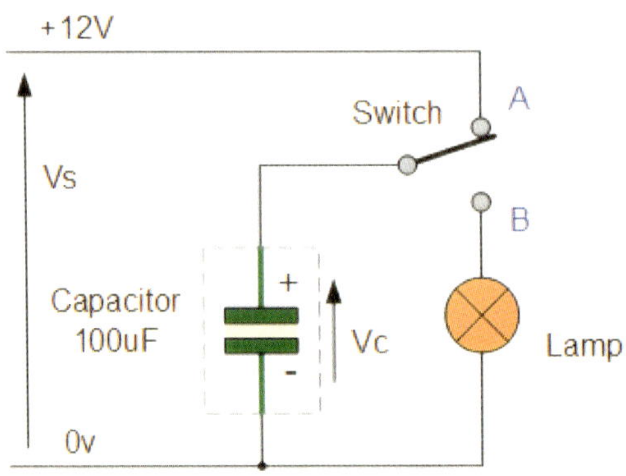

Assume that the capacitor is fully discharged and the switch connected to the capacitor has just been moved to position A. The voltage across the 100uf capacitor is zero at this point and a charging current (i) begins to flow charging up the capacitor until the voltage across the plates is equal to the 12v supply voltage. The charging current stops flowing and the capacitor is said to be "fully-charged". Then, Vc = Vs = 12v.

Once the capacitor is "fully-charged" in theory it will maintain its state of voltage charge even when the supply voltage has been disconnected as they act as a sort of temporary storage device. However, while this may be true of an "ideal" capacitor, a real capacitor will slowly discharge itself over a long period of time due to the internal leakage currents flowing through the dielectric.

This is an important point to remember as large value capacitors connected across high voltage supplies can still maintain a significant amount of charge even when the supply voltage is switched "OFF".

If the switch was disconnected at this point, the capacitor would maintain its charge indefinitely, but due to internal leakage currents flowing across its dielectric the capacitor would very slowly begin to discharge itself as the electrons passed through the dielectric. The time taken for the capacitor to discharge down to 37% of its supply voltage is known as its Time Constant.

If the switch is now moved from position A to position B, the fully charged capacitor would start to discharge through the lamp now connected across it, illuminating the lamp until the capacitor was fully discharged as the element of the lamp has a resistive value.

The brightness of the lamp and the duration of illumination would ultimately depend upon the capacitance value of the capacitor and the resistance of the lamp (t = R*C). The larger the value of the capacitor the brighter and longer will be the illumination of the lamp as it could store more charge.

Example 2

Calculate the charge in the above capacitor circuit.

$$Q = C \times V$$

$$Q = 100\mu F \times 12v = 1.2 \times 10^{-3} = 1.2mC$$

then the charge on the capacitor is 1.2 millicoulombs.

Current Through a Capacitor

Electrical current can not actually flow through a capacitor as it does a resistor or inductor due to the insulating properties of the dielectric material between the two plates. However, the charging and discharging of the two plates gives the effect that current is flowing.

The current that flows through a capacitor is directly related to the charge on the plates as current is the rate of flow of charge with respect to time. As the capacitors ability to store charge (Q) between its plates is proportional to the applied voltage (V), the relationship between the current and the voltage that is applied to the plates of a capacitor becomes,

Current-Voltage (I-V) Relationship:

$$i_{(t)} = C\frac{dv}{dt}$$

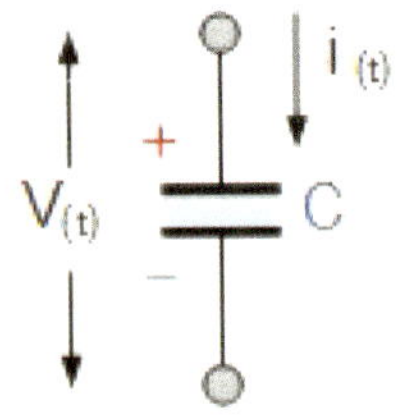

As the voltage across the plates increases (or decreases) over time, the current flowing through the capacitance deposits (or removes) charge from its plates with the amount of charge being proportional to the applied voltage. Then both the current and voltage applied to a capacitance are functions of time and are denoted by the symbols, $i_{(t)}$ and $v_{(t)}$.

However, from the above equation we can also see that if the voltage remains constant, the charge will become constant and therefore the current will be zero!. In other words, no change in voltage, no movement of charge and no flow of current. This is why a capacitor appears to "block" current flow when connected to a steady state DC voltage.

Energy in a Capacitor

When a capacitor charges up from the power supply connected to it, an electrostatic field is established which stores energy in the capacitor. The amount of energy in Joules that is stored in this electrostatic field is equal to the energy the voltage supply exerts to maintain the charge on the plates of the capacitor and is given by the formula:

$$\text{Energy, } W = \frac{1}{2}CV^2 \text{ or } \frac{CV^2}{2} \text{ in Joules, (j)}$$

so the energy stored in the 100uF capacitor circuit above is calculated as:

$$\text{Energy, } W = \frac{CV^2}{2} = \frac{100\times10^{-6}\times12^2}{2} = 7.2\text{mJ}$$

Capacitors in Parallel

Capacitors are connected together in parallel when both of its terminals are connected to each terminal of another capacitor

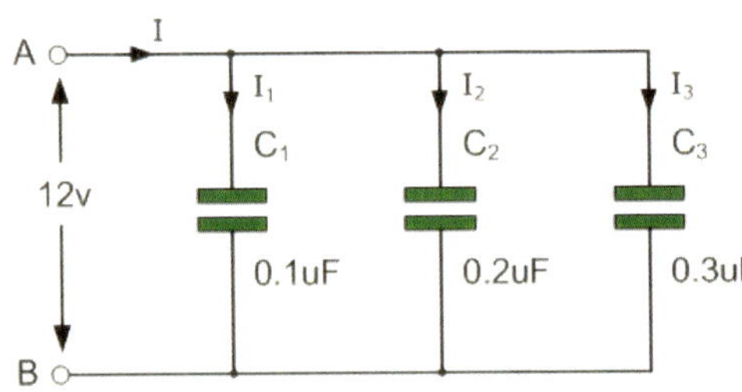

The voltage (V_c) connected across all the capacitors that are connected in parallel is THE SAME. Then, Capacitors in Parallel have a "common voltage" supply across them giving:

$$V_{C1} = V_{C2} = V_{C3} = V_{AB} = 12V$$

In the following circuit the capacitors, C_1, C_2 and C_3 are all connected together in a parallel branch between points A and B as shown.

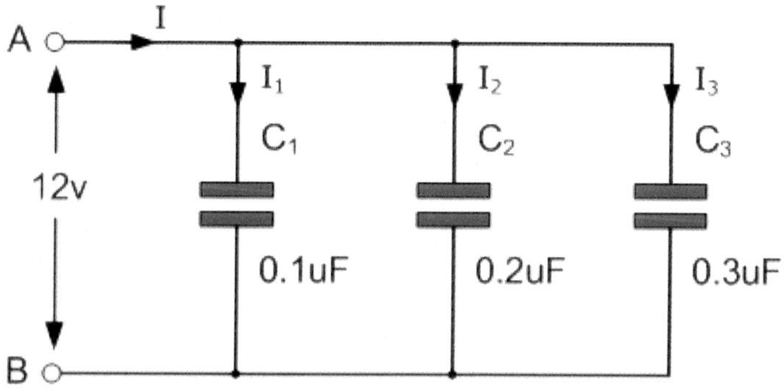

When capacitors are connected together in parallel the total or equivalent capacitance, C_T in the circuit is equal to the sum of all the individual capacitors added together. This is because the top plate of capacitor, C_1 is connected to the top plate of C_2 which is connected to the top plate of C_3 and so on.

The same is also true of the capacitors bottom plates. Then it is the same as if the three sets of plates were touching each other and equal to one large single plate thereby increasing the effective plate area in m^2.

Since capacitance, C is related to plate area ($C = \varepsilon(A/d)$) the capacitance value of the combination will also increase. Then the total capacitance value of the capacitors connected together in parallel is actually calculated by adding the plate area together. In other words, the total capacitance is equal to the sum of all the individual capacitance's in parallel. You may have noticed that the total capacitance of parallel capacitors is found in the same way as the total resistance of series resistors.

The currents flowing through each capacitor and as we saw in the previous tutorial are related to the voltage. Then by applying Kirchoff's Current Law, (KCL) to the above circuit, we have,

$$i_1 = C_1\frac{dv}{dt},\ i_2 = C_2\frac{dv}{dt},\ i_3 = C_3\frac{dv}{dt}$$

$$i_T = i_1 + i_2 + i_3$$

$$\therefore i_T = C_1\frac{dv}{dt} + C_2\frac{dv}{dt} + C_3\frac{dv}{dt}$$

and this can be re-written as:

$$i_T = \left(C_1 + C_2 + C_3\right)\frac{dv}{dt}$$

$$\text{or } i_T = C_T\frac{dv}{dt}$$

Then we can define the total or equivalent circuit capacitance, C_T as being the sum of all the individual capacitance's add together giving us the generalized equation of,

Parallel Capacitors Equation

$$C_T = C_1 + C_2 + C_3 + \ldots\text{etc}$$

When adding together capacitors in parallel, they must all be converted to the same capacitance units, whether it is μF, nF or pF. Also, we can see that the current flowing through the total capacitance value, C_T is the same as the total circuit current, i_T

We can also define the total capacitance of the parallel circuit from the total stored coulomb charge using the Q = CV equation for charge on a capacitors plates. The total charge Q_T stored on all the plates equals the sum of the individual stored charges on each capacitor therefore,

$$Q_T = Q_1 + Q_2 + Q_3 \quad \text{but, } Q=CV$$
$$\therefore Q_T = CV_T = CV_1 + CV_2 + CV_3$$
$$\text{or } C_T = C_1 + C_2 + C_3$$

As the voltage, (V) is common for parallel connected capacitors, we can divide both sides of the above equation through by the voltage leaving just the capacitance and by simply adding together the value of the individual capacitances gives the total capacitance, C_T. Also, this equation is not dependent upon the number of Capacitors in Parallel in the branch, and can therefore be generalized for any number of N parallel capacitors connected together.

Example 1

So by taking the values of the three capacitors from the above example, we can calculate the total equivalent circuit capacitance C_T as being:

$$C_T = C_1 + C_2 + C_3 = 0.1uF + 0.2uF + 0.3uF = 0.6uF$$

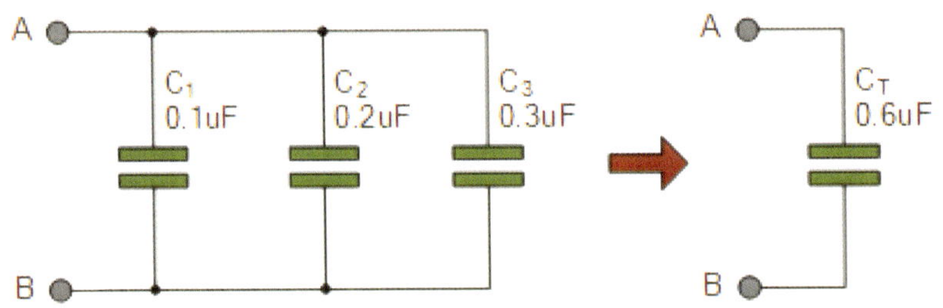

One important point to remember about parallel connected capacitor circuits, the total capacitance (C_T) of any two or more capacitors connected together in parallel will always be GREATER than the value of the largest capacitor in the group as we are adding together values. So in our example above C_T = 0.6μF whereas the largest value capacitor is only 0.3μF.

When 4, 5, 6 or even more capacitors are connected together the total capacitance of the circuit C_T would still be the sum of all the individual capacitors added together and as we know now, the total capacitance of a parallel circuit is always greater than the highest value capacitor.

This is because we have effectively increased the total surface area of the plates. If we do this with two identical capacitors, we have doubled the surface area of the plates which in turn doubles the capacitance of the combination and so on.

Example 2

Calculate the combined capacitance in micro-Farads (μF) of the following capacitors when they are connected together in a parallel combination:

- Two capacitors each with a capacitance of 47nF
- One capacitor of 470nF connected in parallel to a capacitor of 1μF.

a) Total Capacitance,

$$C_T = C_1 + C_2 = 47nF + 47nF = 94nF \text{ or } 0.094\mu F$$

b) Total Capacitance,

$$C_T = C_1 + C_2 = 470nF + 1\mu F$$

therefore, $C_T = 470nF + 1000nF = 1470nF$ or $1.47\mu F$

So, the total or equivalent capacitance, C_T of an electrical circuit containing two or more Capacitors in Parallel is the sum of the all the individual capacitance's added together as the effective area of the plates is increased.

In our next tutorial about capacitors we look at connecting together Capacitors in Series and the affect this combination has on the circuits total capacitance, voltage and current.

Capacitors in Series

Capacitors are connected together in series when they are daisy chained together in a single line.

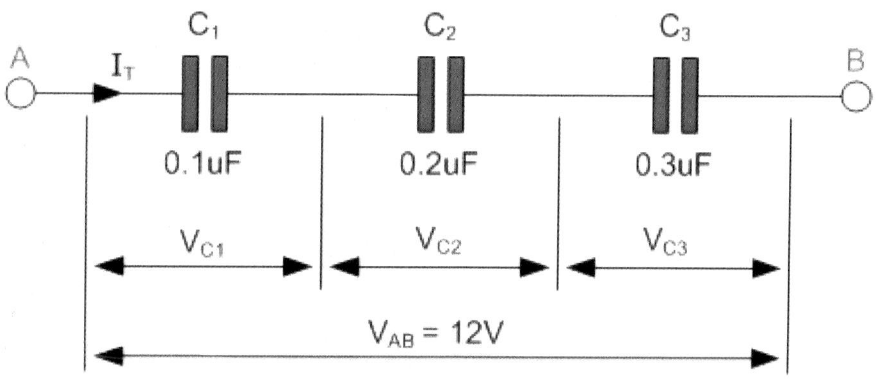

For series connected capacitors, the charging current (i_C) flowing through the capacitors is THE SAME for all capacitors as it only has one path to follow.

Then, Capacitors in Series all have the same current flowing through them as $i_T = i_1 = i_2 = i_3$ etc. Therefore each capacitor will store the same amount of electrical charge, Q on its plates regardless of its capacitance. This is because the charge stored by a plate of any one capacitor must have come from the plate of its adjacent capacitor. Therefore, capacitors connected together in series must have the same charge.

$$Q_T = Q_1 = Q_2 = Q_3 \text{ ....etc}$$

Consider the following circuit in which the three capacitors, C_1, C_2 and C_3 are all connected together in a series branch across a supply voltage between points A and B.

Capacitors in a Series Connection

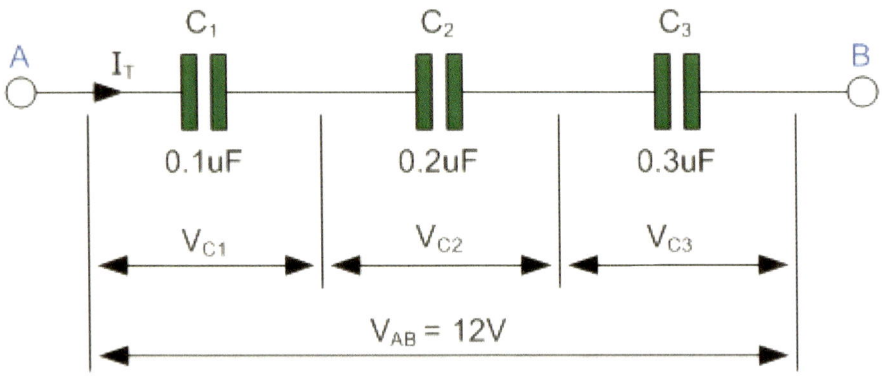

In the previous parallel circuit we saw that the total capacitance, CT of the circuit was equal to the sum of all the individual capacitors added together. In a series connected circuit however, the total or equivalent capacitance CT is calculated differently.

In the series circuit above the right hand plate of the first capacitor, C_1 is connected to the left hand plate of the second capacitor, C_2 whose right hand plate is connected to the left hand plate of the third capacitor, C_3. Then this series connection means that in a DC connected circuit, capacitor C_2 is effectively isolated from the circuit.

The result of this is that the effective plate area has decreased to the smallest individual capacitance connected in the series chain. Therefore the voltage drop across each capacitor will be different depending upon the values of the individual capacitance's.

Then by applying Kirchhoff's Voltage Law, (KVL) to the above circuit, we get:

$$V_{AB} = V_{C1} + V_{C2} + V_{C3} = 12V$$

$$V_{C1} = \frac{Q_T}{C_1}, V_{C2} = \frac{Q_T}{C_2}, V_{C3} = \frac{Q_T}{C_3}$$

Since Q = C*V and rearranging for V = Q/C, substituting Q/C for each capacitor voltage VC in the above KVL equation will give us:

$$V_{AB} = \frac{Q_T}{C_T} = \frac{Q_T}{C_1} + \frac{Q_T}{C_2} + \frac{Q_T}{C_3}$$

dividing each term through by Q gives,

Series capacitors Equation:

$$\frac{1}{C_T} = \frac{1}{C_1} + \frac{1}{C_2} + \frac{1}{C_3} + \text{....etc}$$

When adding together Capacitors in Series, the reciprocal (1/C) of the individual capacitors are all added together (just like resistors in parallel) instead of the capacitance's themselves. Then the total value for capacitors in series equals the reciprocal of the sum of the reciprocals of the individual capacitances.

Example 1

Taking the three capacitor values from the above example, we can calculate the total capacitance, CT for the three capacitors in series as:

$$\frac{1}{C_T} = \frac{1}{0.1\mu F} + \frac{1}{0.2\mu F} + \frac{1}{0.3\mu F} = \frac{1}{18.33 \times 10^6}$$
$$\therefore C_T = 0.055\mu F \text{ or } 55nF$$

One important point to remember about capacitors that are connected together in a series configuration, is that the total circuit capacitance (C_T) of any number of capacitors connected together in series will always be LESS than the value of the smallest capacitor in the series and in our example above C_T = 0.055μF with the value of the smallest capacitor in the series chain is only 0.1μF.

This reciprocal method of calculation can be used for calculating any number of individual capacitors connected together in a single series network. If however, there are only two capacitors in series, then a much simpler and quicker formula can be used and is given as:

$$C_T = \frac{C_1 \times C_2}{C_1 + C_2}$$

If the two series connected capacitors are equal and of the same value, that is: $C_1 = C_2$, we can simplify the above equation further as follows to find the total capacitance of the series combination.

$$C_T = \frac{C^2}{2C} = \frac{C}{2} = \frac{1}{2}C$$

Then we can see that if and only if the two series connected capacitors are the same and equal, then the total capacitance, C_T will be exactly equal to one half of the capacitance value, that is: C/2.

With series connected resistors, the sum of all the voltage drops across the series circuit will be equal to the applied voltage V_S (Kirchhoff's Voltage Law) and this is also true about capacitors in series.

With series connected capacitors, the capacitive reactance of the capacitor acts as an impedance due to the frequency of the supply. This capacitive reactance produces a voltage drop across each capacitor, therefore the series connected capacitors act as a capacitive voltage divider network.

The result is that the voltage divider formula applied to resistors can also be used to find the individual voltages for two capacitors in series. Then:

$$V_{CX} = V_S \frac{C_T}{C_X}$$

Where: C_X is the capacitance of the capacitor in question, V_S is the supply voltage across the series chain and V_{CX} is the voltage drop across the target capacitor.

Example 2

Find the overall capacitance and the individual rms voltage drops across the following sets of two capacitors in series when connected to a 12V AC supply.

- Two capacitors each with a capacitance of 47nF
- One capacitor of 470nF connected in series to a capacitor of 1µF.

a) Total Equal Capacitance,

$$C_T = \frac{C_1 \times C_2}{C_1 + C_2} = \frac{47\text{nF} \times 47\text{nF}}{47\text{nF} + 47\text{nF}} = 23.5\text{nF}$$

Voltage drop across the two identical 47nF capacitors,

$$V_{C1} = \frac{C_T}{C_1} \times V_T = \frac{23.5\text{nF}}{47\text{nF}} \times 12\text{V} = 6\text{volts}$$

$$V_{C2} = \frac{C_T}{C_2} \times V_T = \frac{23.5\text{nF}}{47\text{nF}} \times 12\text{V} = 6\text{volts}$$

b) Total Unequal Capacitance,

$$C_T = \frac{C_1 \times C_2}{C_1 + C_2} = \frac{470\text{nF} \times 47\text{nF}}{470\text{nF} + 47\text{nF}} = 320\text{nF}$$

Voltage drop across the two non-identical Capacitors: C_1 = 470nF and C_2 = 1µF.

$$V_{C1} = \frac{C_T}{C_1} \times V_T = \frac{320\text{nF}}{470\text{nF}} \times 12 = 8.16\text{volts}$$

$$V_{C2} = \frac{C_T}{C_2} \times V_T = \frac{320\text{nF}}{1\text{uF}} \times 12 = 3.84\text{volts}$$

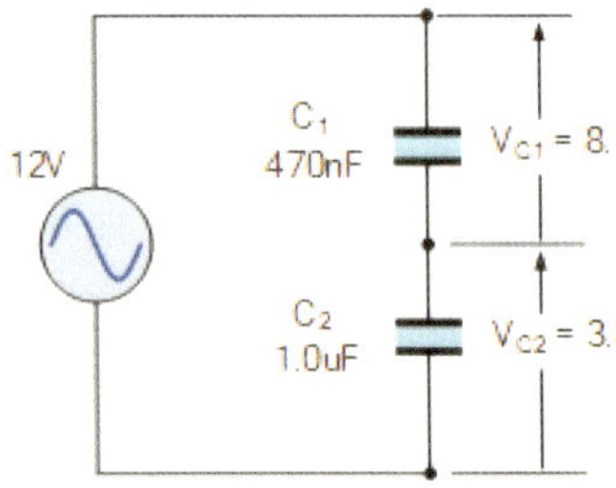

Since Kirchhoff's voltage law applies to this and every series connected circuit, the total sum of the individual voltage drops will be equal in value to the supply voltage, V_S. Then 8.16 + 3.84 = 12V.

Note also that if the capacitor values are the same, 47nF in our first example, the supply voltage will be divided equally across each capacitor as shown. This is because each capacitor in the series chain shares an equal and exact amount of charge (Q = C x V = 0.564μC) and therefore has half (or percentage fraction for more than two capacitors) of the applied voltage, V_S.

However, when the series capacitor values are different, the larger value capacitor will charge itself to a lower voltage and the smaller value capacitor to a higher voltage, and in our second example above this was shown to be 3.84 and 8.16 volts respectively. This difference in voltage allows the capacitors to maintain the same amount of charge, Q on the plates of each capacitors as shown.

$$Q_{C1} = V_{C1} \times C_1 = 8.16\text{V} \times 470\text{nF} = 3.84\mu\text{C}$$
$$Q_{C2} = V_{C2} \times C_2 = 3.84\text{V} \times 1\text{uF} = 3.84\mu\text{C}$$

Note that the ratios of the voltage drops across the two capacitors connected in series will always remain the same regardless of the supply frequency as their reactance, X_C will remain proportionally the same.

Then the two voltage drops of 8.16 volts and 3.84 volts above in our simple example will remain the same even if the supply frequency is increased from 100Hz to 100kHz.

Although the voltage drops across each capacitor will be different for different values of capacitance, the coulomb charge across the plates will be equal because the same amount of current flow exists throughout a series circuit as all the capacitors are being supplied with the same number or quantity of electrons.

In other words, if the charge across each capacitors plates is the same, as Q is constant, then as its capacitance decreases the voltage drop across the capacitors plates increases, because the charge is large with respect to the capacitance. Likewise, a larger capacitance will result in a smaller voltage drop across its plates because the charge is small with respect to the capacitance.

Capacitor Types

There is a large variety of different types of capacitor available in the market place and each one has its own set of characteristics and applications.

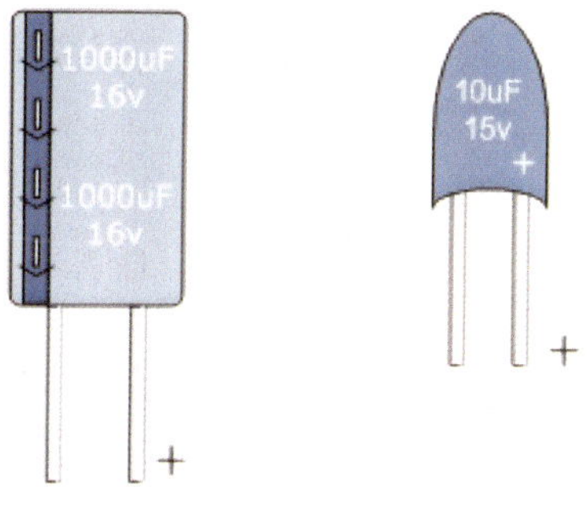

Aluminium Electrolytic   Tantalum Electrolytic

The types of capacitors available range from very small delicate trimming capacitors using in oscillator or radio circuits, up to large power metal-can type capacitors used in high voltage power correction and smoothing circuits.

The comparisons between the the different types of capacitor is generally made with regards to the dielectric used between the plates. Like resistors, there are also variable types of capacitors which allow us to vary their capacitance value for use in radio or "frequency tuning" type circuits.

Commercial types of capacitors are made from metallic foil interlaced with thin sheets of either paraffin-impregnated paper or Mylar as the dielectric material. Some capacitors look like tubes, this is because the metal foil plates are rolled up into a cylinder to form a small package with the insulating dielectric material sandwiched in between them.

Small capacitors are often constructed from ceramic materials and then dipped into an epoxy resin to seal them. Either way, capacitors play an important part in electronic circuits so here are a few of the more "common" types of capacitor available.

Ceramic Capacitor

A ceramic capacitor uses a ceramic material as the dielectric. Ceramics were one of the first materials to be used in the producion of capacitors, as it was a known insulator.

Ceramic capacitors are most commonly found in every electrical device and it uses a ceramic material as the dielectric. The ceramic capacitor is a non-polarity device, that means they do no have polarities. So we can connect it in any direction on a circuit board.

For this reason, they are generally much safer than electrolytic capacitors. Here is the symbol for a non-polarised capacitor given below. Many types of capacitors, such as the tantalum bead do not have a polarity.

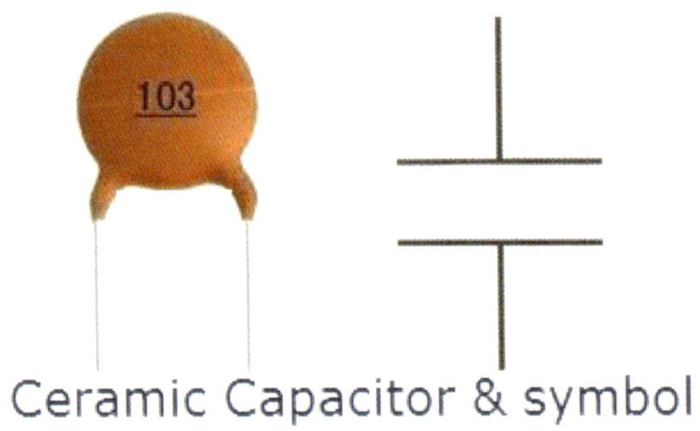

Ceramic Capacitor Polarity and Symbol

Characteristics

Precision and Tolerances

There are two classes of ceramic capacitors available today: class 1 and class 2. Class 1 ceramic capacitors are used where high stability and low losses are required. They are very accurate and the capacitance value is stable in regard to applied voltage, temperature and frequency. The NP0 series of capacitors has a capacitance thermal stability of ±0.54% within the total temperature range of -55 to +125° C. Tolerances of the nominal capacitance value can be as low as 1%.

Class 2 capacitors have a high capacitance per volume and are used for less sensitive applications. Their thermal stability is typically ±15% in the operating temperature range, and the nominal value tolerances are around 20%.

Size Advantages

When high component packing densities are required, as is the case in most modern printed circuit boards (PCBs), MLCC devices offer a great advantage compared to other capacitors. To illustrate this point, the "0402" multi-layer ceramic capacitor package measures just 0.4 mm x 0.2 mm. In such a package, there are 500 or more ceramic and metal layers. The minimum ceramic thickness as of 2010 is on the order of 0.5 microns.

High Voltage and High Power

Physically larger ceramic capacitors can be made to withstand much higher voltages and these are called power ceramic capacitors. These are physically much larger than those used on PCBs and have specialized terminals for safe connection to a high voltage supply. Power ceramic capacitors can be made to withstand voltages in the range of 2kV up to 100 kV, with a power specified at much higher than 200 volt-amperes.

Smaller MLCCs used in printed circuit boards are rated to voltages from only a few volts up to several hundreds of volts, depending on the application.

Construction of Ceramic Capacitor

In a ceramic capacitor, ceramic material is used to construct the dielectric and conductive metals are used to construct the electrodes. Dielectric is the insulating material placed between the electrodes of a capacitor.

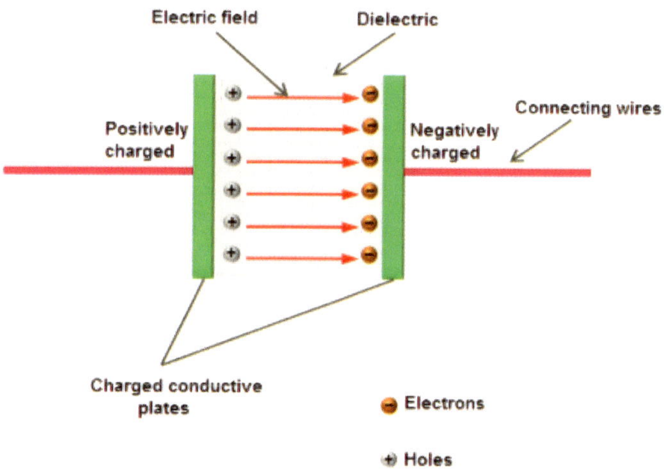

Ceramic material is chosen as dielectric because of its great ability to allow electrostatic attraction and repulsion. Also, ceramic materials are poor conductors of electricity. Therefore, they do not allow flow of electric charges through them. On the other hand, electrodes of the ceramic capacitor are good conductors of electricity. Therefore, they easily allow electric current through them.

Different shapes of Ceramic Capacitor

Ceramic capacitors are available in the following shapes or forms:

- Ceramic disc capacitor
- Ceramic tubular capacitor
- Multilayer ceramic capacitor (MLCC).

Ceramic Disc Capacitor

The ceramic disc capacitors are made by coating silver on both sides of the ceramic disc. The ceramic disc acts as the dielectric and silver coated on both sides of the disc acts as electrodes. For low capacitance, a single ceramic disc coated with silver is used whereas for high capacitance multiple layers are used.

Ceramic disc capacitor

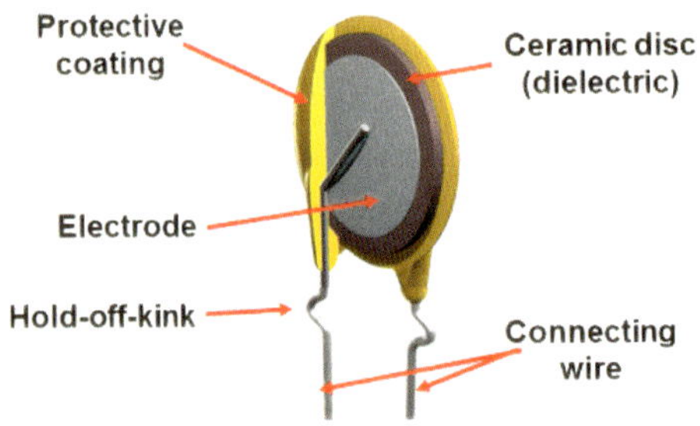

The leads made of copper are attached to the ceramic disc by using soldering technique. A protective coating is applied to the ceramic disc capacitor to protect it from heat.

The area of a ceramic disc or dielectric and spacing between the silver electrodes also determines the capacitance of the ceramic disc capacitor. The main disadvantage of using ceramic disc capacitor is its high capacitance change with slight change in temperature.

Ceramic Tubular Capacitor

As the name suggests, ceramic tubular capacitor is a hollow cylindrical ceramic material. The inner and outer surfaces of the hollow cylindrical ceramic material are coated with the silver ink. The hollow cylindrical ceramic material acts as the dielectric and the silver ink coated on inner and outer surfaces acts as electrodes.

Multilayer Ceramic Capacitor (MLCC)

The multilayer ceramic capacitor (MLCC) is made of multiple layers of ceramic material and conductive electrodes placed one above the other. The conductive electrodes are placed between each layer of ceramic material. In MLCC, the multiple layers of ceramic material acts as dielectric.

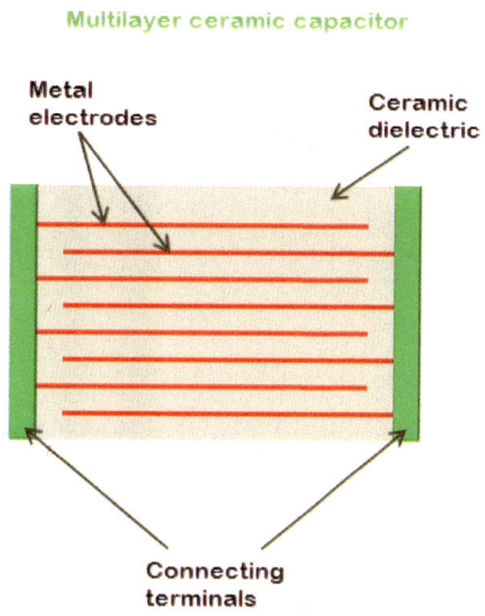

The multiple layers of ceramic and electrodes are contacted via the terminal surfaces. In this way, multiple layers of ceramic material and electrodes are constructed. Some MLCCs contain hundreds of ceramic layers and electrodes, each layer only a few micrometers thick.

The capacitance of each layer in MLCC is same. The total capacitance of MLCC is obtained by multiplying the capacitance of one layer by total number of layers.

For example, if each layer has capacitance of 2 pF and total layers in the MLCC are 50. The total capacitance is equal to 100 pF.

Total capacitance of MLCC = 2 pF × 50 = 100 pF

The capacitance of the MLCCs increases by reducing the space between the ceramic dielectric layers.

Based on the working temperature range, temperature drift, tolerance different ceramic capacitor classes is defined.

Class 1 Ceramic Capacitors

With respect to temperature, these are the most stable capacitors. They have nearly linear characteristics.

The most common compounds used as the dielectrics are:

- Magnesium Titanate for a positive temperature coefficient.
- Calcium Titanate for capacitors with a negative temperature coefficient.

Class 2 Ceramic Capacitors

Class 2 capacitors exhibit better performance with respect to volumetric efficiency, but this is at the cost of lower accuracy and stability. As a result, they are normally used for decoupling, coupling and bypass applications where accuracy is not of prime importance.

- Temperature range: -50C to +85C.
- Dissipation factor: 2.5%.
- Accuracy: average to poor.

Class 3 Ceramic Capacitors

Class 3 ceramic capacitors offer high volumetric efficiency with poor accuracy and a low dissipation factor. It cannot withstand high voltages. The dielectric used is often Barium Titanate.

- Class 3 capacitor will change its capacitance by -22% to +50%.
- Temperature range of +10C to +55C.
- Dissipation factor: 3 to 5%.
- It will have a fairly poor accuracy (commonly, 20%, or -20/+80%).

Class 3 type are typically used for decoupling or in other power supply applications where accuracy is not an issue.

Applications of Ceramic Capacitors

- Ceramic capacitors are majorly used in the resonant circuit in transmitter stations.
- Class 2 high-power capacitors are used in high voltage laser power supplies, power circuit breakers, induction furnaces etc.
- Surface mount capacitors are often used in printed circuit boards and high-density applications.
- Ceramic capacitors can also be used as a general purpose capacitor, because of their non-polarity and are available in a large variety of capacitances, voltage ratings, and sizes.
- Ceramic disc capacitors are used across brush DC motors to minimize RF noise.
- MLCC used in printed circuit boards (PCB) are rated for voltages from only a few volts up to several hundreds of volts, depending on the application.

Advantages

- Any size or shape is available in the market.
- At the same time, ceramic capacitors are inexpensive.
- They are light in weight, too.
- They can be designed to withstand up to sufficient high voltage (up to 100V).
- Their performance is reliable.
- They are suitable for use in hybrid integrated circuits.

Disadvantages

- Very high-voltage ceramic capacitors are not available.
- High capacitance values are not possible.

Film Capacitor

Film capacitors are capacitors which use a thin plastic film as the dielectric. This film is made extremely thin using a sophisticated film drawing process. Once the film is manufactured, it may be metallized or left untreated, depending on the needed properties of the capacitor. Electrodes are then added and the assembly is mounted into a case which protects it from environmental factors. They are used in many applications because of their stability, low inductance and low cost.

Film capacitors also come in an assortment of shapes and case styles which include:

- Wrap & Fill (Oval & Round) – where the capacitor is wrapped in a tight plastic tape and have the ends filled with epoxy to seal them.
- Epoxy Case (Rectangular & Round) – where the capacitor is encased in a moulded plastic shell which is then filled with epoxy.
- Metal Hermetically Sealed (Rectangular & Round) – where the capacitor is encased in a metal tube or can and again sealed with epoxy.

with all the above case styles available in both Axial and Radial Leads.

Film Capacitors which use polystyrene, polycarbonate or Teflon as their dielectrics are sometimes called "Plastic capacitors". The construction of plastic film capacitors is similar to that for paper film capacitors but use a plastic film instead of paper. The main advantage of plastic film capacitors compared to impregnated-paper types is that they operate well under conditions of high temperature, have smaller tolerances, a very long service life and high reliability. Examples of film capacitors are the rectangular metalised film and cylindrical film & foil types as shown below:

Radial Lead Type

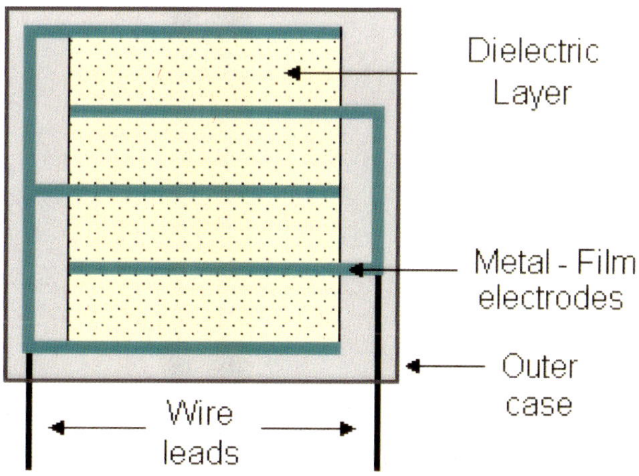

Axial Lead Type

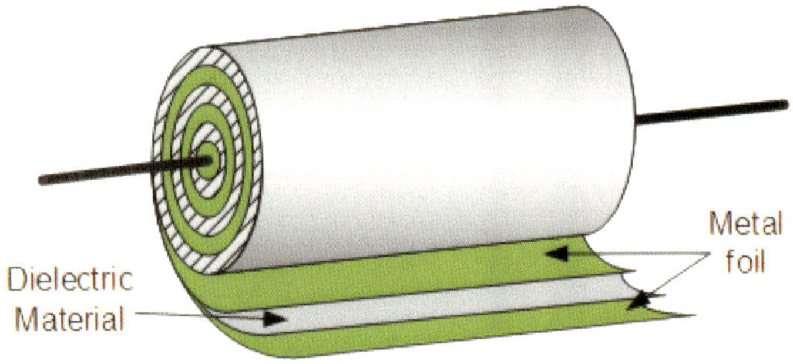

The film and foil types of capacitors are made from long thin strips of thin metal foil with the dielectric material sandwiched together which are wound into a tight roll and then sealed in paper or metal tubes.

Film Capacitor

These film types require a much thicker dielectric film to reduce the risk of tears or punctures in the film, and is therefore more suited to lower capacitance values and larger case sizes.

Metalised foil capacitors have the conductive film metalised sprayed directly onto each side of the dielectric which gives the capacitor self-healing properties and can therefore use much thinner dielectric films. This allows for higher capacitance values and smaller case sizes for a given capacitance. Film and foil capacitors are generally used for higher power and more precise applications.

Types of Plastic Film Capacitors

Film capacitors are classified into two types:

- Film-foil capacitors.
- Metallized film capacitors.

Film-foil Capacitors

The film-foil capacitor is made of two plastic films or sheets; each is layered with thin aluminum metal foil or sheet. The plastic sheets and aluminum sheets are then rolled in the form of a cylinder and wire leads are attached to the both ends of aluminum sheets. Polyester, polypropylene, polyethylene terephthalate, and polyphenylene sulfide are commonly used as dielectric in film capacitors. In plastic film capacitors, aluminum sheets acts as electrodes and plastic sheets acts as dielectric.

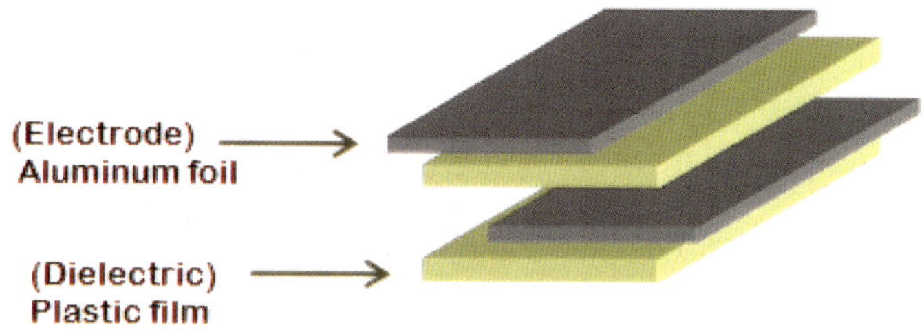

FILM/FOIL

Metallized Film Capacitors

In metallized film capacitors, the aluminum sheet or foil is replaced by a layer of metal vacuum deposited on the film layer. The most commonly used metal layer is aluminum or zinc that is extremely thin.

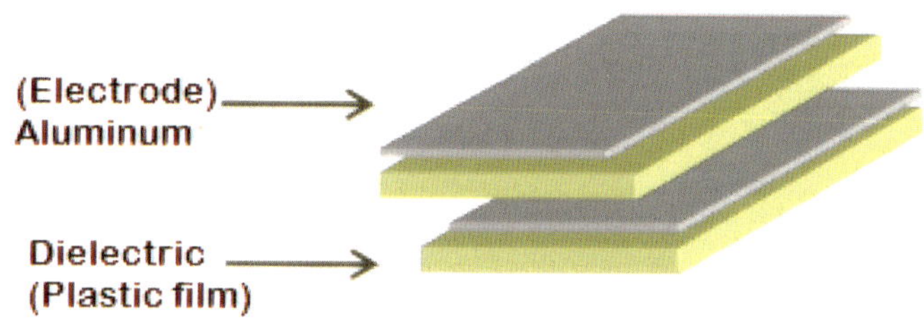

METALLIZED

The plastic film layers made of synthetic material act as dielectric and the aluminum layers act as electrodes. The major advantage of film dielectric capacitors over natural dielectric capacitors is that the plastic film is synthetic or artificial. Therefore, we can able to increase thickness and heat resistance of the dielectric. In other words, we can change the thickness and heat resistance of the plastic film capacitor.

Applications for Film Capacitors

Power film capacitors are used in power electronics devices, phase shifters, X-ray flashes and pulsed lasers, while the low power variants are used as decoupling capacitors, filters and in A/D convertors. Other notable applications are safety capacitors, electromagnetic interference suppression, fluorescent light ballasts and snubber capacitors.

Lighting ballasts are used for proper starting and operation of fluorescent lights. When a ballast is faulty, the light will flicker or fail to start properly. Older ballasts used only an inductor, a solution which provides a poor power factor. New designs use a switched power supply which relies on film capacitors for power factor correction.

Snubber capacitors are protective devices which damp or "snub" inductive kickback voltage spikes. These circuits often use film capacitors because of their low self-inductance, high peak current and low ESR, which are all critical factors in a snubber design. Polypropylene film capacitors are most often used in this type of circuit. Snubbers are used in many areas of electronics, especially power electronics in devices such as flyback DC-DC converters and others.

Film capacitors can also be used in a more conventional way as voltage smoothing capacitors, in filters, audio crossovers. They can be used to store energy and release it in a high-current pulse when needed. High-current electrical pulses are used to power pulsed lasers or generate lighting discharges.

Advantages of Film Capacitors

- High stability.
- Low cost.
- Low losses even at high frequencies.

Electrolytic Capacitor

An electrolytic capacitor is a type of capacitor that uses an electrolyte to achieve a larger capacitance than other capacitor types.

Electrolytic capacitors are mainly used when high charge storage in a small volume is required. In electrolytic capacitors, the liquid electrolyte acts as one of the electrodes (mostly act as cathode). To better understand the concept of electrolytic capacitor first we need to know the working of a general capacitor.

A capacitor is an electronic device that stores electric charge. It consists of two conductive plates separated by an insulating material called dielectric. Different types of insulating materials are used for constructing the dielectric based on the usage.

The conductive plates of the capacitor are good conductors of electricity. Therefore, they easily allow electric current through them. On the other hand, a dielectric medium or material is poor conductor of electricity. Therefore, it does not allow electric current through it.

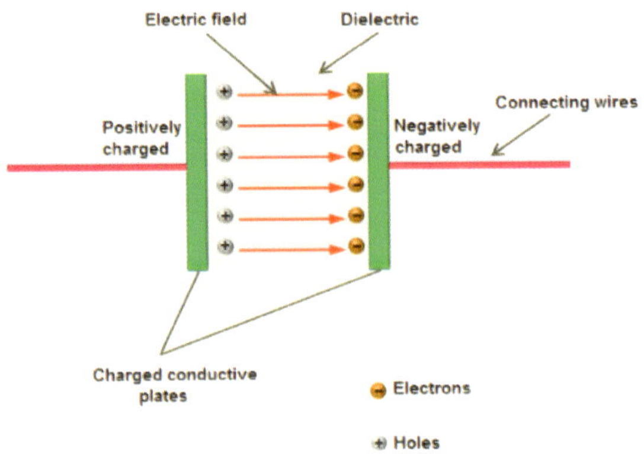

When voltage is applied to the capacitor in such a way that, the negative terminal of the battery is connected to the right side plate and the positive terminal of the battery is connected to the left side plate, the capacitor starts charging.

Because of this supply voltage, the electrons start flowing from negative terminal of the battery and reach the right side plate. After reaching right side plate, the electrons experience strong opposition from dielectric material because dielectric material is poor conductor of electricity.

As a result, a large number of electrons are trapped at the right side plate capacitor. However, these large numbers of electrons exerts force or electric field towards the left side plate. Therefore, electrons at the left side plate experience a repulsive force from excess electrons at right side plate. As a result, electrons move away from left side plate and attracted towards the positive terminal of the battery.

Therefore, the right side plate becomes more negatively charged (negative charge is build) because of the gaining of excess electrons. On the other hand, the left side plate becomes more positively charged (positive charge is build) because of the loss of electrons. As a result, voltage is established between the plates. This is how the normal capacitor works.

The electrolytic capacitor also charges mostly in the similar manner. However, the material used in constructing the electrolytic capacitor is different.

Types

Electrolytic capacitors are classified in to three types based on the material used to construct the dielectric:

- Aluminum electrolytic capacitors.
- Tantalum electrolytic capacitors.
- Niobium electrolytic capacitors.

In this, aluminum electrolytic capacitor is explained. The aluminum, tantalum, and niobium electrolytic capacitors work in similar manner. However, the material used for constructing electrodes is different.

Aluminum Electrolytic Capacitor

Aluminum electrolytic capacitor is made of two aluminum foils, aluminum oxide layer, an electrolytic paper or paper spacer soaked in electrolytic liquid or solutions and liquid or solid electrolyte. Electrolytic liquid contains atoms or molecules that have lost or gained electrons.

In aluminum electrolytic capacitor, the anode (+) and cathode (-) are constructed using pure aluminum foil. The anode aluminum foil is coated with a thin layer of insulating aluminum oxide (aluminum element with oxygen element). This insulating aluminum foil acts as dielectric of the electrolytic capacitor which blocks electric current flow. The cathode and oxide coated anode is separated by an electrolytic paper (which is soaked in an electrolytic liquid).

The cathode aluminum foil also covered with very thin insulating oxide layer or dielectric naturally formed by air. However, this oxide layer is very thin compared to the oxide layer formed on anode.

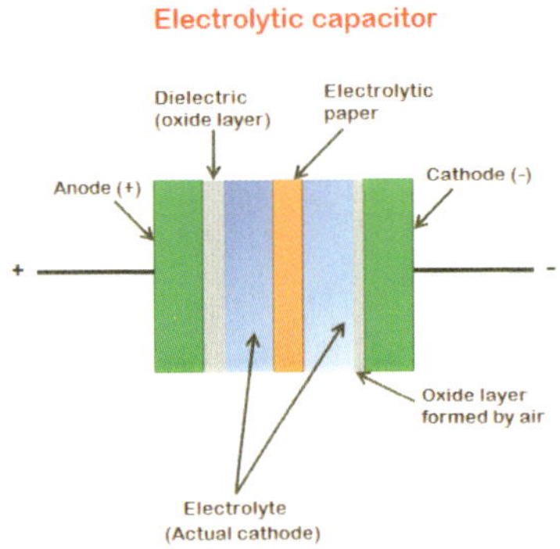

Therefore, the construction of aluminum electrolytic capacitor looks like two capacitors connected in series with anode capacitance CA and cathode capacitance CK.

The total capacitance of the capacitor is thus obtained from the formula of the series connection of two capacitors.

$$C_{ecap} = \frac{C_{A.}C_{K}}{C_{A} + C_{K}}$$

Where, CA = Capacitance of anode

CK = Capacitance of cathode

C_{ecap} = Total capacitance of electrolytic capacitor.

We know that the capacitance or charge storage of the capacitor is directly proportional to the surface area of the conductive plates or electrodes and inversely proportional to the thickness of dielectric. In other words, the capacitors with large electrodes stores large amount of charge whereas the capacitors with small electrodes stores small amount of charge. In the similar way, the capacitors with very thick dielectric stores small amount of charge whereas the capacitors with very thin dielectric stores very large amount of charge.

In normal capacitors, the dielectric is very thick which results in low capacitance per volume. In electrolytic capacitors, the electrolyte acts as the real cathode which has large surface area and the dielectric is very thin. Therefore, because of the large surface area of the electrode and thin dielectric, a large amount of charge storage is achieved in electrolytic capacitors.

The electrical conductivity of the electrolytic capacitor increases as the temperature increases and decreases as the temperature decreases. As a result, the capacitance or charge storage of the aluminum electrolytic capacitor also increases as the temperature increases and decreases as the temperatures decreases. Therefore, the capacitance of the aluminum electrolytic capacitor is largely affected by change in temperature.

Most of the electrolytic capacitors are polarized, that is the voltage applied to the terminals must be in correct polarity (positive connected to positive terminal and negative connected to negative terminal). If it is connected in reverse or wrong direction, the capacitor may be short circuited, that is a large electric current flows through the capacitor and that can permanently damage the capacitor.

In polarized capacitors, a minus sign (-) or plus sign (+) is clearly marked on any one of the two leads. This polarity must be followed.

Symbol of Electrolytic Capacitor

The symbol of electrolytic capacitor is shown in the below figure. The electrolytic capacitor is represented by two parallel straight lines or one straight line and one curved line.

European standard symbol

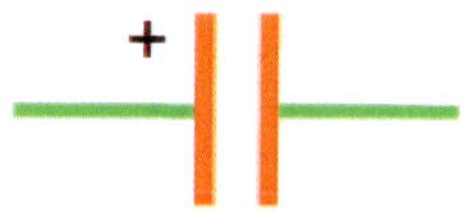

American standard symbol

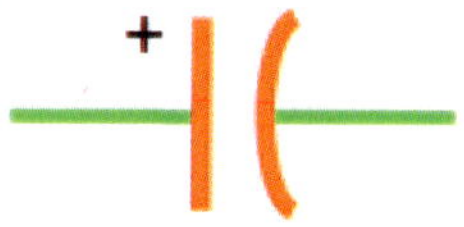

A plus or minus sign is written near any one of the lines to represent whether it is positive or negative terminal (anode or cathode). The voltage must be applied to the correct terminal. Otherwise, the capacitor may be damaged.

Applications

There are many applications which do not need tight tolerances and AC polarization, but require large capacitance values. They are commonly used as filtering devices in various power supplies to reduce the voltage ripple. When used in switching power supplies, they are often the critical component limiting the usable life of the power supply, so high quality capacitors are used in this application.

They may also be used in input and output smoothing as a low pass filter if the signal is a DC signal with a weak AC component. However, electrolytic capacitors do not work well with large amplitude and high frequency signals due to the power dissipated at the parasitic internal resistance called equivalent series resistance (ESR). In such applications, low-ESR capacitors must be used to reduce losses and avoid overheating.

A practical example is the use of electrolytic capacitors as filters in audio amplifiers whose main goal is to reduce mains hum. Mains hum is a 50Hz or 60Hz electrical noise induced from the mains supply which would be audible if amplified.

Supercapacitor

Supercapacitors are electronic devices which are used to store extremely large amounts of electrical charge. They are also known as double-layer capacitors or ultracapacitors. Instead of using a conventional dielectric, supercapacitors use two mechanisms to store electrical energy: double-layer capacitance and pseudocapacitance. Double layer capacitance is electrostatic in origin, while pseudocapacitance is electrochemical, which means that supercapacitors combine the workings of normal capacitors with the workings of an ordinary battery. Capacitances achieved using this technology can be as high as 12000 F. In comparison, the self-capacitance of the entire planet Earth is only about 710 μF, more than 15 million times less than the capacitance of a supercapacitor. While an ordinary electrostatic capacitor may have a high maximum operating voltage, the typical maximum charge voltage of a supercapacitor lies between 2.5 and 2.7 volts. Supercapacitors are polar devices, meaning they have to be connected to the circuit the right way, just like electrolyte capacitors. The electrical properties of these devices, especially their fast charge and discharge times, are very interesting for some applications, where supercapacitors may completely replace batteries.

Characteristics

Charge Time

Supercapacitors have charge and discharge times comparable to those of ordinary capacitors. It is possible to achieve high charge and discharge currents due to their low internal resistance. Batteries usually take up to several hours to reach a fully charged state – a good example is a cell phone battery, while supercapacitors can be brought to the same charge state in less than two minutes.

Specific Power

The specific power of a battery or supercapacitor is a measure used to compare different technologies in terms of maximum power output divided by total mass of the device. Supercapacitors have a specific power 5 to 10 times greater than that of batteries. For example, while Li-ion batteries have a specific power of 1 – 3 kW/kg, the specific power of a typical supercapacitor is around 10 kW/kg. This property is especially important in applications that require quick bursts of energy to be released from the storage device.

Cycle Life and Safety

Supercapacitor batteries are safer than ordinary batteries when mistreated. While batteries are known to explode due to excessive heating when short circuited, supercapacitors do not heat as much due to their low internal resistance. Shorting a fully charged supercapacitor will cause a quick release of the stored energy which can cause electrical arcing, and might cause damage to the device, but unlike batteries, the generated heat is not a concern.

Supercapacitors can be charged and discharged millions of times and have a virtually unlimited cycle life, while batteries only have a cycle life of 500 times and higher. This makes supercapacitors very useful in applications where frequent storage and release of energy is required.

Construction and Properties of Supercapacitors

The construction of supercapacitor is similar to the construction of electrolytic capacitors in that they consist of two foil electrodes, an electrolyte and a foil separator. The separator is sandwiched between the electrodes and the foil is rolled or folded into a shape, usually cylindrical or rectangular. This folded form is placed into a housing, impregnated with electrolyte and hermetically sealed. The electrolyte used in the construction of supercapacitors as well as the electrodes, are different from those used in ordinary electrolytic capacitors.

In order to store electrical charge, a supercapacitor uses porous materials as separators in order to store ions in those pores at an atomic level. The most commonly used material in modern supercapacitors is activated charcoal. The fact that carbon is not a good insulator results in a maximum operating voltage limited to under 3 V. Activated charcoal is not the perfect material for another reason: the charge carriers are comparable in size to the pores in the material and some of them cannot fit into the smaller pores, resulting in a reduced storage capacity.

One of the most exciting materials used in supercapacitor research is graphene. Graphene is a substance consisted of pure carbon, arranged in a planar sheet only one atom thick. It is extremely porous, acting as an ion "sponge". Energy densities achievable using graphene in supercapacitors are comparable to energy densities found in batteries. However, even though prototypes of graphene supercapacitors have been made as a proof of concept, graphene is difficult and expensive to produce in industrial quantities, which postpones the use of this technology. Even so, graphene supercapacitors are the most promising candidate for future supercapacitor technology advances.

Applications

Since supercapacitors bridge the gap between batteries and capacitors, they may be used in a wide variety of applications. One interesting application is the storage of energy in KERS, or dynamic braking systems (Kinetic Energy Recovery System) in automotive industry. The main problem in such systems is building an energy storage device capable of rapidly storing large amounts of energy. One approach is to use an electrical generator which will convert kinetic energy to electrical energy and store it in a supercapacitor. This energy can later be reused to provide power for acceleration.

Another example are low-power applications where a high capacity is not imperative, but a high life cycle or quick recharging is important. Such applications are photographic flash, MP3 players, static memories (SRAM) which need a low power constant voltage source to retain information and so on.

Possible future supercapacitor applications are in cell phones, laptops, electric cars and all other devices that currently run on batteries. The most exciting advantage from a practical perspective is their very fast recharge rate, which would mean that plugging an electric car into a charger for a few minutes would be enough to fully charge the battery.

Disadvantages

Supercapacitors come with some disadvantages as well. One disadvantage is a relatively low specific energy. The specific energy is a measure of total amount of energy stored in the device divided by its weight. While Li-ion batteries commonly used in cell phones have a specific energy of 100-200 Wh/kg, supercapacitors may only store typically 5 Wh/kg. This means that a supercapacitor that has the same capacity (not capacitance) as a regular battery would weigh up to 40 times as much. The specific energy is not to be confused with the specific power, which is a measure of maximum output power of a device per weight.

Another disadvantage is a linear discharge voltage. For example, a battery rated at 2.7V, when at 50% charge would still output a voltage close to 2.7V, while a supercapacitor rated at 2.7V at 50%

charge would output exactly half of its maximum charge voltage – 1.35V. This means that the output voltage would fall below the minimal operating voltage of the device running on a supercapacitor, for example a cellphone, and the device would have to shut down before using all the charge in the capacitor. A solution to this problem is using DC-DC converters. This approach introduces new difficulties, such as efficiency and power noise.

Cost is the third major disadvantage of currently available supercapacitors. The cost per Wh of a supercapacitor is more than 20 times higher than that of Li-ion batteries. However, cost can be reduced through new technologies and mass production of supercapacitor batteries.

Low specific energy, linear discharge voltage and high cost are the main reasons preventing supercapacitors from replacing batteries in most applications.

Variable Capacitor

variable capacitors whose value alters when you vary, either electrically or mechanically. Variable capacitors in general consists of interwoven sets of metallic plates in which one is fixed and the other is variable. These capacitors provide the capacitance values so as to vary between 10 to 500pF.

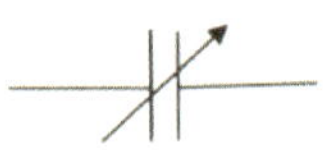

Symbol of a variable capacitor

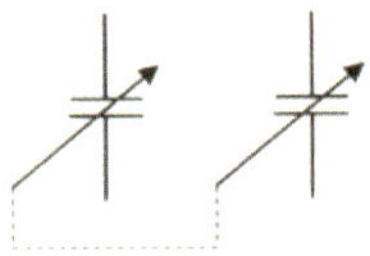

Symbol of a variable ganged capacitor

The ganged capacitor shown here is a combination of two capacitors connected together. A single shaft is used to rotate the variable ends of these capacitors which are combined as one. The dotted line indicates that they are connected internally.

There are many uses of these variable resistors such as for tuning in LC circuits of radio receivers, for impedance matching in antennas etc. The main types of variable capacitors are Tuning capacitors and Trimmer capacitors.

Tuning Capacitors

Tuning capacitors are popular type of variable capacitors. They contain a stator, a rotor, a frame to support the stator and a mica capacitor. The constructional details of a tuning capacitor are shown in the following figure.

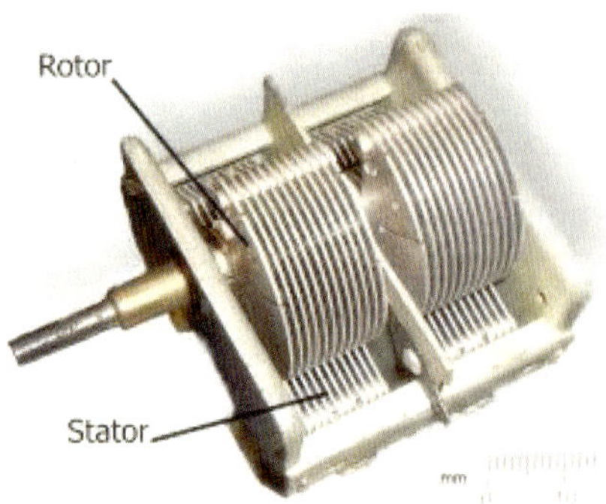

The stator is a stationary part and rotor rotates by the movement of a movable shaft. The rotor plates when moved into the slots of stator, they come close to form plates of a capacitor. When the rotor plates sit completely in the slots of the stator then the capacitance value is maximum and when they don't, the capacitance value is minimum.

The above figure shows a ganged tuning capacitor having two tuning capacitors connected in a gang. This is how a tuning capacitor works. These capacitors generally have capacitance values from few Pico Farads to few tens of Pico Farads. These are mostly used in LC circuits in radio receivers. These are also called as Tuning Condensers.

Trimmer Capacitors

Trimmer capacitors are varied using a screwdriver. Trimmer capacitors are usually fixed in such a place where there is no need to change the value of capacitance, once fixed.

There are three leads of a trimmer capacitor, one connected to stationary plate, one to rotary and the other one is common. The movable disc is a semi-circular shaped one. A trimmer capacitor would look like the ones in the following figure.

Image of a Trimmer Capacitor

There are two parallel conducting plates present with a dielectric in the middle. Depending upon this dielectric used, there are air trimmer capacitors and ceramic trimmer capacitors. The constructional details of a trimmer capacitor are as shown below.

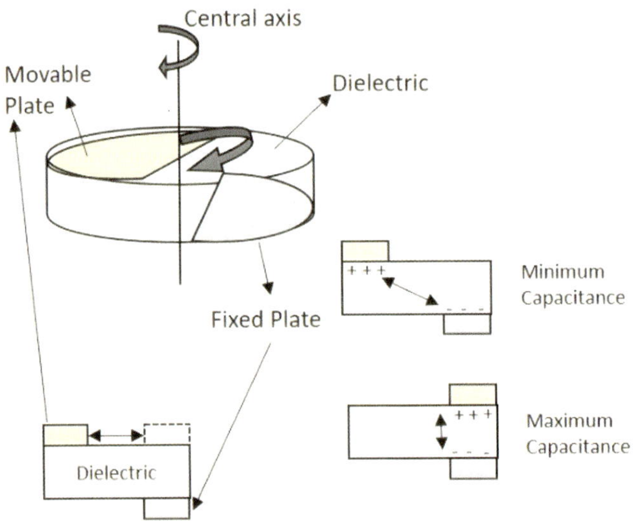

One of the two plates is movable, while the other is fixed. The dielectric material is fixed. When the movable plate is moved, opposite to the area between movable and fixed electrode, then the capacitance can be changed. The capacitance will be higher if the opposite area gets bigger, as both the electrodes act as two plates of a capacitor.

The Trimmer Capacitors are easily fixed on a PCB (Printed Circuit Board) and they are mostly used for calibration of equipment.

References

- Capacitors-capacitance-basics-tutorial, capacitance: radio-electronics.com, Retrieved 21 June 2018
- Self-and-mutual-capacitance, capacitive-touch-cap-touch-9777: renesasrulz.com, Retrieved 19 July 2018
- Mutual-capacitance-vs-self-capacitance-touchscreen-technology: en-touch.com, Retrieved 10 May 2018
- Decreasing-parasitic-capacitance-in-IC-layouts-4426479: edn.com, Retrieved 17 April 2018
- Ceramic-capacitor-working-construction-applications: elprocus.com, Retrieved 21 March 2018
- Basic-electronics-variable-capacitors: tutorialspoint.com, Retrieved 26 July 2018

5

Magnetism and Electromagnetism

Magnetism refers to the class of physical phenomena that occurs as a result of magnetic fields. Electromagnetism is a branch of physics, which is concerned with the study of the electromagnetic force that is produced between charged particles. In order to completely understand the phenomena of magnetism and electromagnetism, it is necessary to understand the concepts of magnetic moment, Gauss's law for magnetism, Oersted's law, electromagnetic induction, Faraday's law of induction and Maxwell's equations, which have been carefully analyzed in this chapter.

Magnetism

Magnetism plays an important role in Electrical and Electronic Engineering because without it components such as relays, solenoids, inductors, chokes, coils, loudspeakers, motors, generators, transformers, and electricity meters etc, would not work if magnetism did not exist.

Then every coil of wire uses the effect of electromagnetism when an electrical current flows through it. But before we can look at Magnetism and especially Electromagnetismin more detail we need to remember back to our physics classes of how magnets and magnetism works.

Nature of Magnetism

Magnets can be found in a natural state in the form of a magnetic ore, with the two main types being Magnetite also called "iron oxide", (Fe_3O_4) and Lodestone, also called "leading stone". If these two natural magnets are suspended from a piece of string, they will take up a position in-line with the Earth's magnetic field always pointing north.

A good example of this effect is the needle of a compass. For most practical applications these natural occurring magnets can be disregarded as their magnetism is very low and because nowadays, man-made artificial magnets can be produced in many different shapes, sizes and magnetic strengths.

There are basically two forms of magnetism, "Permanent Magnets" and "Temporary Magnets", with the type being used dependant upon its application. There are many different types of materials available to make magnets such as iron, nickel, nickel alloys, chromium and cobalt and in their natural state some of these elements such as nickel and cobalt show very poor magnetic quantities on their own.

However, when mixed or "alloyed" together with other materials such as iron or aluminium peroxide they become very strong magnets producing unusual names such as "alcomax", "hycomax", "alni" and "alnico".

Magnetic material in the non-magnetic state has its molecular structure in the form of loose magnetic chains or individual tiny magnets loosely arranged in a random pattern. The overall effect of this type of arrangement results in zero or very weak magnetism as this haphazard arrangement of each molecular magnet tends to neutralise its neighbour.

When the material is Magnetised this random arrangement of the molecules changes and the tiny unaligned and random molecular magnets become "lined-up" in such a way that they produce a series magnetic arrangement. This idea of the molecular alignment of ferromagnetic materials is known as Weber's Theory and is illustrated below.

Magnetic Molecule Alignment of a Piece of Iron and a Magnet

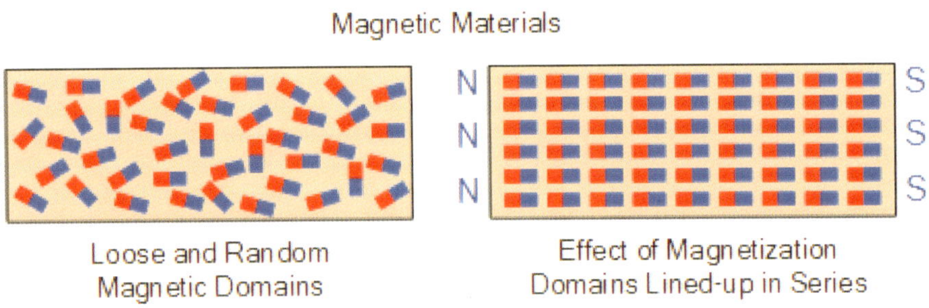

Weber's theory is based on the fact that all atoms have magnetic properties due to the spinning action of the atoms electrons. Groups of atoms join together so that their magnetic fields are all rotating in the same direction. Magnetic materials are composed of groups of tiny magnets at a molecular level around the atoms, and a magnetised material will have most of its tiny magnets lined up in one direction only to produce a north pole in one direction and a south pole in the other direction.

Likewise, a material that has its tiny molecular magnets pointing in all directions will have its molecular magnets neutralised by its neighbouring magnet, thereby neutralising any magnetic effect. These areas of molecular magnets are called "domains".

Any magnetic material will produce a magnetic field itself which depends on the degree of alignment of magnetic domains in the material set up by orbital and spinning electrons. This degree of alignment can be specified by a quantity known as magnetisation, M.

In an unmagnetised material, $M = 0$, but some of the domains remain aligned over small regions in the material once the magnetic field is removed. The effect of applying a magnetising force to the material is to align some of the domains to produce a non-zero magnetisation value.

Once the magnetising force has been removed, the magnetism within the material will either remain or decay away quiet quickly depending on the magnetic material being used. This ability of a material to retain its magnetism is called Retentivity.

Materials which are required to retain their magnetism will have a fairly high retentivity and as such are used to make permanent magnets, while those materials required to lose their magnetism quickly such as soft iron cores for relays and solenoids will have a very low retentivity.

Magnetic fields are always shown visually as lines of force that give a definite pole at each end of the material where the flux lines are more dense and concentrated. The lines which go to make up a magnetic field showing the direction and intensity are called Lines of Force or more commonly "Magnetic Flux" and are given the Greek symbol, Phi (Φ) as shown below.

Lines of Force from a Bar Magnets Magnetic Field

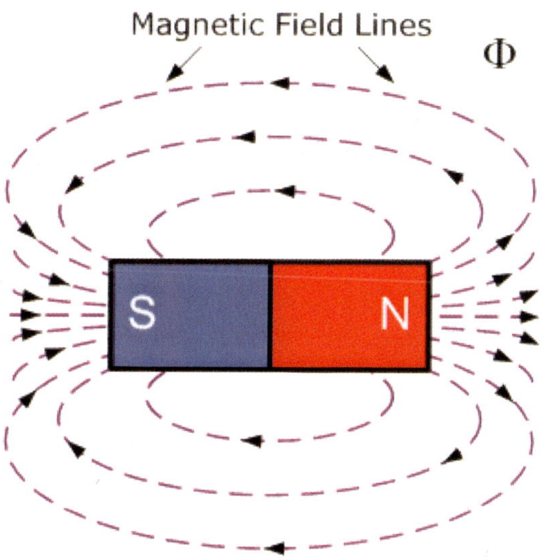

As shown above, the magnetic field is strongest near to the poles of the magnet were the lines of flux are more closely spaced. The general direction for the magnetic flux flow is from the North (N) to the South (S) pole. In addition, these magnetic lines form closed loops that leave at the north pole of the magnet and enter at the south pole. Magnetic poles are always in pairs.

However, magnetic flux does not actually flow from the north to the south pole or flow anywhere for that matter as magnetic flux is a static region around a magnet in which the magnetic force exists. In other words magnetic flux does not flow or move it is just there and is not influenced by gravity. Some important facts emerge when plotting lines of force:

- Lines of force NEVER cross.
- Lines of force are CONTINUOUS.
- Lines of force always form individual CLOSED LOOPS around the magnet.
- Lines of force have a definite DIRECTION from North to South.
- Lines of force that are close together indicate a STRONG magnetic field.
- Lines of force that are farther apart indicate a WEAK magnetic field.

Magnetic forces attract and repel like electric forces and when two lines of force are brought close together the interaction between the two magnetic fields causes one of two things to occur:

1. When adjacent poles are the same, (north-north or south-south) they REPEL each other.
2. When adjacent poles are not the same, (north-south or south-north) they ATTRACT each other.

This effect is easily remembered by the famous expression that "opposites attract" and this interaction of magnetic fields can be easily demonstrated using iron fillings to show the lines of force around a magnet. The effect upon the magnetic fields of the various combinations of poles as like poles repel and unlike poles attract can be seen below.

Magnetic Field of Like and Unlike Poles

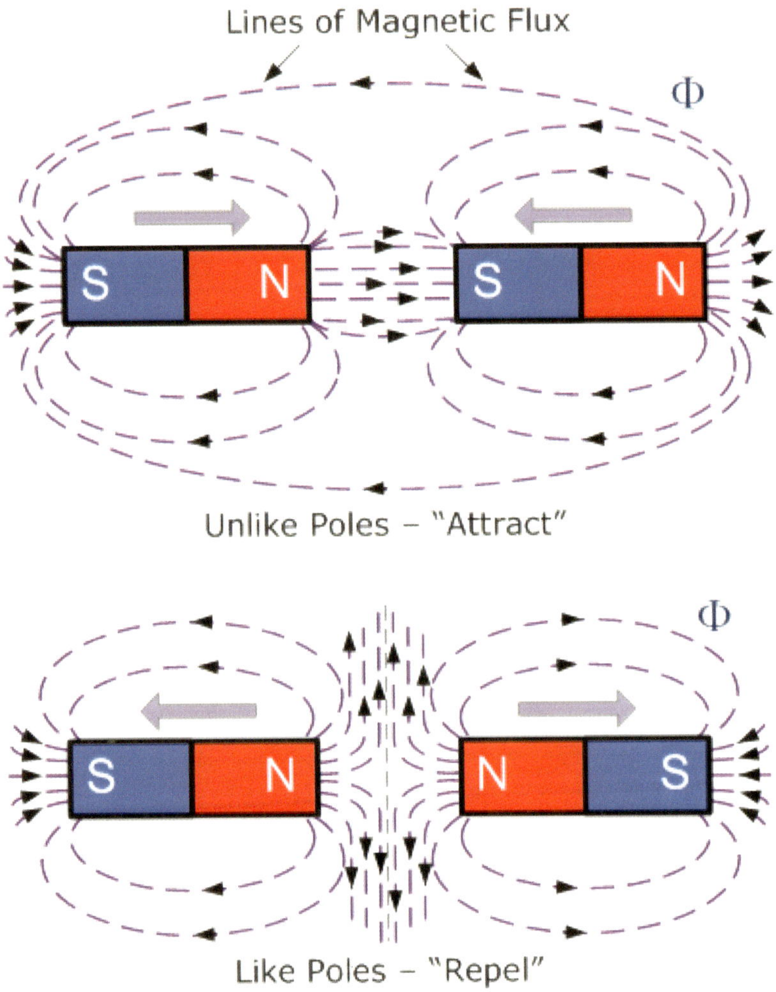

When plotting magnetic field lines with a compass it will be seen that the lines of force are produced in such a way as to give a definite pole at each end of the magnet where the lines of force leave the North pole and re-enter at the South pole. Magnetism can be destroyed by heating or hammering the magnetic material, but cannot be destroyed or isolated by simply breaking the magnet into two pieces.

So if you take a normal bar magnet and break it into two pieces, you do not have two halves of a magnet but instead each broken piece will somehow have its own North pole and a South pole. If you take one of those pieces and break it into two again, each of the smaller pieces will have a North pole and a South pole and so on. No matter how small the pieces of the magnet become, each piece will still have a North pole and a South pole, crazy.

Then in order for us to make use of magnetism in electrical or electronic calculations, it is necessary to define what are the various aspects of magnetism.

The Magnitude of Magnetism

We now know that the lines of force or more commonly the magnetic flux around a magnetic material is given the Greek symbol, Phi, (Φ) with the unit of flux being the Weber, (Wb) after Wilhelm Eduard Weber. But the number of lines of force within a given unit area is called the "Flux Density" and since flux (Φ) is measured in (Wb) and area (A) in metres squared, (m^2), flux density is therefore measured in Webers/Metre^2 or (Wb/m^2) and is given the symbol B.

However, when referring to flux density in magnetism, flux density is given the unit of the Tesla after Nikola Tesla so therefore one Wb/m^2 is equal to one Tesla, $1Wb/m^2 = 1T$. Flux density is proportional to the lines of force and inversely proportional to area so we can define Flux Density as:

Magnetic Flux Density

$$\text{Magnetic Flux Density, (tesla)} = \frac{\text{Magnetic Flux, (weber)}}{\text{Area, } (m^2)}$$

The symbol for magnetic flux density is B and the unit of magnetic flux density is the Tesla, T.

$$B = \frac{\Phi}{A} \quad \text{in Teslas}$$

It is important to remember that all calculations for flux density are done in the same units, e.g., flux in webers, area in m^2 and flux density in Teslas.

Example

The amount of flux present in a round magnetic bar was measured at 0.013 webers. If the material has a diameter of 12cm, calculate the flux density.

The cross sectional area of the magnetic material in m_2 is given as:

$$\text{Diameter} = 12\text{cm}$$

$$\therefore \text{Area} = \pi r^2$$

$$A = 3.142 \times 0.06^2 = 0.0113m^2$$

The magnetic flux is given as 0.013 webers, therefore the flux density can be calculated as:

$$B = \frac{\Phi}{A} = \frac{0.013}{0.0113} = 1.15T$$

When dealing with magnetism in electrical circuits it must be remembered that one Tesla is the density of a magnetic field such that a conductor carrying 1 ampere at right angles to the magnetic field experiences a force of one newton-metre length on it.

Magnetic Field

Magnetic fields are produced by electric currents. Magnetic fields may be represented by continuous lines of force or magnetic flux that emerge from north-seeking magnetic poles and enter south-seeking magnetic poles. The density of the lines indicates the magnitude of the magnetic field. At the poles of a magnet, for example, where the magnetic field is strong, the field lines are crowded together, or more dense. Farther away, where the magnetic field is weak, they fan out, becoming less dense. A uniform magnetic field is represented by equally spaced parallel straight lines. The direction of the flux is the direction in which the north-seeking pole of a small magnet points. The lines of flux are continuous, forming closed loops. For a bar magnet, they emerge from the north-seeking pole, fan out and around, enter the magnet at the south-seeking pole, and continue through the magnet to the north pole, where they again emerge.

The magnetic field B is defined in terms of force on moving charge in the Lorentz force law. Magnetic field sources are essentially dipolar in nature, having a north and south magnetic pole. The SI unit for magnetic field is the Tesla, which can be seen from the magnetic part of the Lorentz force law.

Fmagnetic = qvB to be composed of (Newton × second)/(Coulomb × meter). A smaller magnetic field unit is the Gauss (1 Tesla = 10,000 Gauss).

$$B = v \times \frac{1}{c^2} E$$

Where,

V-is velocity of the electric charge, measured in metres per second

X indicates a vector cross product

c is the speed of light measured in metres per second

E is the electric field measured in newtons per coulomb or volts per metre.

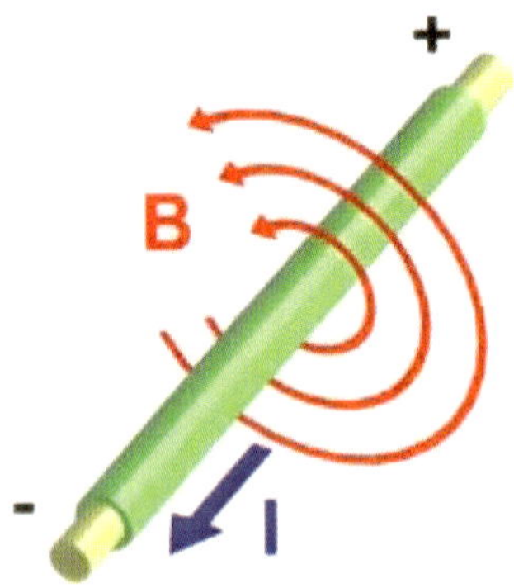

Magnetic Flux

Magnetic flux is a measurement of the total magnetic field which passes through a given area. It is a useful tool for helping describe the effects of the magnetic force on something occupying a given area. The measurement of magnetic flux is tied to the particular area chosen. We can choose to make the area any size we want and orient it in any way relative to the magnetic field.

If we use the field-line picture of a magnetic field then every field line passing through the given area contributes some magnetic flux. The angle at which the field line intersects the area is also important. A field line passing through at a glancing angle will only contribute a small component of the field to the magnetic flux. When calculating the magnetic flux we include only the component of the magnetic field vector which is normal to our test area.

If we choose a simple flat surface with area A as our test area and there is an angle θ between the normal to the surface and a magnetic field vector (magnitude B) then the magnetic flux is:

More generally, the magnetic flux can be found using the vector dot product. If $\vec{B}$, with, vector, on top is a magnetic field vector and $\vec{A}$, with, vector, on top is the surface-normal vector to the test area then $\Phi = \vec{B} \cdot \vec{A}$.

$$\Phi = BAcos\theta$$

In the case that the surface is perpendicular to the field then the angle is zero and the magnetic flux is simply BA. Figure shows an example of a flat test area at two different angles to a magnetic field and the resulting magnetic flux.

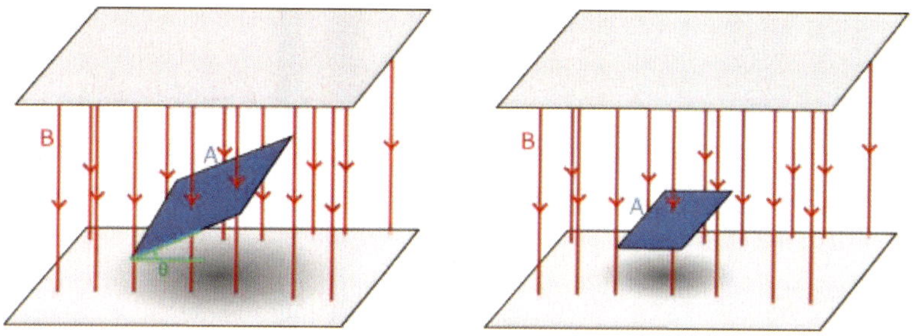

Figure: Magnetic flux through given areas (blue) oriented at an angle (left) and normal to (right) the magnetic field.

Exercise

If the blue surfaces shown in figure both have equal area and the angle θ is 25°, degree, how much smaller is the flux through the area in figure-left vs figure-right?

Solution

Looking at the equation for magnetic flux through a surface, we can see that if the area and field remain the same then the only remaining factor is the angle. In this case the diagram shows the angle between the normal to the magnetic field and the blue surface. This is the same as the angle between the magnetic field and a vector normal to the blue surface. So as,

$$cos25^{\circ} \simeq 0.91$$

the magnetic flux through the tilted area is about 9% smaller than through the area normal to the field.

If the magnetic field is non-uniform and at different parts of the surface, the magnetic field is different in magnitude and direction, then the total magnetic flux through the given surface can be given as the summation of the product of all such area elements and their corresponding magnetic field. Mathematically,

$$\phi_B = B_1.dA_1 + B_2.dA_2 + B_3.dA_3 + \ldots = \Sigma_{all} B_i.dA_i$$

Magnetic Flux Unit

SI unit of magnetic flux is Weber, as explained below,

$$\varphi = BAcos\theta$$

$$\varphi = BA \quad \therefore\ surface\,area\ perpendicular\,to\,B,\ so\,Cos\theta = 1$$

$$\varphi = \frac{Newton}{Ampere - meter} * meter^2$$

$$\varphi = \frac{N}{A - m} * m^2$$

$$\varphi = \frac{Nm}{A}$$

$$\varphi = W\,eber$$

Hence magnetic flux unit is NmA^{-1} or weber. Whereas,

$$1\,Weber = 10^8\,lines\,of\ \,force$$

Example

A certain magnet has a flux φ = 1.5 Weber. How many lines of force does flux represent?

Solution

By definition,

$$1\,Weber = 10^8\,lines\,of\ \,force$$

So,

$$1.5\,Weber = 1.5 * 10^8\,lines\,.$$

Magnetic Flux Density

Magnetic flux density (B) is defined as the force acting per unit current per unit length on a wire placed at right angles to the magnetic field.

- Units of B is Tesla (T) or $Kgs^{-2}A^{-1}$

- B is a vector quantity,

$$B = \frac{F}{Il}$$

Where,

l=length of wire

F=total force acting on the wire

I=current flowing through the wire.

Magnetic Moment

Magnetic moment is a determination of its tendency to get arranged through a magnetic field. As we all know, a magnet has two poles, i.e., North and South. The distance between these two poles of a magnetic or a magnetic dipole is named as the magnet length and is given as the 2 ıı. If m is the power of any magnetic pole then the magnets magnetic dipole moment is provided by the vector M and it is articulated as,

$$M = \frac{m}{2l}$$

Where, m = Strength of any magnetic dipole

ıı = Magnet length.

Magnetic Moment Units

In the definition for current loop, the Magnetic moment is produce of current flowing and the area,

M = I A

So the unit conferring to this definition is articulated by Amp-m^2.

It can also be suggested in terms of torque and moment. Conferring to that, the torque is measured in Joules (J) and the magnetic field is measured in tesla (T) and thus the unit is J T^{-1}.

So, these two units are equivalent of each other and is provided by : 1 Amp-m^2 = 1 J T^{-1}.

Magnetic Dipole Moment

A Magnetic Dipole comprises of two unlike poles of equivalent strength and parted by a small distance.

For instance : The needle of a compass, a bar magnet, etc. are the magnetic dipoles. We shall show that a current loop works as a magnetic dipole.

Magnetic Dipole Moment is described as the product of pole strength and the distance amidst the two poles. The distance between the two poles of a magnetic or a magnetic dipole is named as the magnet length and is given as the 2 ι.

If m is the power of any magnetic pole then the magnetic dipole moment of the magnet is signified by the vector M and it is represented as,

$$M = \frac{m}{2l}$$

The Magnetic dipole moment is a vector as termed above and it has direction from the South Pole of the magnet to the north pole of the magnet as presented in the figure below.

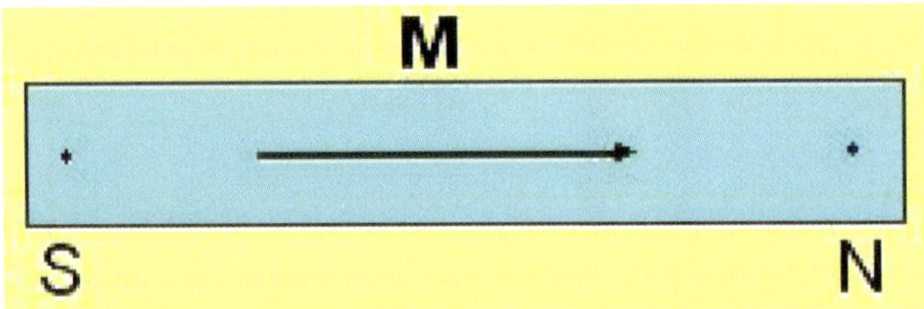

Expression for Magnetic Dipole Force

The force on a magnetic dipole is because of both the poles of the magnet and we consider the magnetic dipole of a bar magnet and assume that the magnet is kept in a unbroken magnetic field B. In that situation, the force on the separate poles is articulated as,

m_B which is along the magnetic field B = Force on the N-pole

m_B and this is opposite to magnetic field B = Force on the S-pole.

These forces are equivalent in magnitude, but opposite in direction and they form a parallel couple which rotates the magnet clockwise and creates a net torque on the magnet because of the individual force in a couple thus we have torque acting on the bar magnet.

τ = Moment of the couple.

$\tau = m_B \times 2L \sin\theta$

Where θ is the angle amid the magnet and the magnetic field therefore we have,

M = m x 2L

Thus, Magnetic dipole moment is articulated by,

$\tau\tau = MB \sin\theta$

In vector form, it can be rephrased as,

$\tau = M \times B$.

This is the required expression for the magnetic dipole force.

Orbital Moment

According to Ampere's law, when a current flows through a coil, it gives rise to a magnetic field in a direction perpendicular to the plane of the loop. The field is related to the current by,

$$H = NI$$

where N is the number of loops per unit length of the coil, I is the current in Amperes and H is the induced magnetic field in Ampere per m i.e. $A.m^{-1}$.

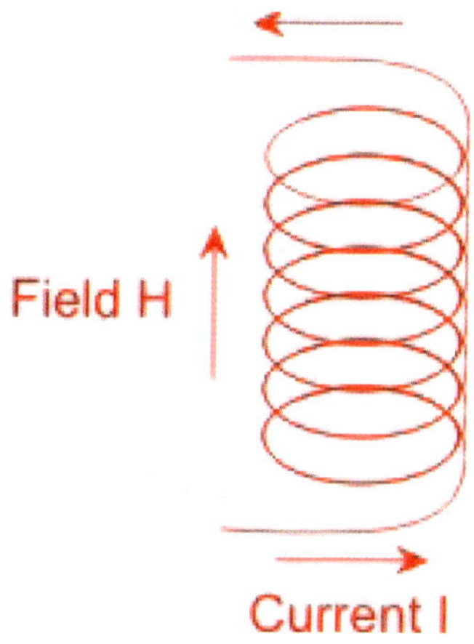

A current carrying coil

Now based on classical physics, the current, *I*, is nothing but charge per unit time and is expressed as,

$$I = \frac{q}{t}$$

The time is nothing but the circumference divided by the velocity of electron, v, *i.e.* $t = 2\pi r/v$ which can be substituted in equation to yield,

$$I = q\frac{v}{2\pi r}$$

where r is the radius of the orbital.

Therefore, the orbital magnetic moment, μ_{orbit}, which is a product of current, *I*, and area, $A = \pi r^2$ can be expressed as,

$$\mu_{orbit} = A.I = \pi r^2 . q . \frac{v}{2\pi r} = \frac{qvr}{2}$$

or

$$\mu_{orbit} = -\frac{e}{2m} . mvr = -\frac{e}{2m} . J_{orbit}$$

where *q* is the charge of electron which equals to $-e$; *v* is velocity of electron; *m* is the mass of electron and *J* is the angular momentum which is equal to mvr.

The equation above shows that orbital moment can be expressed in the multiple of angular momentum.

Spin Moment

As we saw earlier, spinning of electron about its own axis also gives rise to a moment. For purely quantum mechanical reasons, for an electron spinning around its own axis, the ratio of spin magnetic moment to angular moment is twice as large as for orbital moment *i.e.*

$$\mu_{spin} = -\frac{e}{m} . J_{spin}$$

Net Magnetic Moment

So the net magnetic moment for an atom containing several atoms is basically net sum of these two contributions.

$$\mu_{net} = \mu_{orbit}^{net} + \mu_{spine}^{net}$$

Although it is not possibly to explain classically, from quantum mechanics, the ratio of magnetic moment to the angular momentum always works out to be between 1 and 2, related by a factor called a Landé g-factor (g) purely because net moment is the sum of these two and can be expressed as,

$$\mu_{net} = -g.\left(\frac{e}{2m}\right).J$$

Value of g would be 1 for purely orbital moment and 2 for purely spin moment. The latter is true for many ferrites where orbital magnetic moment is thought as quenched and the dominating contribution to magnetic moment arises from spin magnetization.

Again, quantum mechanics argues that the angular momentum can take only certain values defined by orbital and spin quantum number, I and s respectively. Hence the angular moment for an electron with a characteristic number, *j* is expressed as,

$$J = \sqrt{j(j+1)}.\hbar$$

Where $\hbar$ is equal to h/2π;; h being Planck's constant (=6.6×10^{-34} $J.s^{1}$). The characteristics number *j* can be orbital quantum number, *l*, for orbital moment and spin quantum number, *s*, for spin moment. Since magnetic moment is now quantized as a result of angular momentum being quantized, for an electron, we can define a basic quantum unit of magnetic moment as Bohr magneton, μ_B, as

$$\mu_B = \frac{e\hbar}{2m}$$

The value of one Bohr magneton is $9.274\times10^{-24}A.m^2$.

Nuclear magnetic moment is only of the order of $10^{-3}\,\mu_B$ and hence is neglected for most cases.

Multipole Expansion

Actually, considering only magnetic dipoles is necessarily an approximation, as a current loop produces a "pure" dipole field only in the limit of small loop area. In reality, for a macroscopic current distribution, there will be additional terms in the expression for the magnetic field.

The multipole expansion of the vector potential in terms of Legendre polynomials yield,

$$A(r) = \frac{\mu_0 I}{4\pi} \sum_{n=0}^{\infty} \frac{1}{r^{n+1}} \int r'^n P_n(\cos\theta') dI',$$

where P_n denotes the nth Legendre polynomial. The first few terms of the sum are,

$$A_{monopole}(r) = \frac{\mu_0 I}{4\pi} \frac{1}{r} dI'$$
$$= 0$$

$$A_{dipole}(r) = \frac{\mu_0 I}{4\pi} \frac{1}{r^2} \int r' \cos\theta' dI'$$

$$A_{quadrupole}(r) = \frac{\mu_0 I}{4\pi} \frac{1}{r^3} \int \frac{r'^2}{2} (3\cos^3\theta' - 1) dI'.$$

Appropriately, the so-called monopole term vanishes, and Gauss's law for magnetism is satisfied.

Clearly, the magnetic dipole term reappears in the multipole expansion, but additional terms (quadrupole, octupole, etc.) appear. To show that the dipole term is consistent with the previous discussion, one can rewrite the cosine term as a dot product:

$$A_{dipole}(r) = \frac{\mu_0 I}{4\pi} \frac{1}{r^3} \int (r'.\hat{r}) dI'.$$

One can show that the integral evaluates to $-\hat{r} \times a$, in which case the dipole term becomes

$$A_{dipole}(r) = \frac{\mu_0 I}{4\pi r^2} \frac{m \times \hat{r}}{r^2},$$

where $m = Ia$ is simply the magnetic dipole moment defined previously. The remaining terms all represent higher-order contributions to the field. In many practical situations, if non-vanishing, the dipole term tends to dominate over all others, and the higher moments may be ignored.

Gauss's Law for Magnetism

Gauss's law for magnetism is a physical application of Gauss's theorem (also known as the divergence theorem) in calculus, which was independently discovered by Lagrange in 1762, Gauss in 1813, Ostrogradsky in 1826, and Green in 1828.

Gauss's law for magnetism states that no magnetic monopoles exists and that the total flux through a closed surface must be zero. This topic describes the time-domain integral and differential forms of Gauss's law for magnetism and how the law can be derived. The frequency-domain equation is also given.

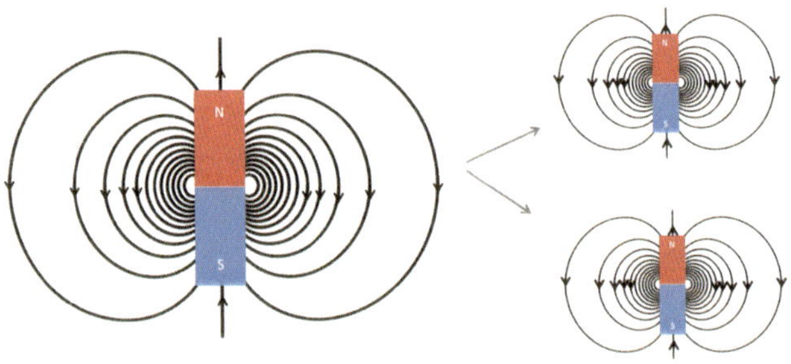

When a bar magnet is cut in two, you get two bar magnets

Integral Equation

The Gauss's law for magnetic fields in integral form is given by:

$$\oint_S \mathrm{b.da} = 0,$$

where:

b is the magnetic flux

The equation states that there is no net magnetic flux b (which can be thought of as the number of magnetic field lines through an area) that passes through an arbitrary closed surface S. This means the number of magnetic field lines that enter and exit through this closed surface S is the same. This is explained by the concept of a magnet that has a north and a south pole, where the strength of the north pole is equal to the strength of the south pole. This is equivalent to saying that a magnetic monopole, meaning a solitary north or south pole, does not exist because for every positive magnetic pole, there must be an equal amount of negative magnetic poles.

Differential Equation

Gauss's law for magnetic fields in the differential form can be derived using the divergence theorem. The divergence theorem states:

$$\int_V (\nabla.\mathrm{f})\,dv = \oint_S \mathrm{f.da},$$

where f is a vector. The right-hand side looks very similar to first equation. Using the divergence theorem, Equation first is rewritten as follows:

$$0 = \oint_S \mathrm{d.da} = \int_V (\nabla.\mathrm{f})\,dv.$$

Because the expression is set to zero, the integrand (∇·b) must be zero also. Thus the differential form of Gauss's law becomes:

$$\nabla \cdot b = 0.$$

Derivation using Biot-Savart Law

Gauss's law can be derived using the Biot-Savart law, which is defined as:

$$b(r) = \frac{\mu_0}{4\pi} \int_V \frac{(\mathrm{j}(r')dv) \times \hat{\underline{r}}}{|r - r'|^2},$$

where:

- b(r) is the magnetic flux at the point r.
- j(r′) is the current density at the point r′.
- μ_o is the magnetic permeability of free space.

Taking the divergence of both sides of equation above yields:

$$\nabla \cdot b(r) = \frac{\mu_0}{4\pi} \int_V \nabla . \frac{(\mathrm{j}(r')dv) \times \hat{\underline{r}}}{|r - r'|^2}.$$

To carry through the divergence of the integrand in equation above, the following vector identity is used:

$$\nabla \cdot (A \times B) = B \cdot (\nabla \times A) - A \cdot (\nabla \times B).$$

Thus, the integrand becomes:

$$\left[\mathrm{j}(r') . \left(\nabla \times \frac{\hat{\underline{r}}}{|r - r'|^2} \right) \right] - \left[\frac{\hat{\underline{r}}}{|r - r'|^2} . (\nabla \times \mathrm{j}(r')) \right]$$

The first part of above equation is zero as the curl of $\frac{\hat{\underline{r}}}{|r - r'|^2}$ is zero. The second part of above equation becomes zero because j depends on r' and ∇ depends only on r. Plugging this back into the one mentioned before the above the above two equations, the right-hand side of the expression becomes zero. Thus, we see that:

$$\nabla \cdot b(r) = 0,$$

which is Gauss's law for magnetism in differential form.

Differential Equation in the Frequency-domain

The equation can also be written in the frequency-domain as:

$$\nabla \cdot B = 0$$

Units

Magnetic flux	b	T	tesla
Electric current density	j	$\frac{A}{m^2}$	ampere per square meter

Constants

Magnetic constant	$\mu_0 = 4\pi \times 10^{-7} \frac{N}{A^2} \approx 1.2566370614... \times 10^{-6} \frac{T \cdot m}{A}$

Electromagnetism

Electromagnetism can be considered as the science of electromagnetic fields. When an electric current passes through a conductor, a circular electromagnetic field is created around it. The direction of the current decides the direction of rotation of the created magnetic field. The current strength as well as length of the conductor decides the force of electromagnetism developed. Change of the magnetic field can produce electricity.

Electromagnetism is governed by the basic law known as "Faraday's law of induction." According to Faraday's law, the induced electromotive force in any closed circuit is equal to the negative of the time rate of change of the magnetic flux enclosed by the circuit. The law shows how a magnetic field interacts with an electric circuit. Electromagnetism helped in establishing the relationship between electricity and magnetism. The electromagnetism created can be turned on and off by cutting off the electrical current flowing in the conductor. Electromagnetism also causes radiation, which shows up in an electromagnetic spectrum. Electromagnetism has been extended to quantum physics. It has helped in findings related to the nature of light and in the discovery of the electromagnetic spectrum. Electromagnetism is also used in devices such as speakers, solenoids, electric motors and magnetic disks.

Magnetic Field Created from Flowing Electrons

When DC electricity is passed through a wire, a magnetic field rotates around the wire in a specific direction.

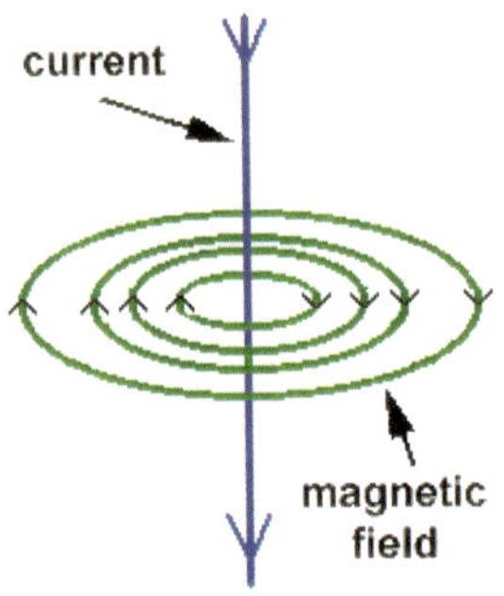

Electromagnetism Around a Loop

To understand the concept of electromagnetism around a loop we need to understand two concepts mentioned below.

Right Hand Screw Rule

When current passes through a conductor we can tell the direction of magnetic field around the conductor using Right Hand Screw Rule. Which states that if the current is entering the conductor (the + side) then the field when looked from above is South pole and if it is exciting (the • sign) then it is north pole.

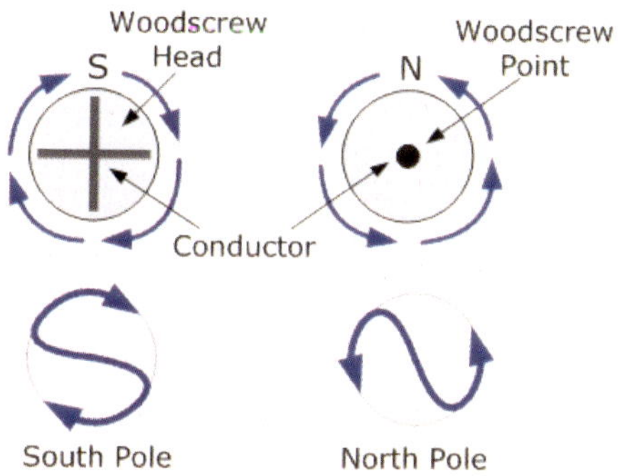

Here as we can see the first one is clockwise hence it is South pole and the second one is anti-clockwise so it is north pole.

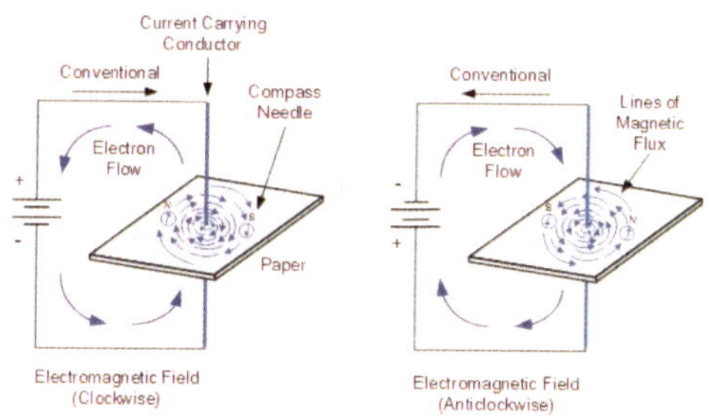

Left Hand Rule

The second concept needed is called Left Hand Rule which states that if the thumb of the left-hand points towards the direction of current flow then the direction in which the fingers rotate is the corresponding pole.

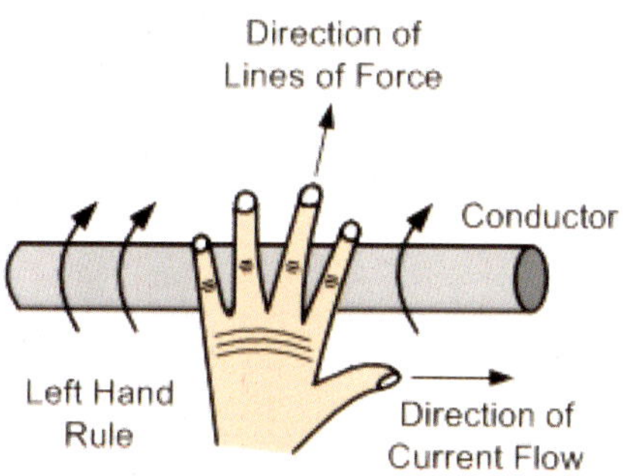

Now we are ready to understand the concept of Electromagnetism around a Loop.

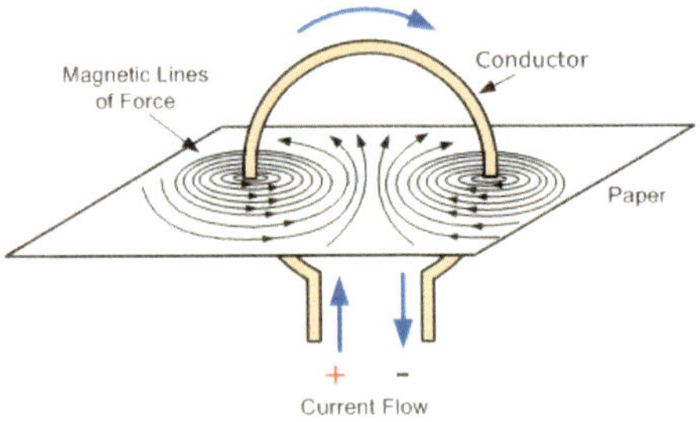

If we apply the above rules we can say that the current flowing through the two parallel conductors of the loop are in opposite directions as the current through the loop exits the left-hand side and returns on the right-hand side. This results in the magnetic field around each conductor inside the loop are in the "SAME" direction to each other.

The resulting lines of force generated by the current flowing through the loop oppose each other in the space between the two conductors where the two like poles meet thereby deforming the lines of force around each conductor. However, as the conductors cannot move, the two magnetic fields, therefore, help each other by generating a north and a south pole along this line of interaction. This results in the magnetic field are strongest in the middle between the two conductors. The intensity of the magnetic field around the conductor is proportional to the distance from the conductor and by the amount of current flowing through it.

The magnetic field generated around a straight length of current-carrying wire is very weak even with a high current passing through it. However, if several loops of the wire are wound together along the same axis producing a coil of wire, the resultant magnetic field will become even more concentrated and stronger than that of just a single loop. This produces an electromagnetic coil more commonly called a Solenoid.

Electromagnetic Induction

Electromagnetic induction is creating an electric current from a changing magnetic field. When a wire moves through a magnetic field, the electromagnetic effect takes place, creating an electric current through the wire. Current created in the wire moving through magnetic field Current created in the wire moving through the magnetic field.

Likewise, when a changing magnetic field affects a metal wire, the current is changed. One application is to change the electrical voltage, through the use of a transformer.

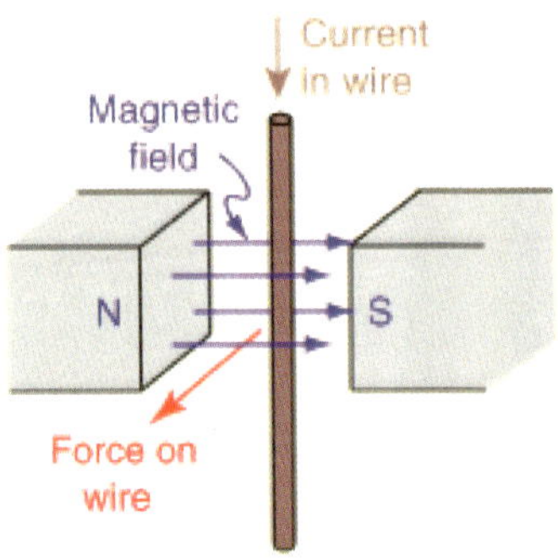

Example

Permanent Magnetic speakers commonly used in TV's and Radios are perfect examples of Electromagnetic devices. Let's see the operation of these devices which are based on the principle of electromagnetism. See the picture below.

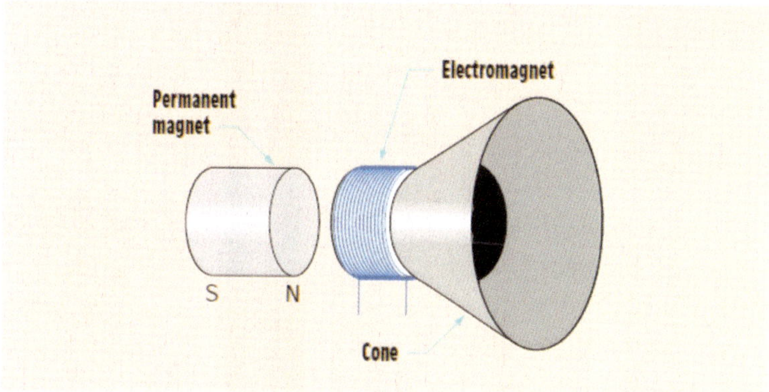

In order to convert electrical waves into audible sound, the speakers are designed. A metal coil is attached to a permanent magnet and when current passes through the coil it generates a magnetic field. The newly formed magnetic field is repelled by the permanent magnetic field resulting in the vibrations. These vibrations are amplified by the cone-like structure causing the sound.

Applications of Electromagnetism

Electromagnetism has numerous applications in today's world of science and physics.

- The very basic application of electromagnetism is in the use of motors. The motor has a switch that continuously switches the polarity of the outside of the motor. An electromagnet does the same thing. We can change the direction by simply reversing the current. The inside of the motor has an electromagnet, but the current is controlled in such a way that the outside magnet repels it.
- One of the most common uses of the properties of electromagnetism is in common speakers. In a speaker, there are a few main parts including a cone, a voice coil, and a permanent magnet. The ends of the voice coil are connected to the back of a panel receiver or stereo and the receiver sends a signal through the voice coil which interacts with the magnetic field of the magnet which produces an alternating force that pushes and pulls on the cone. This pushing and pulling create sound waves.
- Television screens and computer monitors use solenoids to produce images by exerting magnetic forces on moving electrons also. In a television, there is an electron gun that fires a narrow beam of electrons at the screen of a tube.
- Another very useful application of electromagnetism is the "CAT scan machine." This machine is usually used in hospitals to diagnose a disease. As we know that current is present in our body and the stronger the current, the strong is the magnetic field. This scanning technology is able to pick up the magnetic fields, and it can be easily identified where there is a great amount of electrical activity inside the body.

- The work of the human brain is based on electromagnetism. Electrical impulses cause the operations of the brain and it has some magnetic field. When two magnetic fields cross each other inside the brain, interference occurs which is not healthy for the brain.

Oersted's Law

In electromagnetism, Oersted's law, also spelled Oersted's law, is the physical law stating that an electric current creates a magnetic field.

The magnetic effect of current was discovered by H.C. Oersted in 1820. He observed that flow of electric current through a conductor produce a magnetic field around it. The magnetic effect of current can be demonstrated by doing following experiment.

Experiment to Demonstrate the Magnetic Effect of Current

First of all make a simple electrical circuit by joining a long straight wire with a battery and a plug. Now, take a magnetic compass needle and place the straight wire parallel to and over the compass needle. Then switch on the circuit so that current flows through the wire from south to north directions. Now, you will found that the north pole of compass needle gets deflected towards the west.

But, if the direction of current in the wire is reversed then the north pole of the compass needle gets deflected towards the east.

The direction of deflection of magnetic needle can be easily determined by SNOW rule. According to SNOW rule, when current flows through a conductor from South to North direction and the wire is held over magnetic needle then the north pole of the magnetic needle deflects towards west.

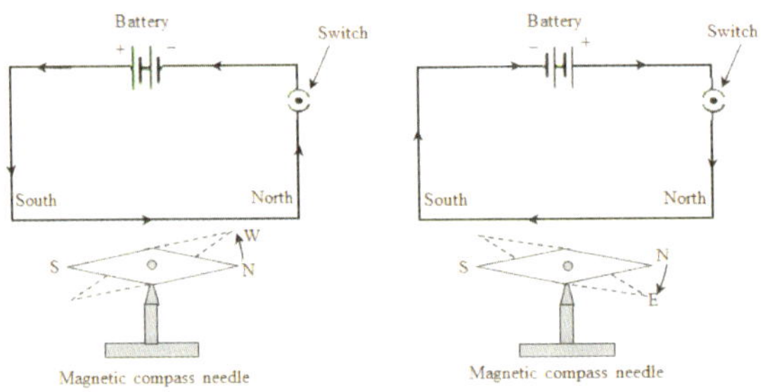

Electromagnetic Induction

Electromagnetic induction refers to the production of a voltage or a potential difference across a conductor when it is exposed to a changing magnetic field. The name of Faraday is generally acknowledged with the discovery of induction.

Faraday discovered that a voltage would be generated across a length of wire if that wire was exposed to a perpendicular magnetic field flux of changing intensity.

An easy way to create a magnetic field of changing intensity is to move a permanent magnet next to a wire or coil of wire. Remember: the magnetic field must increase or decrease in intensity perpendicular to the wire (so that the lines of flux "cut across" the conductor), or else no voltage will be induced.

Electromagnetic induction

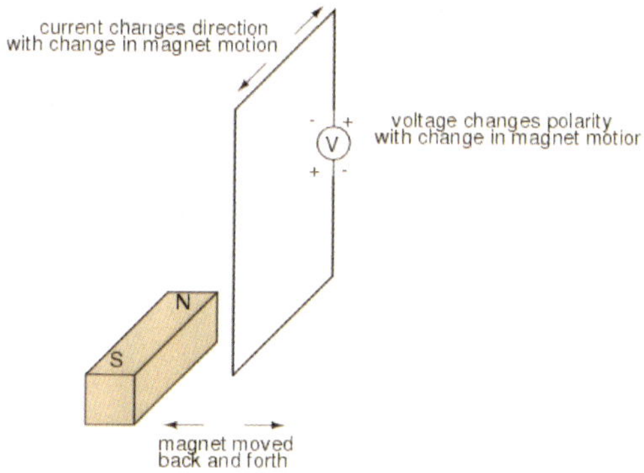

Faraday was able to mathematically relate the rate of change of the magnetic field flux with induced voltage (note the use of a lower-case letter "e" for voltage. This refers to instantaneous voltage, or voltage at a specific point in time, rather than a steady, stable voltage.):

$$e = N\frac{d\Phi}{dt}$$

Where,

$e = (\text{Instantaneous})$ induced voltage in volts

$N =$ Number of turns in wire coil (straight wire $= 1$)

$\Phi =$ Magnetic flux in Webers

$t =$ Time in seconds

The "d" terms are standard calculus notation, representing rate-of-change of flux over time. "N" stands for the number of turns, or wraps, in the wire coil (assuming that the wire is formed in the shape of a coil for maximum electromagnetic efficiency).

This phenomenon is put into obvious practical use in the construction of electrical generators, which use mechanical power to move a magnetic field past coils of wire to generate voltage. However, this is by no means the only practical use for this principle.

If we recall that the magnetic field produced by a current-carrying wire was always perpendicular to that wire, and that the flux intensity of that magnetic field varied with the amount of current through it, we can see that a wire is capable of inducing a voltage along its own length simply due to a change in current through it. This effect is called self-induction: a changing magnetic field produced by changes in current through a wire inducing voltage along the length of that same wire. If the magnetic field flux is enhanced by bending the wire into the shape of a coil, and/or wrapping that coil around a material of high permeability, this effect of self-induced voltage will be more intense. A device constructed to take advantage of this effect is called an inductor.

Faraday's Motional Emf Expression

$$\varepsilon = -\beta.l.v \quad volts$$

If the conductor does not move at right angles (90°) to the magnetic field then the angle θ° will be added to the above expression giving a reduced output as the angle increases:

$$\varepsilon = -\beta.l.v\sin\theta \quad volts$$

Lenz's Law of Electromagnetic Induction

Faraday's Law tells us that inducing a voltage into a conductor can be done by either passing it through a magnetic field, or by moving the magnetic field past the conductor and that if this conductor is part of a closed circuit, an electric current will flow. This voltage is called an induced emf as it has been induced into the conductor by a changing magnetic field due to electromagnetic induction with the negative sign in Faraday's law telling us the direction of the induced current (or polarity of the induced emf). But a changing magnetic flux produces a varying current through the coil which itself will produce its own magnetic field. This self-induced emf opposes the change that is causing it and the faster the rate of change of current the greater is the opposing emf. This self-induced emf will, by Lenz's law oppose the change in current in the coil and because of its direction this self-induced emf is generally called a back-emf.

Lenz's Law states that, "the direction of an induced emf is such that it will always opposes the change that is causing it". In other words, an induced current will always oppose the motion or change which started the induced current in the first place and this idea is found in the analysis of Inductance.

Likewise, if the magnetic flux is decreased then the induced emf will oppose this decrease by generating and induced magnetic flux that adds to the original flux.

Lenz's law is one of the basic laws in electromagnetic induction for determining the direction of flow of induced currents and is related to the law of conservation of energy.

According to the law of conservation of energy which states that the total amount of energy in the universe will always remain constant as energy can not be created nor destroyed. Lenz's law is derived from Michael Faraday's law of induction.

One final comment about Lenz's Law regarding electromagnetic induction. We now know that when a relative motion exists between a conductor and a magnetic field, an emf is induced within the conductor.

But the conductor may not actually be part of the coils electrical circuit, but may be the coils iron core or some other metallic part of the system, for example, a transformer. The induced emf within this metallic part of the system causes a circulating current to flow around it and this type of core current is known as an Eddy Current.

Eddy currents generated by electromagnetic induction circulate around the coils core or any connecting metallic components inside the magnetic field because for the magnetic flux they are acting like a single loop of wire. Eddy currents do not contribute anything towards the usefulness of the system but instead they oppose the flow of the induced current by acting like a negative force generating resistive heating and power loss within the core. However, there are electromagnetic induction furnace applications in which only eddy currents are used to heat and melt ferromagnetic metals.

Eddy Currents Circulating in a Transformer

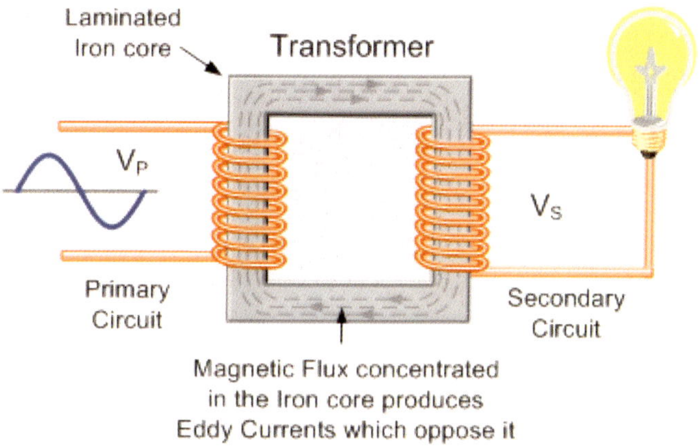

The changing magnetic flux in the iron core of a transformer above will induce an emf, not only in the primary and secondary windings, but also in the iron core. The iron core is a good conductor, so the currents induced in a solid iron core will be large. Furthermore, the eddy currents flow in a direction which, by Lenz's law, acts to weaken the flux created by the primary coil. Consequently, the current in the primary coil required to produce a given B field is increased, so the hysteresis curves are fatter along the H axis.

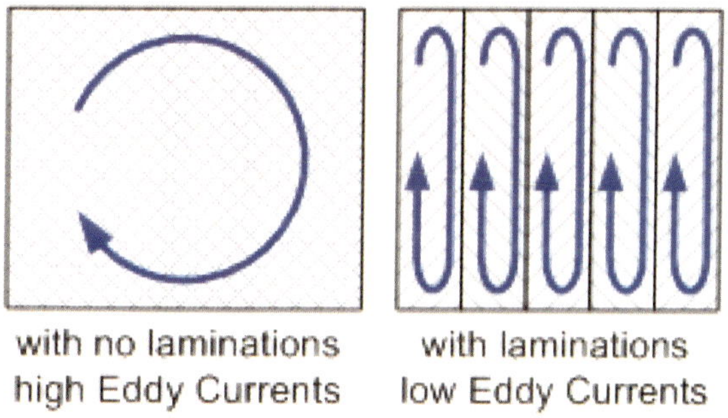

Laminating the Iron Core

Eddy current and hysteresis losses can not be eliminated completely, but they can be greatly reduced. Instead of having a solid iron core as the magnetic core material of the transformer or coil, the magnetic path is "laminated".

These laminations are very thin strips of insulated (usually with varnish) metal joined together to produce a solid core. The laminations increase the resistance of the iron-core thereby increasing the overall resistance to the flow of the eddy currents, so the induced eddy current power-loss in the core is reduced, and it is for this reason why the magnetic iron circuit of transformers and electrical machines are all laminated.

Faraday's Law of Induction

In 1831, Michael Faraday, an English physicist gave one of the most basic laws of electromagnetism called Faraday's law of electromagnetic induction. This law explains the working principle of most of the electrical motors, generators, electrical transformers and inductors. This law shows the relationship between electric circuit and magnetic field. Faraday performs an experiment with a magnet and a coil. During his experiment, he found how emf is induced in the coil when flux linked with it changes.

Faraday's Experiment

Relationship between Induced Emf and Flux

In this experiment, Faraday takes a magnet and a coil and connects a galvanometer across the coil. At starting, the magnet is at rest, so there is no deflection in the galvanometer i.e needle of galvanometer is at the center or zero position. When the magnet is moved towards the coil, the needle of galvanometer deflects in one direction.

When the magnet is held stationary at that position, the needle of galvanometer returns to zero position. Now when the magnet moves away from the coil, there is some deflection in the needle but opposite direction, and again when the magnet becomes stationary, at that point respect to the coil, the needle of the galvanometer returns to the zero position. Similarly, if the magnet is held stationary and the coil moves away, and towards the magnet, the galvanometer similarly shows deflection. It is also seen that the faster the change in the magnetic field, the greater will be the induced emf or voltage in the coil.

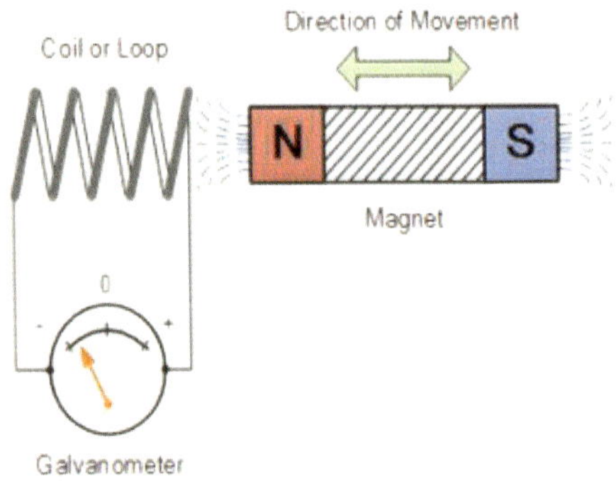

Conclusion: From this experiment, Faraday concluded that whenever there is relative motion between conductor and a magnetic field, the flux linkage with a coil changes and this change in flux induces a voltage across a coil.

Michael Faraday formulated two laws on the basis of above experiments. These laws are called Faraday's laws of electromagnetic induction.

Faraday's Laws

Faraday's First Law

Any change in the magnetic field of a coil of wire will cause an emf to be induced in the coil. This emf induced is called induced emf and if the conductor circuit is closed, the current will also circulate through the circuit and this current is called induced current.

Method to change magnetic field:

1. By moving a magnet towards or away from the coil.
2. By moving the coil into or out of the magnetic field.
3. By changing the area of a coil placed in the magnetic field.
4. By rotating the coil relative to the magnet.

Faraday's Second Law

It states that the magnitude of emf induced in the coil is equal to the rate of change of flux that linkages with the coil. The flux linkage of the coil is the product of the number of turns in the coil and flux associated with the coil.

Faraday Law Formula

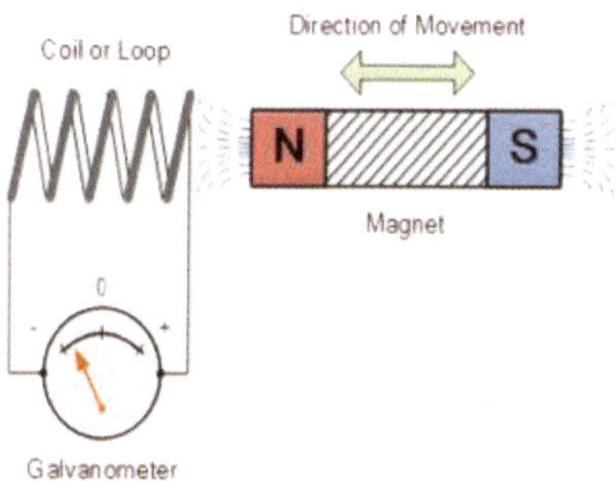

Consider, a magnet is approaching towards a coil. Here we consider two instants at time T1 and time T_2.

Flux linkage with the coil at time,

$$T_1 = N\phi_1 \, wb$$

Flux linkage with the coil at time,

$$T_2 = N\phi_2 \, wb$$

Change in flux linkage,

$$N(\phi_2 - \phi_1)$$

Let this change in flux linkage be,

$$\phi = (\phi_2 - \phi_1)$$

So, the Change in flux linkage,

$$N\phi$$

Now the rate of change of flux linkage,

$$\frac{N\phi}{t}$$

Take derivative on right hand side we will get,

The rate of change of flux linkage,

$$N\frac{d\phi}{dt}$$

But according to Faraday's law of electromagnetic induction, the rate of change of flux linkage is equal to induced emf.

$$E = N\frac{d\phi}{dt}$$

Considering Lenz's Law.

$$E = -N\frac{d\phi}{dt}$$

Where,

flux Φ in Wb = B.A

B = magnetic field strength

A = area of the coil.

Ways to Increase Emf Induced in A Coil

- By increasing the number of turns in the coil i.e N, from the formulae derived above it is easily seen that if number of turns in a coil is increased, the induced emf also gets increased.

- By increasing magnetic field strength i.e B surrounding the coil- Mathematically, if magnetic field increases, flux increases and if flux increases emf induced will also get increased. Theoretically, if the coil is passed through a stronger magnetic field, there will be more lines of force for coil to cut and hence there will be more emf induced.
- By increasing the speed of the relative motion between the coil and the magnet - If the relative speed between the coil and magnet is increased from its previous value, the coil will cut the lines of flux at a faster rate, so more induced emf would be produced.

Integral form in the Time-domain

Faraday's law in integral form can be expressed using the following equation:

$$\oint_C e.\mathrm{dI} = -\int_S \frac{\partial b}{\partial t}.\hat{n}\,da,$$

where:

- e is the electric field defined around a closed path C.
- b is the magnetic flux density defined over a closed surface A contoured by C.
- $\hat{n}$ is an outward normal unit vector perpendicular to da.
- dl is a vector element of length along contour C.

Eq. above states that the time-dependent rate of change in magnetic flux, through a surface bounded by a closed path, is negatively proportional to the line integral of the electric field it induces over that path.

Differential form in the Time-domain

By applying Stokes's theorem to left-hand side of Eq. above, we can obtain the differential form of Faraday's law:

$$\nabla \times e = -\frac{\partial b}{\partial t}$$

Eq. above states that time varying magnetic fields will induce rotational electric fields. Furthermore, the curl of the induced electric fields opposes time-dependent changes in the inducing magnetic field.

Faraday's Law in the Frequency-domain

The frequency-domain representation of Faraday's law can be obtained by applying a Fourier transform to above two equations. The integral form of Faraday's law in the frequency domain is:

$$\oint_C E.\mathrm{dI} = -i\omega\int_A B.\hat{n}\,da$$

Similarly using Stokes' theorem, the differential form of Faraday's law is:

$$\nabla \times E = -i\omega B$$

where ω is the angular frequency, E is the frequency-dependent electric field and B is the frequency-dependent magnetic flux density.

From equation above, we can infer two things:

1. Induced rotational electric fields are proportional to the angular frequency; this implies that electromagnetic induction is larger at higher frequencies.
2. Induced rotational electric fields, and the frequency-dependent magnetic fields responsible for them, are 90 degrees out of phase.

Applications of Faraday's Law

Faraday law is one of the most basic and important laws of electromagnetism. This law finds its application in most of the electrical machines, industries, and the medical field, etc.

- Electrical Transformers work on Faraday's law.
- The basic working principle of electrical generator is Faraday's law of mutual induction.
- The Induction cooker is the fastest way of cooking. It also works on the principle of mutual induction. When current flows through the coil of copper wire placed below a cooking container, it produces a changing magnetic field. This alternating or changing magnetic field induces an emf and hence the current in the conductive container, and we know that the flow of current always produces heat in it.
- Electromagnetic Flow Meter is used to measure the velocity of certain fluids. When a magnetic field is applied to electrically insulated pipe in which conducting fluids are flowing, then according to Faraday's law, an electromotive force is induced in it. This induced emf is proportional to the velocity of fluid flowing.
- Form bases of Electromagnetic theory, Faraday's idea of lines of force is used in well known Maxwell's equations. According to Faraday's law, change in magnetic field gives rise to change in electric field and the converse of this is used in Maxwell's equations.
- It is also used in musical instruments like electric guitar, electric violin etc.

Maxwell's Equations

Maxwell's equations are of great generality. They apply to all charge and current densities, whether static or time-dependent. Together, they describe the dynamical behaviour of the electromagnetic field. Each of Maxwell's equations is a local equation, relating field quantities at each point in space and at each instant in time, so all trace of instantaneous action at a distance has been eliminated. The revolutionary nature of this description was recognised by Einstein, who wrote:

'The formulation of Maxwell's equations is the most important event in physics since Newton's time, not only because of their wealth of content, but also because they form a pattern for a new type of law. Maxwell's equations are laws representing the structure of the field. All space is the scene of these laws and not, as for mechanical laws, only points in which matter or charges are present.'

Maxwell's equations are partial differential equations. They link the spatial and temporal rates of change of electric and magnetic fields, and they show how these rates of change are related to the sources of the fields – charge and current densities. The spatial rates of change of the fields are neatly bundled up as div E, div B, curl E and curl B – divergences and curls. This, in itself, is an immense simplification. Each field has three components, which can be partially differentiated with respect to three coordinates, so there are 18 first-order spatial partial derivatives of the electric and magnetic fields. The divergences and curls collect these partial derivatives together, focusing attention on only eight quantities of interest (a scalar for each divergence and three components for each curl). Moreover, divergences and curls have clear physical interpretations, telling us how the fields spread out and circulate at each point.

Where do the electric and magnetic fields come from? The modern answer is that they come from the terms in Maxwell's equations that describe matter – the charge and current densities, ρ and J. To be explicit about this, we can re-order and rearrange Maxwell's equations so that the two source terms appear on the right-hand sides of the first two equations:

$$div\ E = \frac{\rho}{\varepsilon_0}$$

$$curl\ B - \varepsilon_0 \mu_0 \frac{\partial E}{\partial t} = \mu_0\ \mathrm{J}$$

$$div\ B = 0$$

$$curl\ E + \frac{\partial B}{\partial t} = 0.$$

In regions where there are no charges or currents, all four equations have zero on the right-hand sides. They then tell us the conditions that electric fields and magnetic fields must satisfy in empty space. These conditions describe the internal structure and dynamics of the electromagnetic field.

In regions where there are charges and currents, the first two equations have an additional role. They tell us how the electromagnetic field is coupled to matter, and the left-hand sides of these equations describe the response of the electromagnetic field to the local charge and current densities. The last two equations do not have this role, so Maxwell's equations are asymmetrical. The absence of source terms in the last two equations arises because magnetic monopoles, and monopole currents, are assumed to be non-existent.

When Maxwell introduced his equations, he expected them to apply in a special frame of reference – the frame of the stationary ether. This is not the modern view. We now believe equations apply in all inertial frames of reference – that is, all frames in which free particles move uniformly, with no acceleration. Maxwell's equations are also unaffected by time-reversal and by reflections in space.

Only one caveat need be mentioned. Maxwell's equations do not apply in non-inertial frames. In a rotating frame of reference, for example, the electric flux over a closed surface can be non-zero even though the surface encloses no net charge – a clear violation of Gauss's law. This should not alarm you. Most laws of physics, including the laws of conservation of energy and momentum, are restricted to inertial frames of reference, and Maxwell's equations are no exception.

Maxwell's 4 Equation

- Gauss' law: Electric charges produce an electric field. The electric flux across a closed surface is proportional to the charge enclosed.
- Gauss' law for magnetism: There are no magnetic monopoles. The magnetic flux-and-faradays-law-quantitative across a closed surface is zero.
- Faraday's law: Time-varying magnetic fields produce an electric field.
- Ampère's law: Steady currents and time-varying electric fields (the latter due to Maxwell's correction) produce a magnetic field.

First assembled together by James Clerk 'Jimmy' Maxwell in the 1860s, Maxwell's equations specify the electric and magnetic fields and their time evolution for a given configuration

Integral Form

Where q and v are respectively the electric charge and velocity of a particle, the Lorentz law defines the electric field E and magnetic field B by specifying the total electromagnetic force F as,

$$F = qE + qv \times B.$$

In essence, one takes the part of the electromagnetic force that arises from interaction with moving charge (qv) as the magnetic field and the other part to be the electric field.

Gauss' Law

The earliest of the four Maxwell's equations to have been discovered (in the equivalent form of Coulomb's law) was Gauss' law. In its integral form in SI units, it states that the total charge contained within a closed surface is proportional to the total electric flux (sum of the normal component of the field) across the surface:

$$\int_S E \cdot da = \frac{1}{\varepsilon_0} \int \rho dV,$$

where the constant of proportionality is $1/\varepsilon_0$ the reciprocal of the electric constant. The total charge is expressed as the charge density integrated over a region.

Gauss' Law for Magnetism

Although magnetic dipoles can produce an analogous magnetic flux, which carries a similar mathematical form, there exist no equivalent magnetic monopoles, and therefore the total "magnetic charge" over all space must sum to zero. Therefore, Gauss' law for magnetism reads simply,

$$\int_S \mathbf{B} \cdot da = 0.$$

Faraday's Law

The electric and magnetic fields become intertwined when the fields undergo time evolution. In the 1820s, Faraday discovered that a change in magnetic flux produces an electric field over a closed loop. This relation is now called Faraday's law:

$$\int_{loop} E \cdot ds = -\frac{d}{dt} \int_S \mathbf{B} \cdot da.$$

With the orientation of the loop defined according to the right hand rule, the negative sign reflects Lenz's law.

Ampère's Law

Finally, Ampère's law suggests that steady current across a surface leads to a magnetic field (expressed in terms of flux). In addition, Maxwell determined that that rapid changes in the electric flux $(d/dt)E \cdot da$ can also lead to changes in magnetic flux. Altogether, Ampère's law with Maxwell's correction holds that,

$$\int_{loop} B \cdot ds = \mu_0 \int_S \mathrm{J} \cdot da + \mu_0 \varepsilon_0 \frac{d}{dt} \int_S E \cdot da.$$

In summary,

- Gauss' law

$$\int_S E \cdot da = \frac{1}{\varepsilon_0} \int \rho dV$$

- Gauss' law for magnetism

$$\int_S \mathbf{B} \cdot da = 0.$$

- Faraday's law

$$\int_{loop} E \cdot ds = -\frac{d}{dt} \int_S B \cdot da$$

- Ampère's law

$$\int_{loop} B \cdot ds = \mu_0 \int_S \mathrm{J} \cdot da + \varepsilon_0 \mu_0 \frac{d}{dt} \int_S E \cdot da$$

In their integral form, Maxwell's equations can be used to make statements about a region of charge or current.

Differential Form

To make local statements and evaluate Maxwell's equations at individual points in space, one can recast Maxwell's equations in their differential form, which use the differential operators div and curl.

Differential Form of Gauss' Law

The divergence theorem holds that a surface integral over a closed surface can be written as a volume integral over the divergence inside the region. Thus,

$$\frac{1}{\varepsilon_0}\iiint \rho dV = \int_S E \cdot da = \iiint \nabla .E dV.$$

Since the statement is true for all closed surfaces, it must be the case that the integrands are equal, and thus,

$$\nabla \cdot \mathrm{E} = \frac{\rho}{\varepsilon_0}.$$

(The derivation of the differential form of Gauss' law for magnetism is identical.)

Differential Form of Ampère's Law

One can use Stokes' theorem to rewrite the line integral $\int B \cdot ds$ in terms of the surface integral of the curl of B:

$$\int_{loop} B \cdot ds = \int_{surface} \nabla \times B \cdot da.$$

Ampère's law says that,

$$\int_{loop} B \cdot ds = \mu_0 \int_S \mathrm{J} \cdot da + \mu_0 \varepsilon_0 \frac{d}{dt}\int_S E \cdot da$$

Of course, the surface integral in both equations can be taken over any chosen closed surface, so the integrands must be equal:

$$\nabla \times B = \mu_0 \mathrm{J} + \mu_0 \varepsilon_0 \frac{\partial E}{\partial t}.$$

Differential form of Faraday's Law

It follows from the integral form of Faraday's law that,

$$\int_{loop} E \cdot ds = -\frac{d}{dt}\int_S B \cdot da.$$

As was done with Ampère's law, one can invoke Stokes' theorem on the left side to equate the two integrands:

$$\int_S \nabla \times E \cdot da = -\frac{d}{dt}\int_S B \cdot da.$$

Again, one argue that since the relationship must hold true for any arbitrary surface S, it must be the case that the two integrands are equal, and therefore,

$$\nabla \times E = -\frac{dB}{dt}.$$

In all, the result is as follows:

- Gauss' law

$$\nabla \cdot \mathrm{E} = \frac{\rho}{\varepsilon_0}.$$

- Gauss' law for magnetism

$$\nabla \cdot B = 0$$

- Faraday's law

$$\nabla \times E = -\frac{dB}{dt}.$$

- Ampère's law

$$\nabla \times B = \mu_0 \mathrm{J} + \mu_0 \varepsilon_0 \frac{\partial E}{\partial t}.$$

Electromagnetic Waves

By assembling all four of Maxwell's equations together and providing the correction to Ampère's law, Maxwell was able to show that electromagnetic fields could propagate as traveling waves. In other words, Maxwell's equations could be combined to form a wave equation. Maxwell's insight stands as one of the great theoretical triumphs of physics. A simple sketch of this result is as follows.

For simplicitly, suppose there is some region of space in which the electric field $E(x)$ is non-zero only along the z axis and the magnetic field $B(x)$ is non-zero only along the y axis, such that both are functions of x only. Then Faraday's law gives,

$$\frac{\partial E}{\partial x} = -\frac{\partial B}{\partial t}.$$

Even though $\mathrm{J} = 0$, with the additional term, Ampere's law now gives,

$$\frac{\partial B}{\partial x} = -\frac{1}{c^2}\frac{\partial E}{\partial t}.$$

Taking the partial derivative of the first equation with respect to x and the second with respect to t yields,

$$\frac{\partial^2 E}{\partial x^2} = -\frac{\partial^2 B}{\partial x \partial t}$$

$$\frac{\partial^2 B}{\partial t \partial x} = -\frac{1}{c^2}\frac{\partial^2 E}{\partial t^2}.$$

Therefore,

$$\frac{\partial^2 E}{\partial x^2} = -\frac{1}{c^2}\frac{\partial^2 E}{\partial t^2}.$$

This equation has solutions for $E(x)$ (and corresponding solutions for $B(x)$) that represent traveling electromagnetic waves. In fact, the equation that has just been derived is in fact in the same form as the classical wave equation in one dimension. In other words, the laws of electricity and magnetism permit for the electric and magnetic fields to travel as waves, but only if Maxwell's correction is added to Ampère's law. Indeed, Maxwell was the first to provide a theoretical explanation of a classical electromagnetic wave and, in doing so, compute the speed of light.

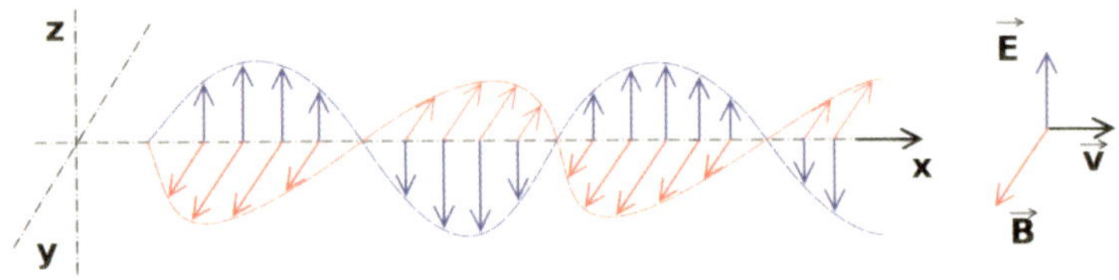

Maxwell's Equations and the Speed of Light

Let's consider light travelling in a vacuum, ie in a region in which there are no electric charges, so that ρ = 0 and J = 0. Again for the vacuum, Maxwell's equations become:

$$\nabla \times E = -\frac{\partial B}{\partial t} \qquad \nabla . E = 0$$

$$\nabla \times B = \mu_0 \varepsilon_0 \frac{\partial E}{\partial t} \qquad \nabla . B = 0$$

To get a wave equation, we want second derivatives in both time and space. First, we take the curl of both sides of $\nabla \times E = -\frac{\partial B}{\partial t}$. Then substitute in $\nabla \times B = \mu_0 \varepsilon_0 \frac{\partial E}{\partial t}$, and we have,

$$\nabla \times (\nabla \times E) = -\frac{\partial(\nabla \times B)}{\partial t} = -\mu_0 \varepsilon_0 \frac{\partial^2}{\partial t^2} E$$

A theorem in vector calculus is that, for any vector a, $\nabla \times (\nabla \times \mathrm{a}) = -\nabla^2 \mathrm{a} + \nabla.(\nabla.\mathrm{a})$ Here, $\nabla . E = 0$, *so*,

$$\nabla^2 E = \mu_0 \varepsilon_0 \frac{\partial^2}{\partial t^2} E$$

This is the wave equation in three dimensions. In one dimension, we could write it,

$$\frac{\partial^2 E}{\partial x^2} = \mu_0 \varepsilon_0 \frac{\partial^2}{\partial t^2}$$

Let's look for a solution that is a sinusoidal wave, with speed v and wavelength λ. One such wave has the expression:

$$E = E_0 \sin\left(2\pi \frac{x - vt}{\lambda}\right)$$

Differentiating, we get,

$$\frac{\partial^2 E}{\partial^2 x} = -E_0 \left(\frac{2\pi}{\lambda}\right)^2 \sin\left(2\pi \frac{x - vt}{\lambda}\right) \text{ and } \frac{\partial^2 E}{\partial^2 t} = -E_0 \left(\frac{2\pi}{\lambda}\right)^2 \sin\left(2\pi \frac{x - vt}{\lambda}\right)$$

Substituting this back into the wave equation, we see that it is a solution, provided that,

$$v^2 = \frac{1}{\mu_0 \varepsilon_0}$$

In other words, the equation has a wave solution and the speed of the waves is $\sqrt{\frac{1}{\mu_0 \varepsilon_0}}$. When Hertz found this solution, he recognized that the value of this quantity was the speed already known for light, which strongly suggested that light was a wave, and part of the electromagnetic spectrum, along with radio.

Other forms of Maxwell's Equations

Note that the values of ε_0 and μ_0 relate our units of mechanics, electricity and magnetism. We could choose our system of units such that $\varepsilon_0 = 1$ and $\mu_0 = 1$. In fact, many theoretical physicists sometimes use these units. They have the effect of making Maxwell's equations even simpler and more symmetrical. (They have the disadvantage that other units then have uncomfortable sizes. With $c = 1$ in these units, either the unit of length is very long or (more usually) that of time very short.)

Another asymmetry is the appearance of ρ and J in the equations: the electric charge density and electric current density. These arise because individual charges can move independently. In magnetism, a North and South pole always appear together. If there were isolated magnetic monopoles (as Dirac had predicted*), we could define a magnetic charge density ρ_{mag} and magnetic current density J_{mag}. Using grey symbols for the magnetic monopole terms that seem to be zero in our universe, this would give us the following set of Maxwell's equations:

Maxwell's equations: symmetric form $(\varepsilon_0 = 1 \; and \; \mu_0 = 1)$

$$\nabla \times E = -\frac{\partial B}{\partial t} + J_{mag} \qquad \nabla . E = \rho_{\text{elec}}$$

$$\nabla \times B = -\frac{\partial E}{\partial t} + J_{\text{elec}} \qquad \nabla . B = \rho_{mag}$$

References

- Electric-current-magnetic-field, electricity-magnetism-sim: livephysics.com, Retrieved 22 March 2018
- Magnetic-flux-definition-unit-formula, electromagnetism: electricalacademia.com, Retrieved 28 June 2018
- Magnetic-moment, physics: byjus.com, Retrieved 31 March 2018
- Electromagnetism-7003: techopedia.com, Retrieved 11 July 2018
- Electromagnetic-induction: allaboutcircuits.com, Retrieved 16 March 2018
- Faraday-law-of-electromagnetic-induction: electrical4u.com, Retrieved 21 April 2018

Permissions

All chapters in this book are published with permission under the Creative Commons Attribution Share Alike License or equivalent. Every chapter published in this book has been scrutinized by our experts. Their significance has been extensively debated. The topics covered herein carry significant information for a comprehensive understanding. They may even be implemented as practical applications or may be referred to as a beginning point for further studies.

We would like to thank the editorial team for lending their expertise to make the book truly unique. They have played a crucial role in the development of this book. Without their invaluable contributions this book wouldn't have been possible. They have made vital efforts to compile up to date information on the varied aspects of this subject to make this book a valuable addition to the collection of many professionals and students.

This book was conceptualized with the vision of imparting up-to-date and integrated information in this field. To ensure the same, a matchless editorial board was set up. Every individual on the board went through rigorous rounds of assessment to prove their worth. After which they invested a large part of their time researching and compiling the most relevant data for our readers.

The editorial board has been involved in producing this book since its inception. They have spent rigorous hours researching and exploring the diverse topics which have resulted in the successful publishing of this book. They have passed on their knowledge of decades through this book. To expedite this challenging task, the publisher supported the team at every step. A small team of assistant editors was also appointed to further simplify the editing procedure and attain best results for the readers.

Apart from the editorial board, the designing team has also invested a significant amount of their time in understanding the subject and creating the most relevant covers. They scrutinized every image to scout for the most suitable representation of the subject and create an appropriate cover for the book.

The publishing team has been an ardent support to the editorial, designing and production team. Their endless efforts to recruit the best for this project, has resulted in the accomplishment of this book. They are a veteran in the field of academics and their pool of knowledge is as vast as their experience in printing. Their expertise and guidance has proved useful at every step. Their uncompromising quality standards have made this book an exceptional effort. Their encouragement from time to time has been an inspiration for everyone.

The publisher and the editorial board hope that this book will prove to be a valuable piece of knowledge for students, practitioners and scholars across the globe.

Index

A

Alternating Current, 4-5, 19-20, 31, 33, 50, 56, 72, 90, 117, 120, 167, 185

Alternator, 20, 31, 33, 38

Avalanche Breakdown, 112-113, 116

B

Band Stop Filter, 122, 126-127

Bipolar Junction Transistor, 107, 142, 146-149, 152

Bridge Rectifier, 107, 117-121

C

Capacitance, 22, 52-54, 71, 141, 166, 170-181, 183-184, 186-212, 214, 216-222

Capacitor, 52, 54-57, 59-61, 71-72, 106, 122-125, 166-173, 176-177, 184-218, 220-222

Ceramic Capacitor, 206-209

Circuit Diagram, 45, 48, 61-63, 92-94, 98, 100-101, 114, 122, 124, 126-127, 132, 166

Circuit Element, 24, 55, 59-61, 79, 83

Clipper With Bias, 131, 133, 135, 137, 139

Closed Circuit, 41-42, 45-47, 85, 238, 244

Complex Impedance, 55-56, 70-72

Control Panel, 32, 36-37

Current Electricity, 1-4

Current Source, 59, 90-95, 97, 100-102, 104, 116

D

Dc Generator, 39

Dielectric, 170-172, 185, 187-197, 206-217, 222

Diode, 18, 60, 107-117, 119-121, 131-141, 143-144, 150-151, 160, 163, 169

Direct Current, 4-5, 19-20, 30, 33, 70, 117-118, 120, 167, 185

E

Electric Circuit, 15, 46-47, 58, 94, 238, 246

Electric Current, 1, 3-4, 15-16, 18-22, 28, 30, 45, 70, 112-113, 115, 118-119, 121, 132-133,

135-136, 153, 186, 208, 214-216, 238, 240, 242, 244, 257

Electric Field, 9, 13-15, 17, 58-59, 113-116, 152, 173, 176, 215, 228, 249-250, 252-253, 255

Electrical Generator, 219, 250

Electricity, 1-9, 15, 25, 31, 33, 42-45, 48, 57-58, 63, 106, 170, 175, 207-208, 214-215, 223, 238, 257-258

Electrolytic Capacitor, 193, 214-217

Electromagnetic Induction, 38, 223, 240, 242, 244-246, 248, 250

Electromagnetism, 223, 238-242, 246, 250, 258

Electromotive Force, 1, 4, 16, 25, 28, 87, 238, 250

Exhaust System, 36

F

Faraday's Law of Induction, 85, 88, 223, 238, 244, 246

Feedback Biasing, 145, 161

Film Capacitor, 211-213

Filter Circuit, 122, 124, 126

Frequency Response, 123, 125-128, 130, 164

G

Gauss's Law, 223, 235-237, 252

Generator, 1, 6, 31-39, 44, 219, 250

H

High Pass Filter, 122, 124, 129-130

I

Impedance, 23-24, 47, 50-51, 55-56, 70-72, 79, 90, 104, 106, 156, 164, 168-169, 177, 181, 194, 203, 220

Inductor, 48-49, 51, 59-60, 71-72, 122-125, 173, 176-177, 197, 213, 243

J

Junction Rule, 85-86, 89

L

Loop Rule, 86-89

Low Pass Filter, 122-123, 128, 217

M

Magnetic Dipole Moment, 231-232, 235

Magnetic Effect, 20-21, 224, 242

Magnetic Field, 1, 14, 21, 33, 35, 37-39, 58-59, 69, 85, 88, 223-233, 235-236, 238-244, 246-250, 252-253, 255

Magnetic Flux, 39, 225, 227-230, 236-238, 244-245, 249-250, 252-253

Magnetic Flux Density, 227, 230, 249-250

Magnetic Moment, 223, 231, 233-234

Magnetism, 5-6, 37, 39, 223-224, 226-227, 235-238, 252-255, 257-258

Maxwell's Equations, 223, 250-255, 257

Mosfet Transistor, 155-157

Mutual Capacitance, 174-176

N

Negative Charge, 2, 9-10, 12, 14-15, 18, 144, 185, 215

Nodal Analysis, 73-74, 76, 100, 106

Norton's Theorem, 45, 94-97

O

Ohm's Law, 27-28, 74-75, 80-84, 96

Open Circuit, 45-47, 95-96, 100, 102-104

Output Waveform, 129, 131, 167

P

Parallel Circuit, 57-58, 72, 96-97, 200, 202

Parallel Plate Capacitor, 185-187, 189-190, 195-196

Parasitic Capacitance, 166, 170, 173-174, 176, 179-181

Peak Inverse Voltage, 111, 119-120

Permittivity, 171, 187-189, 195-196

Phase Angle, 24, 50, 56-57, 64-66, 69

Phase Difference, 56-57, 63, 65

Phasor Diagram, 45, 49, 53, 64-66, 68-69

Positive Charge, 8-9, 11, 14-15, 17-18, 156, 185, 215

Power Factor, 51, 56-57, 185, 213

R

Rc Coupled Amplifier, 164, 166, 168

Reference Node, 73-76

Resistor, 27, 46, 48-49, 51-52, 54-61, 71, 74, 77, 82, 94-95, 97, 99-102, 104-106, 110, 116-119, 124, 141, 145-148, 160-163, 167, 169-170, 173, 197

Rl Circuit, 48, 51

S

Self Capacitance, 170, 173-174

Series Circuit, 48-51, 53, 66, 87-88, 202-203, 205

Series Negative Clipper, 131, 134-136

Short Circuit, 46-47, 95-96, 103-104

Shunt Positive Clipper, 131, 137-138, 140

Sinusoidal Waveform, 64-65

Solar Cell, 6, 42

Source Transformation, 90-91

Static Electricity, 1-3, 9, 25

Supercapacitor, 217-220

T

Thermoelectric Device, 41-42

Time Varying Electric Field, 14-15

Transistor, 59-60, 107, 116, 141-153, 155-169, 178-179

Transistor Amplifier, 145, 152, 163-164, 166-167, 169

Transistor Biasing, 107, 145, 158-163

Trimmer Capacitor, 221-222

V

Variable Capacitor, 220

Voltage Regulator, 32, 34-35, 116

Voltage Source, 54, 56-57, 59, 76-77, 90-94, 97, 99-104, 107, 117, 164, 170, 184, 219

W

Wave Shaping, 117, 121-122, 130-131

Z

Zener Diode, 107, 111-117

Zener Voltage, 112-115, 117